Fungi in the Environment

Fungi are of fundamental importance in the terrestrial environment. They have roles as decomposers, plant pathogens and symbionts, and in elemental cycles. Fungi are often dominant, and in soil can comprise the largest pool of biomass (including other microorganisms and invertebrates). They also play a role in the maintenance of soil structure owing to their filamentous growth habit and exopolymer production. Despite their important roles in the biosphere, fungi are frequently neglected within broader environmental and microbiological spheres. In addition, mycological interests can be somewhat fragmented between traditional subject boundaries.

This multidisciplinary volume explores the roles and importance of fungi in the environment. Particular emphasis is given to major research advances made in recent years as a result of molecular and genomic approaches, and in cell imaging and biology. Drawing together microbiologists, mycologists and environmental scientists, this work is a unique account of modern environmental mycology and a pivotal contribution to the field.

The British Mycological Society promotes all aspects all mycology. It is an international society with members throughout the world. Further details regarding membership, activities and publications can be obtained at www.britmycolsoc.org.uk.

GEOFF GADD is Professor of Microbiology and Head of the Division of Environmental and Applied Biology at the University of Dundee.

SARAH WATKINSON is Research Lecturer in the Department of Plant Sciences, and Tutor in Biology at St Hilda's College, University of Oxford.

PAUL DYER is a Lecturer in the School of Biology at the University of Nottingham.

Fungi in the Environment

EDITED BY

GEOFFREY MICHAEL GADD

SARAH C. WATKINSON AND

PAUL S. DYER

CAMBRIDGE UNIVERSITY PRESS
Cambridge, New York, Melbourne, Madrid, Cape Town, Singapore,
São Paulo, Delhi, Dubai, Tokyo, Mexico City

Cambridge University Press
The Edinburgh Building, Cambridge CB2 8RU, UK

Published in the United States of America by Cambridge University Press, New York

www.cambridge.org
Information on this title: www.cambridge.org/9780521850292

First published 2007

A catalogue record for this publication is available from the British Library

ISBN 978-0-521-85029-2 Hardback

Contents

Contributors

Anne Ashford
School of Biological, Earth and Environmental Sciences
The University of New South Wales
Sydney, NSW 2052
Australia

Anne Beauvais
Aspergillus Unit
Institut Pasteur
25, rue du Dr Roux
75724 Paris Cedex 15
France

Daniel P. Bebber
Department of Plant Sciences
University of Oxford
South Parks Road
Oxford OX1 3RB
UK

Meredith Blackwell
Department of Biological Sciences
Louisiana State University
Baton Rouge
LA 70803
USA

Lynne Boddy
Cardiff School of Biosciences
Cardiff University
Main Building
Park Place
Cardiff CF10 3TL
UK

Euan P. Burford
Division of Environmental and Applied Biology
Biological Sciences Institute
School of Life Sciences
University of Dundee
Dundee DD1 4HN
UK

Zachary Cartwright
School of Biosciences
University of Exeter
Washington Singer Laboratories
Perry Road
Exeter EX4 4QG
UK

Peter R. Darrah
Department of Plant Sciences
University of Oxford
South Parks Road
Oxford OX1 3RB
UK

Fordyce A. Davidson
Division of Mathematics
School of Engineering and Physical Sciences
University of Dundee
Dundee DD1 4HN
UK

Amy E. Davies
School of Biosciences
University of Exeter
Washington Singer Laboratories
Perry Road
Exeter EX4 4QG
UK

Jeremy Dettman
Department of Botany
University of Toronto
Mississauga
Ontario
Canada L5L 1C6

Paul S. Dyer
School of Biology
University of Nottingham
University Park
Nottingham NG7 2RD
UK

Matthew C. Fisher
Imperial College Faculty of Medicine
Department of Infectious Diseases and Epidemiology
St Mary's Campus
Norfolk Place
London W2 1PG
UK

Marina Fomina
Division of Environmental and Applied Biology
Biological Sciences Institute
School of Life Sciences
University of Dundee
Dundee DD1 4HN
UK

Petra M. A. Fransson
Department of Forest Mycology and Pathology
Swedish University of Agricultural Sciences
Box 7026
S-75007 Uppsala
Sweden

Mark D. Fricker
Department of Plant Sciences
University of Oxford
South Parks Road
Oxford OX1 3RB
UK

Geoffrey M. Gadd
Division of Environmental and Applied Biology
Biological Sciences Institute
School of Life Sciences
University of Dundee
Dundee DD1 4HN
UK

Maria J. Harrison
Boyce Thompson Institute for Plant Research
Tower Road
Ithaca
NY 14853
USA

Rosmarie Honegger
Institute of Plant Biology
University of Zurich
Zollikerstr. 107
CH-8008 Zurich
Switzerland

Juliet Hynes
Cardiff School of Biosciences
Cardiff University
Biomedical Sciences Building
Museum Avenue
Cardiff CF10 3US
UK

Gregory Jedd
Temasek Life Sciences Laboratory and
Department of Biological Sciences
1 Research Link
National University of Singapore
Singapore 117604

Joanna M. Jenkinson
School of Biosciences
University of Exeter
Washington Singer Laboratories
Perry Road
Exeter EX4 4QG
UK

Hanna Johannesson
Department of Evolutionary Biology
Uppsala University
752 36 Uppsala
Sweden

T. Hefin Jones
Cardiff School of Biosciences
Cardiff University
Main Building
Park Place
Cardiff CF10 3TL
UK

Michael J. Kershaw
School of Biosciences
University of Exeter
Washington Singer Laboratories
Perry Road
Exeter EX4 4QG
UK

Jean Paul Latgé
Aspergillus Unit
Institut Pasteur
25 rue du Dr Roux
75724 Paris Cedex 15
France

Jonathan R. Leake
Department of Animal & Plant Sciences
University of Sheffield
Alfred Denny Building
Weston Bank
Sheffield S10 2TN
UK

Jinyuan Liu
Boyce Thompson Institute for Plant Research
Tower Road
Ithaca
NY 14853
USA

Melina Lopez-Meyer
Boyce Thompson Institute for Plant Research
Tower Road
Ithaca
NY 14853
USA

Justine I. Lyons
The University of Georgia – Marine Institute
Sapelo Island
GA 31327
USA

Ignacio Maldonado-Mendoza
Boyce Thompson Institute for Plant Research
Tower Road
Ithaca
NY 14853
USA

Karrie Melville
Division of Environmental and Applied Biology
Biological Sciences Institute
School of Life Sciences
University of Dundee
Dundee DD1 4HN
UK

Nicholas P. Money
Department of Botany
Miami University
Oxford
OH 45056
USA

Mary Ann Moran
The University of Georgia – Marine Institute
Sapelo Island
GA 31327
USA

James B. Nardi
Department of Entomology
University of Illinois
Urbana-Champaign
IL 61801
USA

Steven Y. Newell
The University of Georgia – Marine Institute
Sapelo Island
GA 31327
USA

David S. Perlin
Public Health Research Institute
Newark, NJ 07103
USA

Anne Pringle
Department of Plant and Microbial Biology
University of California
Berkeley
CA 94720-3102
USA

Nick D. Read
Institute of Cell Biology
University of Edinburgh
Rutherford Building
Edinburgh EH9 3JH
UK

Darren M. Soanes
School of Biosciences
University of Exeter
Washington Singer Laboratories
Perry Road
Exeter EX4 4QG
UK

Sung-Oui Suh
Department of Biological Sciences
Louisiana State University
Baton Rouge
LA 70803
USA

Nicholas J. Talbot
School of Biosciences
University of Exeter
Washington Singer Laboratories
Perry Road
Exeter EX4 4QG
UK

Andy F. S. Taylor
Department of Forest Mycology and Pathology
Swedish University of Agricultural Sciences
Box 7026
S-75007 Uppsala
Sweden

John W. Taylor
Department of Plant and Microbial Biology
University of California
Berkeley
CA 94720-3102
USA

Monika Tlalka
Department of Plant Sciences
University of Oxford
South Parks Road
Oxford OX1 3RB
UK

Anders Tunlid
Department of Microbial Ecology
Lund University
Ecology Building
SE 223 62 Lund
Sweden

Elizabeth Turner
Department of Plant and Microbial Biology
University of California
Berkeley
CA 94720-3102
USA

Sarah C. Watkinson
Department of Plant Sciences
University of Oxford
South Parks Road
Oxford OX1 3RB
UK

Richard Wilson
School of Biosciences
University of Exeter
Washington Singer Laboratories
Perry Road
Exeter EX4 4QG
UK

Preface

Fungi are ubiquitous in the aquatic and terrestrial environments, occurring as unicellular yeasts, polymorphic and filamentous fungi, and as both free-living and symbiotic forms. In the terrestrial environment, that part of the biosphere most closely associated with fungal activities, fungi are of fundamental importance as decomposer organisms, plant pathogens and mutualistic symbionts (mycorrhizas and lichens), playing important roles in carbon, nitrogen and other biogeochemical cycles. In soil they can comprise the largest pool of biomass (even exceeding that of other micro-organisms and invertebrates) and also play a role in maintenance of soil structure owing to their filamentous branching growth habit and exopolymer production. Despite their important roles in the biosphere, fungi are frequently neglected within broader environmental and microbiological spheres, in contrast to bacteria. For example, symbiotic mycorrhizal fungi can be associated with the majority of plant species and are responsible for major transformations and re-distribution of inorganic nutrients as well as carbon flow, while free-living fungi have major roles in decomposition and solubilization of plant and other organic materials, including xenobiotics, and low-solubility phosphate compounds. As well as this general neglect, mycological interests can be somewhat fragmented between traditional microbiological and botanical activities and the fields of cell biology, plant symbiosis, and pathogenesis and genetics. This symposium volume provides a unique account of modern environmental mycology and includes accounts of major recent advances in molecular, imaging and modelling methodologies, which have the potential to draw together these disparate areas of mycology.

The book is divided into six general sections, each dealing with an important area in environmental mycology, but with a considerable degree of complementarity between all the sections. The first section, *Imaging and modelling of fungi in the environment*, includes modern contributions in the areas of cell biology where molecular tools and dyes have revealed unsuspected complexity in fungal cell structure, and allowed key questions to be answered in relation to roles of sub-cellular organelles. A chapter on peroxisomal roles in the ascomycete lifestyle is complemented by chapters on cellular structure and how environmental signals influence hyphal and colony morphogenesis. Complex nutrient dynamics in mycelial networks can be dissected by using novel radiolabelling imaging techniques, and mathematical modeling using sound biological data enables understanding of mycelial form and function, particularly in complex realistic systems. The section on *Functional ecology of saprotrophic fungi* includes modern approaches to study saprotrophic growth in soil that reveal unsuspected fungal roles in mineral transformations and biogeochemical cycles, the adaptability and importance of the mycelial habit being a recurring theme. *Mutualistic interactions in the environment* includes chapters on mycorrhizas, which not only influence cycling of major elements such as carbon and phosphorus but can even determine plant community composition. Lichens represent a unique fungal lifestyle of global importance, not least because of their success as pioneer organisms on rocks and other exposed environments where water relations are an important survival determinant. In contrast, *Pathogenic interactions in the environment*, as well as being a significant part of ecosystem function, may cause serious crop diseases, such as rice blast, potentially affecting millions of human lives. Nematode-trapping fungi are important predators that have relevance to the biocontrol of nematode populations. Aspects of fungal structure can be significant determinants of virulence and this is the case for human as well as plant pathogens: the glucans of *Aspergillus fumigatus* provide an example from the human health perspective. Within the human built environment, contamination, spoilage and potential health effects are well known and are also highlighted in this section. Advances in molecular biology techniques are leading to greater understanding of *Environmental population genetics of fungi*; some of the approaches are highlighted in a case history with *Penicillium marneffei* as the lead player in a complex system of relationships. In complement to this, the final section on *Molecular ecology of fungi in the environment* details examples where molecular approaches provide understanding of the gut mycoflora of beetles, and fungal populations in saltmarsh decomposition systems.

This book arises from the British Mycological Society Symposium on *Fungi in the Environment* held at the University of Nottingham in September 2004 and the editors would like to thank those authors who contributed enthusiastically to this book. Special thanks also go to Diane Purves in Dundee, who greatly assisted communication, collation, editing and formatting of chapters. The prime objective was to produce a volume that would highlight the roles and importance of fungi in the environment together with the modern approaches and tools that are now revealing the importance of fungi in a wider biological context. We think that has been achieved and hope the volume has broad appeal not only to mycologists of all persuasions but also to cell biologists, molecular biologists and geneticists who routinely rely on fungal model systems. It will be interesting to see how the new interdisciplinary and systems approaches influence developments over the next few years.

Geoffrey Michael Gadd
Sarah C. Watkinson
Paul S. Dyer

I

Imaging and modelling of fungi in the environment

1

Imaging complex nutrient dynamics in mycelial networks

DANIEL P. BEBBER[1], MONIKA TLALKA[1], JULIET HYNES[2], PETER R. DARRAH[1], ANNE ASHFORD[3], SARAH C. WATKINSON[1], LYNNE BODDY[2] AND MARK D. FRICKER[1]

[1]*Department of Plant Sciences, University of Oxford*
[2]*Cardiff School of Biosciences, Cardiff University*
[3]*School of Biological, Earth and Environmental Sciences, The University of New South Wales*

Introduction

Basidiomycetes are the major agents of decomposition and nutrient cycling in forest ecosystems, occurring as both saprotrophs and mycorrhizal symbionts (Boddy & Watkinson, 1995; Smith & Read, 1997). The mycelium can scavenge and sequester nutrients from soil, concentrate nutrients from decomposing organic matter, relocate nutrients between different organic resources, and ultimately make nutrients available to plants to maintain primary productivity. Hyphae of both saprotrophic and ectomycorrhizal basidiomycetes that ramify through soil often aggregate to form rapidly extending, persistent, specialized high-conductivity channels termed cords (Rayner *et al.*, 1994, 1999; Boddy, 1999; Watkinson, 1999; Cairney, 2005). These cords form complex networks that can extend for metres or hectares in the natural environment. The distribution of resources is extremely heterogeneous and unpredictable in space and time, and these fungi have developed species-specific strategies to search for new resources and to capitalize on resources landing on their mycelial systems (Chapter 6, this volume). Thus the architecture of the network is not static, but is continuously reconfigured in response to local nutritional or environmental cues, damage or predation, through a combination of growth, branching, fusion or regression (Boddy, 1999; Watkinson, 1999; Chapter 6, this volume). At this stage it is not clear whether specific global mechanisms exist to couple local sensory perception and responses over different length scales specifically to maximize the long-term success of the whole colony, or whether such collective behaviour is an emergent property arising solely from local interactions of individual hyphae.

Fungi in the Environment, ed. G. M. Gadd, S. C. Watkinson & P. S. Dyer. Published by Cambridge University Press.

Embedded within the physical structure is an equally complex set of physiological processes that contribute to uptake, storage and redistribution of nutrients throughout the network in an apparently well coordinated manner (Olsson, 1999; Cairney, 2005). As the colony grows out from a resource base, nutrient translocation would be expected to be predominantly towards the growing margin. If additional resources are found, then re-distribution back to the base can also occur, although not necessarily by the same transport system at the same time. However, we know little about the cellular and sub-cellular anatomy of the pathway, the mechanism of transport and its driving forces, and the nature of the information pathways through mycelium that might contribute to coordinated system-wide responses to localized nutritional stimuli (Cairney, 2005; Watkinson *et al.*, 2006). A detailed understanding of nutrient translocation requires analysis of processes occurring across a range of length scales, from uptake by transporters in individual hyphae, to translocation through corded networks spanning several metres. In this chapter we describe the range of overlapping techniques that we are developing to try to track nutrient flow directly or indirectly in these systems across a range of length scales.

To simplify the initial state of the system, we have chosen to focus on microcosms in which foraging fungi grow from a central resource (agar or wood-block inoculum) over an inert (scintillation screen) or nutrient-depleted (sand or soil–sand mix) surface. Under these conditions, all nutrient transport, and in the first case water movement, is initially from the centre.

Transport at the micrometre to millimetre scale

The precise mechanisms underlying nutrient translocation in fungi are not yet known but are thought to include mass flow, diffusion, generalized cytoplasmic streaming and specific vesicular transport (Cairney, 2005). It has been proposed that the highly dynamic pleiomorphic vacuolar system present in filamentous fungi of all the major fungal taxonomic groups so far examined might have a role in long-distance translocation over millimetres or centimetres (Ashford & Allaway, 2002). The structure of the vacuole within a single hypha develops from a complex reticulum of fine tubes interspersed with small spherical vacuoles at the tip (Fig. 1.1D) to a series of larger, more spherical, adherent vacuoles interconnected with fine tubes (Fig. 1.1A). As there are no convenient fluorescent probes to study N-movement directly, we have adopted an indirect approach to address whether the vacuolar system plays a role in long-distance transport of N. The lumen of the vacuole

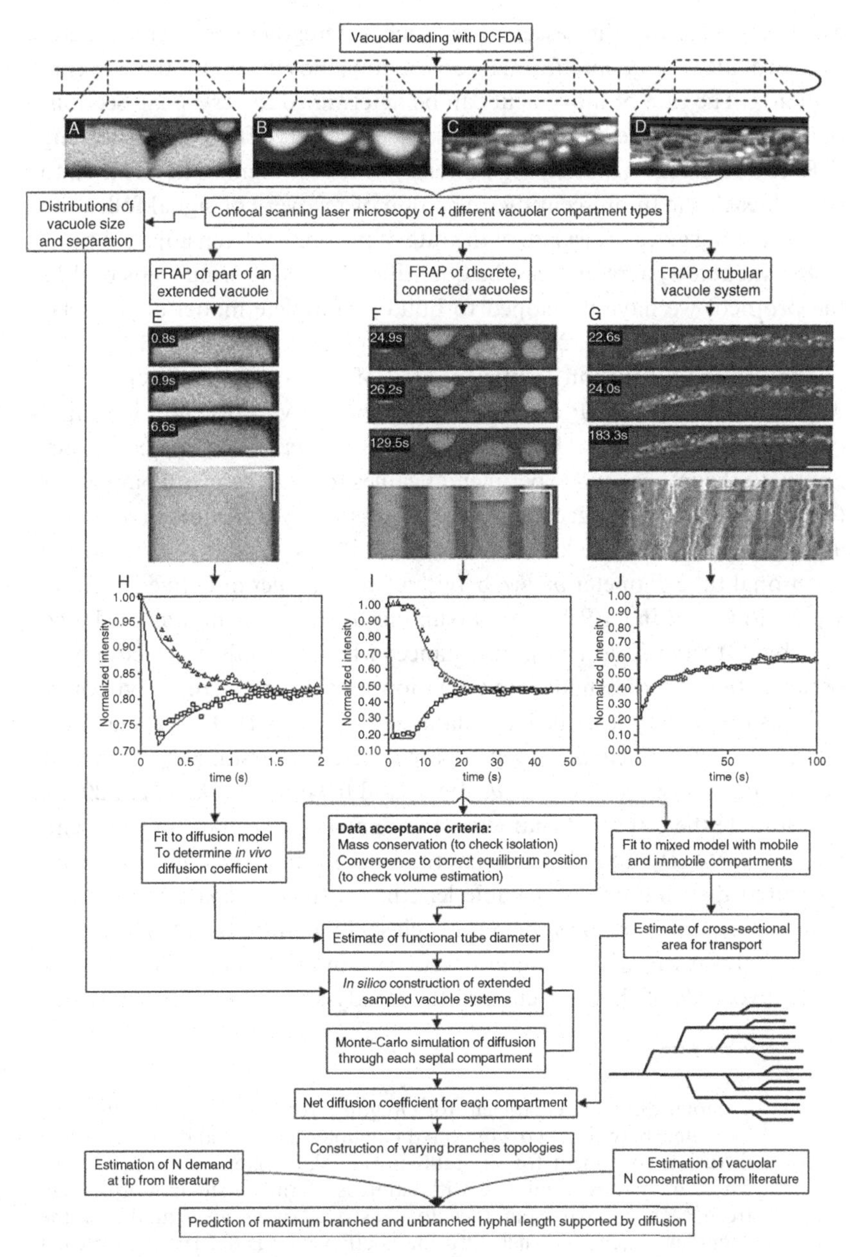

Fig. 1.1. Measurement and modelling of longitudinal vacuolar transport in *Phanerochaete velutina*. Vacuoles were labelled with carboxy-DFFDA (carboxydifluorofluorescein diacetate) and imaged with time-lapse confocal laser scanning microscopy in different septal compartments progressing towards the tip (A–D). Fluorescence recovery after photobleaching (FRAP) of part of single, isolated large vacuoles was used to determine the

can be labelled with fluorescent dyes, such as Oregon Green, which therefore provide a non-specific marker for any movement of the lumenal contents. The rate of movement can be determined in very small sections of individual hyphae by using fluorescence recovery after photobleaching (FRAP). In essence, a brief, high-intensity pulse of illumination is used to photobleach the fluorescent dye. The rate of recovery of signal following this bleaching gives a measure of the rate of movement from adjacent parts of the vacuole system and can be quantified. The following steps outline the protocol we have developed to build a complete model of vacuolar transport (Darrah *et al.*, 2006). A flow diagram is given in Fig. 1.1.

The vacuolar diffusion coefficient (D_v) of Oregon Green (OG) *in vivo* was estimated by FRAP of half a large, isolated vacuole by using rapid confocal imaging (Fig. 1.1E, H). Values measured *in vivo* compared favourably with theoretical and experimental values for fluorescein in pure water (deBeer *et al.*, 1997) suggesting that OG was freely diffusible in a largely aqueous vacuole. With a known value of D_v, it was possible to estimate the functional tube diameter *in vivo* between two vacuoles of defined size and separation by using FRAP and assuming dye movement was mediated only by diffusion (Fig. 1.1F). In instances where the only connection was between the vacuoles under investigation and not adjacent neighbours, a diffusion model described the data well (Fig. 1.1I). Functional tube diameters determined *in vivo* compared well with estimates from EM data of 0.24–0.48 μm (Rees *et al.*, 1994) and 0.3 μm (Uetake *et al.*, 2002). To estimate the transport characteristics of an entire septal compartment, the values of D_v and the median tube diameter were combined with measured distributions of vacuole length, width and separation to construct an *in silico* vacuole system for each septal compartment. The resulting model was run with constant boundary conditions of $C = 0$ and $C = 1$ at the two ends of the filament and the steady-state flux recorded, which

---

Fig 1.1. (cont.)
vacuolar diffusion coefficient for Oregon Green *in vivo* (E, H); the tube diameter between two connected vacuoles (F, I); and the effective diffusion coefficient for the tubular vacuolar region at the tip (G, J). These data were combined with samples of the vacuolar morphology from each compartment type to construct an *in silico* model of the vacuolar system. The net diffusion coefficient was determined for each compartment by Monte Carlo simulation. The maximum hyphal length that could be supported by diffusion was then calculated by using estimates of the N demand at the tip and the maximum likely vacuolar N concentration. Horizontal scale bar, 10 μm. Vertical scale bar, 2 s (E) or 60 s (F, G).

allowed an effective diffusion coefficient, $D_V\alpha$, for the whole compartment to be calculated from Fick's first law, where α has a range from 0–1 and measures the reduction in the vacuolar diffusion coefficient D_V caused by including many vacuoles and tubes of smaller diameter in the string relative to a uniform vacuole of the same length. One thousand Monte Carlo simulations of *in silico* hyphae were conducted to give a mean α for this compartment type. Similar simulations were run for two other categories of vacuolar organization with progressively smaller vacuoles and increasing amounts of tubular network (Darrah *et al.*, 2006).

In contrast to the discrete vacuoles distal from the tip, the tubular vacuole region consisted of a structurally complex reticulate network of predominantly longitudinal, tube-like elements and small vesicles (Fig. 1.1D). Superficially the network appeared quite dynamic, but most of the motion appeared to be short-range micrometre-scale oscillations rather than longer-range translocation of entire structures. Net diffusion in this tubular region was determined by FRAP of a region 40–60 μm long spanning the entire hyphal diameter (Fig. 1.1G). The data were well described by a model that included a well-connected (tubular) component and a smaller immobile (vesicle) phase (Fig. 1.1J).

Diffusion was a sufficient mechanism to explain observed transport in the various regions of the vacuolar organelle. Therefore, we used Fick's first law to estimate the length scales for effective diffusional translocation to maintain tip growth through the system at steady state with literature-based estimates of the concentration gradient for N in the vacuole system, a value for the N demand at the tip, and the effective diffusion coefficient, D_e, for the composite branched structure. This last value was calculated from the composite diffusion coefficients for each type of compartment. Although this analysis is not yet complete, preliminary observations suggest that an unbranched hypha possessing a continuous tubular vacuole system could sustain growth over a transport distance of around 12–24 mm. Conversely, diffusion alone in a maximally branched system would operate over only a few millimetres. This poise between translocation being sufficient or insufficient depending on the amount of branching and status of the vacuolar network suggests that the vacuolar system is an important organ for coordinating and controlling tip growth and branching. For example, the range of simulated effective diffusion coefficients varied by orders of magnitude when we used the sampled vacuole distribution data. It is therefore possible that the vacuolar system could be regulated to change its translocation capacity according to local nutrient conditions. The system could thus shift between increasing transport to tips or preventing unnecessary nutrient mobility by isolating tips. An alternative possibility is that the vacuolar system translocates material acquired by the tips back into the main colony, against the mass flow component needed for tip growth (Darrah *et al.*, 2006).

Transport at the millimetre to centimetre scale

One approach to measuring transport of solutes is to follow the movement of radiolabelled compounds through the mycelium. Typically, the final radiolabel distribution has been visualized by using autoradiography techniques or phosphor-imaging, or analysed by destructive harvesting of the tissues followed by scintillation counting. We have developed a novel non-invasive technique to track movement of ^{14}C-labelled N compounds in foraging mycelial networks in contact with an inert scintillation screen, by using photon-counting scintillation imaging (PCSI) (Tlalka *et al.*, 2002). We have developed two protocols to analyse the distribution patterns of the non-metabolized amino-acid analogue α-amino-isobutyrate (^{14}C-AIB). The first approach focuses on the correlation between local distribution patterns of AIB and local patterns of growth and is designed to accommodate the marked asymmetry in colony development, particularly in the presence of additional resources that provide a highly polarized resource environment. The second approach focuses on mapping the pulsatile component of transport that was observed superimposed on the net AIB translocation pattern (Tlalka *et al.*, 2002, 2003).

In simple microcosms with mycelium growing out from a central inoculum, ^{14}C-AIB was taken up and distributed through the growing colony. The AIB distribution pattern can be characterized by three parameters, the position of the centre of mass of AIB ($CM_{\Delta AIB}$) relative to the centre of the inoculum, the extent to which the AIB was spread evenly around the colony or concentrated in a particular area (measured by the angular concentration of the AIB, $Conc_{\Delta AIB}$) and the alignment of the $CM_{\Delta AIB}$ vector with the new resource, if present. We have no completely independent measure of colony growth during the experiment as (i) PCSI precludes separate bright-field imaging, and (ii) the lack of contrast between a white mycelium and a white screen on which it is growing makes it extremely difficult to characterize growth effectively. As an alternative approach, we determined the area covered from the scintillation image by using contrast-limited adaptive histogram equalization (CLAHE) and automated grey-scale thresholding algorithms (Otsu, 1979). We have validated this approach by comparison of the segmented boundary with a bright-field image of a colony grown across Mylar film 1.5 µm thick to facilitate imaging against a black background at the end of the PCSI experiment (Fig. 1.2). There was very good correspondence between the automatically segmented area (Fig. 1.2A) and the visible distribution pattern of the colony (Fig. 1.2B). In a similar manner to the analysis of AIB distribution, we used difference values for the change in area, averaged over a sliding 12 or 24 h window, and calculated the displacement of the centre of mass of the change in area ($CM_{\Delta area}$), the angular concentration ($Conc_{\Delta area}$) and the alignment of the $\Delta area$ vector to quantify the dynamics of the underlying processes (Fig. 1.3). In the first phase of colony development, the change in area and transport were almost symmetrical (Fig. 1.3A, B). This was followed by a transition to sparser, more asymmetric growth. The duration of the first growth phase depended on nutrient availability and developmental age of the colony. The different growth phases can be described by two superimposed logistic equations, which allow normalization of all data sets to a common developmental stage of the colony.

Time series data for $CM_{\Delta AIB}$, $Conc_{\Delta AIB}$, $CM_{\Delta area}$ and $Conc_{\Delta area}$ were normalized to the start of the second growth phase and fitted by using

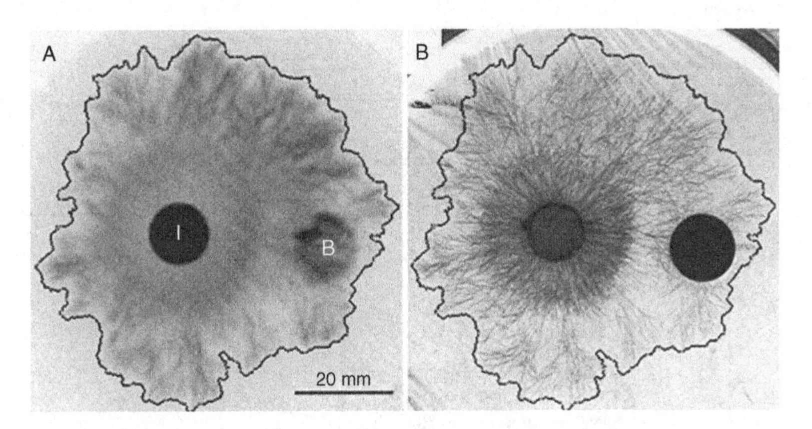

Fig. 1.2. Validation of automated colony area segmentation from photon-counting scintillation images. The area of a growing colony was estimated from photon-counting scintillation imaging (PCSI) by a combination of contrast-limited adaptive histogram equalization (CLAHE) and automated grey-scale segmentation. The perimeter determined by this approach is shown superimposed on the PCSI image (A) and the corresponding bright-field image (B) taken at the end of the experiment for a colony growing from an agar inoculum (I) with a filter-paper bait (B). The difference between the first phase of symmetrical growth and the second phase of sparser, more asymmetric growth is also clearly visible in these images.

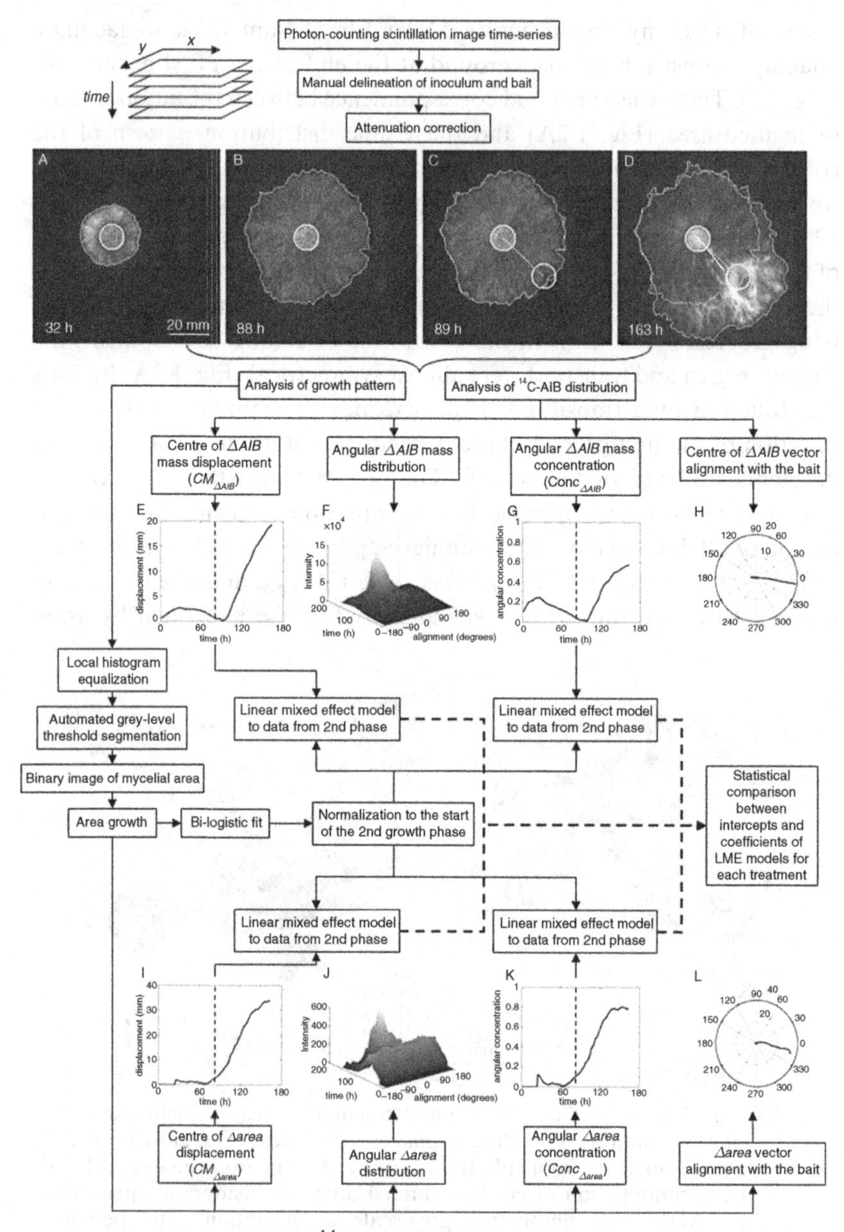

Fig. 1.3. Analysis of ^{14}C-AIB distribution and growth by using photon-counting scintillation imaging. Time-lapse photon-counting scintillation imaging (PCSI) was used to map the distribution of N in *Phanerochaete velutina* mycelium during growth from a central inoculum with or without the addition of either a glass-fibre or a filter-paper 'resource'. Representative PCSI images are shown (A–D) with the automatically

Linear Mixed Effects models (Pinheiro & Bates, 2000). Controls with no additions spontaneously switched to asymmetric growth and selectively allocated resources to broad sectors of the colony. Canalized flow patterns in cords also emerged with the transition to the second phase. Added damp cellulosic resources induced a change in internal nitrogen allocation, promoting marked N accumulation (Fig. 1.3E–G) and asymmetric growth (Fig. 1.3I–K) tightly focused on the new resource (Fig. 1.3H, L). The effect of a damp glass-fibre 'resource' was more variable, often with a transient response to perturbation that was not sustained in comparison with filter-paper resources.

Analysis of the pulsatile component

In addition to the evolution of the longer-term trends described above, a strong pulsatile component was found to be associated with rapid transport, particularly through corded systems (Tlalka *et al.*, 2002, 2003). We have previously analysed this pulsatile component by using Fourier techniques from discrete regions of interest manually defined on the image (Tlalka *et al.*, 2002, 2003). For example, results for three regions shown in Fig. 1.4A are given in Fig. 1.4B–E. This revealed that the assimilatory mycelium on the inoculum and the foraging mycelium both pulsed, but were out of phase with each other. To determine whether there were other

---

Fig. 1.3. (cont.)
segmented colony margin shown as a dotted line around the periphery. A filter-paper resource was added at 89 h (C) and resulted in localized increased transport of ^{14}C-AIB and localized growth (D). The magnitude of these effects was determined from analysis of the displacement of the centre of mass of AIB ($CM_{\Delta AIB}$, E) and displacement of the centre of mass of the colony area ($CM_{\Delta area}$, I), in which zero displacement represents symmetrical N distribution and growth. The displacement following resource addition ($CM_{\Delta AIB}$) is plotted as a solid line from the centre of the inoculum in (D). The total angular distribution in 12° sectors of AIB (F) and growth (J) indicates how tightly focused these changes were. The angular concentration ($Conc_{\Delta AIB}$ and $Conc_{\Delta area}$) provides a quantitative measure of these distributions (G, K), in which values of zero represent completely even distribution or growth and values approaching 1 indicate a very tightly focused distribution. The alignment between the centre of the mass displacement vector and the vector between the inoculum and bait provides a measure of the degree to which resource allocation is directed specifically towards the bait. A value of zero represents perfect alignment. Linear mixed effects models were fitted to the data for the $CM_{\Delta AIB}$, $CM_{\Delta area}$, $Conc_{\Delta AIB}$ and $Conc_{\Delta area}$ to allow statistical comparison between the control and the treatments.

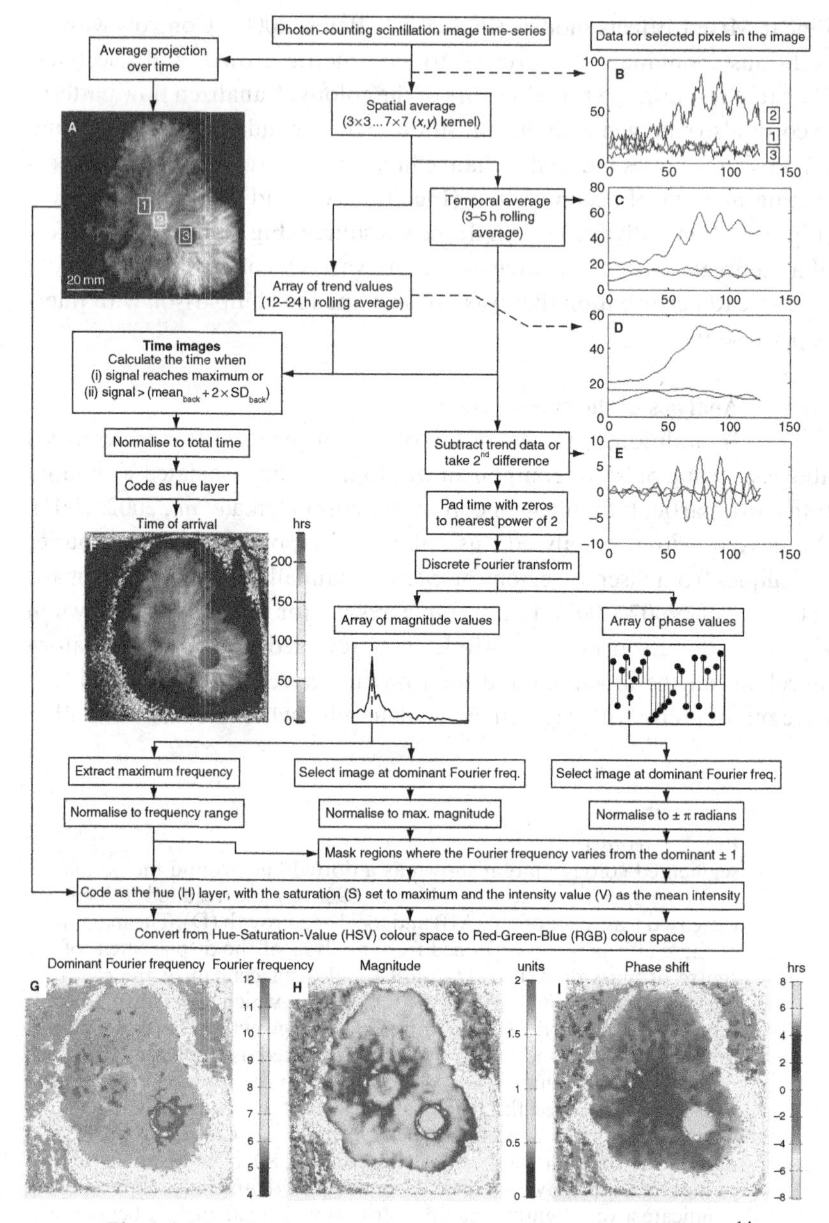

Fig. 1.4. Mapping frequency, amplitude and phase of the pulsatile ^{14}C-AIB transport by using Fourier analysis. Time-series of ^{14}C-AIB movement were recorded by PCSI, smoothed and detrended to provide a stationary time-series suitable for Fourier analysis. Data from three individual pixels shown in A are presented in B–E to illustrate the effect of these

phase domains with locally synchronized behaviour within the colony, we have extended the Fourier analysis to provide colour-coded pixel-by-pixel maps of frequency (Fig. 1.4G), amplitude (Fig. 1.4H) and phase (Fig. 1.4I) according to the scheme outlined in Fig. 1.4.

These maps revealed that the signals from the assimilatory hyphae in the inoculum and new resource, and from the foraging hyphae, all oscillate, but were out of phase with each other. The phase differences became established as distinct domains that were locally synchronized. In several colonies, the amplitude of the pulsing centre also shifted towards the bait.

Pulsatile behaviour in larger microcosms

We have now modified the PCSI approach to allow measurements from more realistic microcosms with wood-block inocula and sand or soil substrata overlaid with a translucent scintillation screen. With this system we can continuously image ^{14}C-AIB dynamics for extended periods (in excess of 6 weeks) and have observed a complex sequence of shifts in N distribution and transport priority throughout the network as it develops over time.

For example, Fig. 1.5 shows analysis of AIB transport in *Phanerochaete velutina* allowed to grow across compressed sand from a wood inoculum. Contact with a new wood resource after a few weeks triggered an increase in local branching and proliferation. After 8 weeks' growth, the microcosm was overlaid with a translucent scintillation screen and ^{14}C-AIB added to the initial inoculum. Within 1 h of loading, the ^{14}C-AIB had travelled 250 mm along one of the major cords (Fig. 1.5A). After 4 h, signal was present in most of the growing mycelium subtended by this cord (Fig. 1.5B). The signal from regions 1–3 along this cord (Fig. 1.5D) showed pronounced oscillations, superimposed on the longer-term trend, that continued for around 5–7 d (Fig. 1.5E). The overall level of signal decreased in the cords as the growing mycelial margin advanced out of this region. Not all the cords transported simultaneously. For example, the

Fig. 1.4. (cont.)
manipulations. The time when the signal was detected above background was extracted from the trend images and pseudo-colour-coded as a concise summary of the long-term translocation pattern of ^{14}C-AIB around the colony. The discrete Fast Fourier transform (FFT) was calculated pixel by pixel for the detrended oscillations. Results were displayed as pseudo-colour-coded images of the dominant Fourier frequency (G), the amplitude (H) and the phase (I) at this frequency (note only black and white versions are reproduced here).

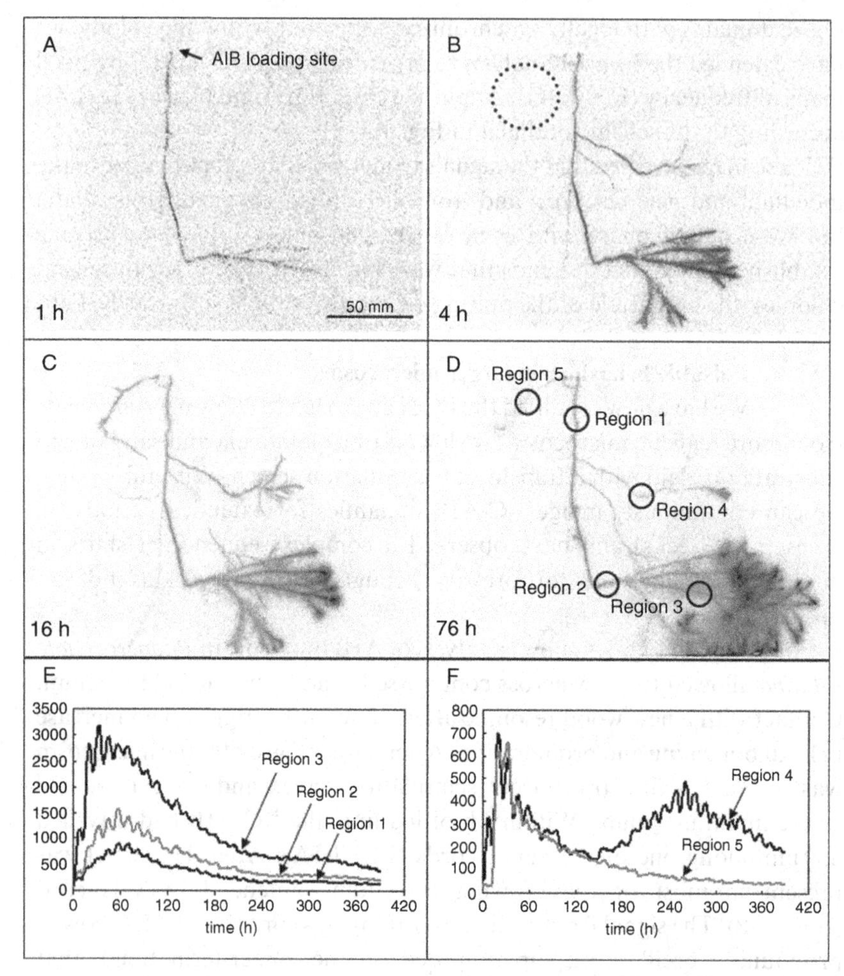

Fig. 1.5. Photon-counting scintillation imaging of ^{14}C-AIB transport in *Phanerochaete velutina* growing from a wood-block inoculum across a sand microcosm. ^{14}C-AIB was added at the wood inoculum site of a mycelium that had been growing on sand for 2 months. The movement of AIB was imaged for the subsequent 420 h. The initial 76 h are shown in the images. Rapid transport was visible 1 h after loading (A). Within 4 h, the signal reached the mycelial front and additional cords became labelled (B). At 16 h a pre-existing cord from the inoculum became transiently labelled (C) but signal had disappeared again by 76 h (D). The signals averaged over a number of regions of interest are shown in E and F. Synchronous pulsation was observed from regions 1, 2 and 3 on the main cord between the inoculum and the new resource (E). The filling and subsequent emptying of the cord in region 5 is shown in (F); the cord in region 4 shows two filling phases. The first is initiated at the start of the experiment and the second starts around 120 h.

pre-existing cord in the area highlighted by the dotted ellipse (Fig. 1.5B) showed no ^{14}C-AIB movement until around 12 h, then filled at a similar rate to the primary cord (region 5 in Fig. 5D, F), a process we term 'route-switching'. This cord appeared to act as a transport route only transiently, as the signal declined after around 30 h. Likewise, one of the other subsidiary cords showed two phases of transport, one initiated almost synchronously with the main cord and the second starting at around 150 h (region 4 in Fig. 1.5D, F).

Transport at the centimetre to metre scale

The dynamics of N-fluxes observed in these networks have prompted us to ask what routes are actually available for transport, how they are formed and how direction and switching are controlled. Different species exhibit different foraging strategies and we presume that these represent a balance between efficient resource capture, tolerance to damage or predation, and cost. In other areas of biology, networked systems such as ecological food webs or metabolic and genetic networks have been successfully analysed by using tools originally developed for graph theory (Albert & Barabási, 2001; Strogatz, 2001; Dorogovtsev & Mendes, 2002; Newman, 2003; Amaral & Ottino, 2004). To explore whether it is possible to apply network analysis tools to fungal mycelia, it is necessary to translate the morphological structures observed into a form appropriate for network modelling. Our starting assumption is that the fungal mycelium forms a (planar) spatial network that can be represented as a graph comprising a set of nodes (or vertices, V) connected by links (or edges, E).

As our first approximation we have focused on cords as the most convenient spatial scale, as these are readily identifiable discrete structures that represent the major transport pathways through the mycelium (Fig. 1.6A–C). Each branch point or junction is represented as a node and the persistent cords connecting them form the links (Fig. 1.6D–F). The number of links associated with any node is termed the degree of the node (k). Thus tips will have a degree of 1 as they are only connected to the previous node; branch points will typically have a degree of 3, because the growth processes forming the network tend usually to give a single branch or a single fusion (anastomosis) at each point. It is unlikely that there will be any loops where a link curls back round on itself to rejoin the same node, although multiple parallel links between two nodes are possible. As the fine structure of the mycelium within a food resource (agar or wood block) cannot be resolved, each of these is represented as a node with

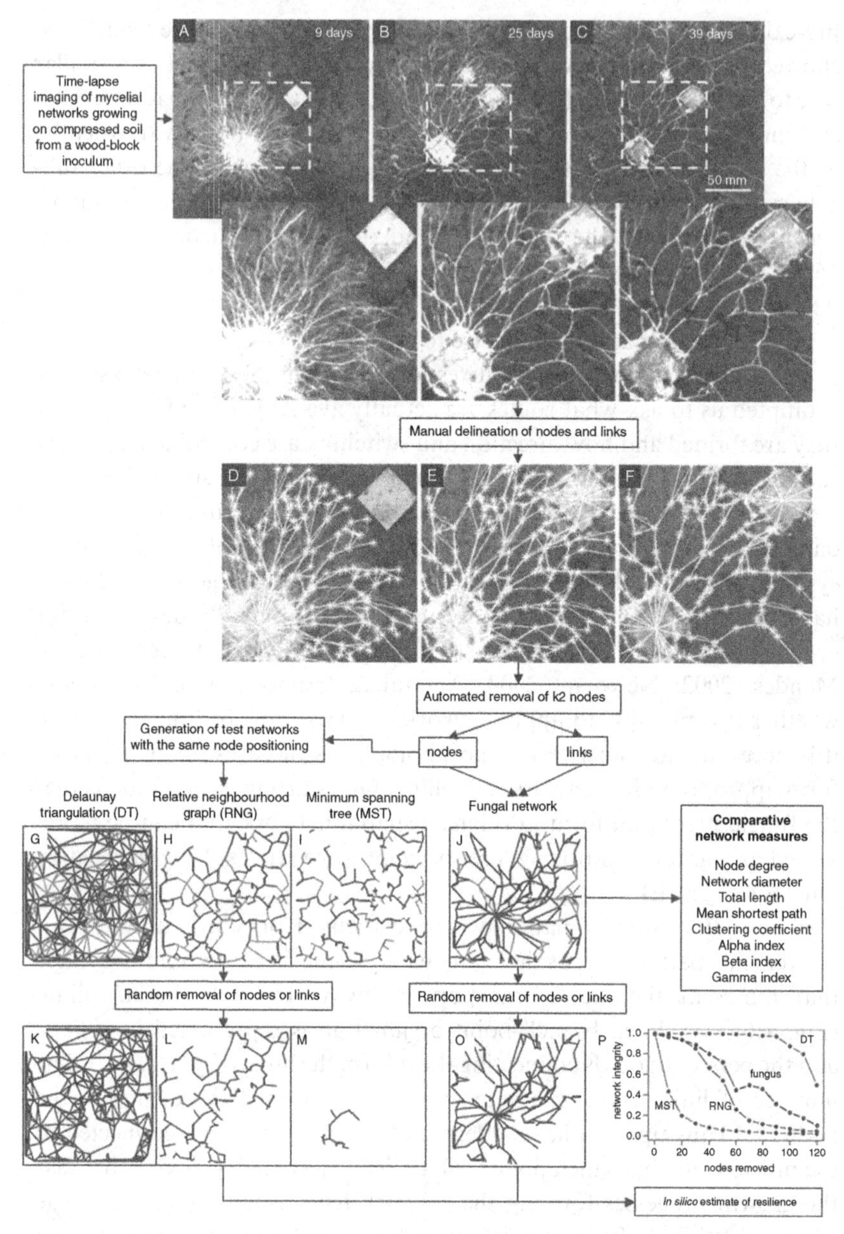

Fig. 1.6. Network analysis of corded mycelial development in *Phanerochaete velutina*. Bright-field images of *P. velutina* growing from a beech-block inoculum over compressed soil were collected at 9 (A), 25 (B) and 39 days (C). Branch points and anastomoses were manually coded as nodes connected by links (D–F) to construct a network graph (J). The experimental node positions were also connected

many links, resembling a hub in other network systems (Fig. 1.6J). A number of different quantities are typically measured for a network to try to understand its properties better, including:

1. the minimum path length that must be traversed between two nodes;
2. the maximum diameter of the network, in terms of both physical distance and the number of nodes traversed;
3. the degree distribution, which is given by the frequency of nodes with different numbers of links;
4. various indices derived from physical geography to describe both natural and artificial networks that estimate (i) the actual number of closed paths compared to the maximum possible number of closed paths (alpha index), (ii) the actual number of links compared with the number of nodes (beta index) and (iii) the actual number of links compared with the maximum possible number of links (gamma index);
5. the local clustering or transitivity, which measures the probability that if a node is connected to two other nodes, they will also be connected to each other;
6. the network resilience, which estimates the extent to which the network properties change as nodes or links are removed from the network.

These quantities can also be derived for model networks that are connected according to well-defined rules exemplifying different classes of network organization (Fig. 1.6G–I), which can then provide a useful reference point against which to test the behaviour of the real network (Strogatz, 2001; Dorogovtsev & Mendes, 2002; Newman, 2003; Amaral & Ottino, 2004). Most network analysis has so far focused on the network topology rather than the spatial relations of the nodes. However, it is clear that most networks have a spatial context and this must impose constraints on the probability of interconnection between different nodes in the

---

Fig. 1.6. (cont.)
by using different algorithms to give 'toy' networks for comparison, ranging from highly connected Delaunay triangulation (G), through the relative neighbourhood graph (H) to the minimum spanning tree (I). A range of network parameters were measured for the fungal network and the 'toy' networks. To assess the resilience of the network, nodes (120 in the illustration) were removed at random (K–O) and the network integrity measured as the fraction of the remaining nodes still connected to the initial inoculum (P).

network (Artzy-Randrup *et al.*, 2004; Gorman & Kulkarni, 2004). Thus, in a spatial network, nodes are likely to have a much higher probability of connecting to their physical neighbours and, depending on the way the network is built, a low or even zero (in 2-D planar networks) probability of links crossing over each other without forming a new node. This makes it difficult or impossible to create topological equivalents to 'short cuts' between physically remote parts of the network. Equally, it is interesting to speculate that inclusion of weighting by transport speed and/or capacity may have the equivalent effect of long-range communication, bringing distant parts of the network into closer contact than expected from their spatial separation or unweighted path length.

In *P. velutina* grown in a microcosm 24 cm square, the size of the corded experimental networks is around 300–500 main nodes, depending on growth conditions. If the finest hyphae readily discernible are included, the number of nodes increases to around 3000. It might be appropriate to consider the links to be directed on the basis of their initial growth direction. In practice, the physiological direction of nutrient fluxes is more important and does not have to follow the developmental connection sequence. Unfortunately, however, we cannot predict *a priori* in which direction the flux may move; indeed, we expect it to vary depending on the source–sink relations within the network. In the future, the techniques to map fluxes described above may provide this information, but at this stage it is simpler to assume that links are bidirectional.

The network architecture was not static, but continuously evolved. For example, growth from the inoculum to connect to a new food source progressed through an initial proliferation phase with many links forming (Fig. 1.6A), followed by selection and reinforcement of a sub-set of paths to create a more limited number of strong links (Fig. 1.6B), eventually with regression of the remainder of the links to leave a sparser network (Fig. 1.6C). Nodes were (manually) assigned to each tip, branch or fusion during colony development to try to understand the network topology that developed (Fig. 1.6D–F). In the early stages of growth a substantial number of cords were forming. However, there was also a considerable number of fine foraging hyphae that could not be resolved clearly. By 18 d, the mycelium had contacted the new food resource and established more obvious cords, and much of the fine mycelium had receded. By 39 d, the number of interconnected cords had reduced further, although the history of the previous connections remained as a series of nodes with degree 2 left on the main connecting cords. If these were excluded from the analysis, the average degree for each node stabilized at around 3.5.

Although efficient transport is likely to shape fungal mycelial networks, resilience to accidental damage and predation is also likely to be important in determining the network architecture. Resilience was assessed *in silico* by measuring how the network properties changed as individual nodes or links were removed at random or in a targeted manner (Fig. 1.6O). The same approach was used on the model networks (Fig. 1.6K–M) to provide a basis for comparison (Fig. 1.6P). In a real mycelial network, the probability of node or edge removal is unlikely to be random and may also show a high degree of correlation between adjacent nodes. For example, grazing by soil invertebrates may be in specific locations in the network (Tordoff *et al.*, 2006). This is because some regions are more palatable than others. Part of the resilience of such a biological network may not be just the architecture of the network prior to damage, but the ease and efficiency with which the network can reconnect itself. In this respect, a self-organizing spatial network may have considerable advantages over a random network in the cost, consistency and efficacy of the rewiring process needed to re-establish a functioning system.

Conclusions

The imaging approaches described here appear to provide a rich source of new information on the dynamic behaviour of nutrient translocation in saprotrophic basidiomycetes. So far most of our results have been derived for *P. velutina*. We are now expanding the range of species examined to determine whether these results are of general significance. Despite the exciting progress made, we still do not have techniques that can easily span all the length scales of interest. For example, we do not yet have a clear anatomical description of how the transport pathways map onto the individual hyphae in a differentiating corded system. These structures have so far proved difficult to investigate, even by using confocal microscopy, when they are growing in their natural dry state, but might be susceptible to serial EM sectioning and 3-D reconstruction. We are also only just beginning to develop analysis routines that can accommodate the immense plasticity of a mycelial network, which makes quantitative comparisons between notionally replicate experiments quite difficult. Despite these challenges, simply being able to visualize the beauty and sophistication of solute translocation has already proved very rewarding.

Acknowledgements

Research in the authors' laboratories has been supported by BBSRC (43/P19284), NERC (GR3/12946 & NER/A/S/2002/882),

EPSRC (GR/S63090/01), EU Framework 6 (STREP No. 12999), Oxford University Research Infrastructure Fund and the University Dunston Bequest.

References

Albert, R. & Barabási, A.-L. (2001). Statistical mechanics of complex networks. *Reviews of Modern Physics* **74**, 47–97.

Amaral, L. A. N. & Ottino, J. M. (2004). Complex networks: augmenting the framework for the study of complex systems. *European Physical Journal* **B38**, 147–62.

Artzy-Randrup, Y., Fleishman, S. J., Ben-Tal, N. & Stone, L. (2004). Comment on "Network motifs: simple building blocks of complex networks" and "superfamilies of evolved and designed networks". *Science* **305**, 1107c.

Ashford, A. E. & Allaway, W. G. (2002). The role of the motile tubular vacuole system in mycorrhizal fungi. *Plant and Soil* **244**, 177–87.

Boddy, L. (1999). Saprotrophic cord-forming fungi: meeting the challenge of heterogeneous environments. *Mycologia* **91**, 13–32.

Boddy, L. & Watkinson, S. C. (1995). Wood decomposition, higher fungi, and their role in nutrient redistribution. *Canadian Journal of Botany* **73** (suppl.1), S1377–S1383.

Cairney, J. W. G. (2005). Basidiomycete mycelia in forest soils: dimensions, dynamics and roles in nutrient distribution. *Mycological Research* **109**, 7–20.

Darrah, P. R., Tlalka, M., Ashford, A., Watkinson, S. C. & Fricker, M. D. (2006). The vacuole system is a significant intracellular pathway for longitudinal solute transport in basidiomycete fungi. *Eukaryotic Cell* **5**, 1111–25.

deBeer, D., Stoodley, P. & Lewandowski, Z. (1997). Measurement of local diffusion coefficients in biofilms by microinjection and confocal microscopy. *Biotechnology and Bioengineering* **53**, 151–8.

Dorogovtsev, S. N. & Mendes, J. F. F. (2002). Evolution of networks. *Advances in Physics* **51**, 1079–87.

Gorman, S. P. & Kulkarni, R. (2004). Spatial small worlds: New geographic patterns for an information economy. *Environmental Planning* **B31**, 273–96.

Newman, M. E. J. (2003). The structure and function of complex networks. *SIAM Review* **45**, 167–256.

Olsson, S. (1999). Nutrient translocation and electrical signalling in mycelia. In *The Growing Fungus*, ed. N. A. R. Gow & G. M. Gadd, pp. 25–48. London: Chapman and Hall.

Otsu, N. (1979). A threshold selection method from gray-level histograms. *IEEE Trans. SMC* **9**, 62–6.

Pinheiro, J. & Bates, D. M. (2000). *Mixed Effects Models in S and S-PLUS*. New York: Springer-Verlag.

Rayner, A. D. M., Griffith, G. S. & Ainsworth, A. M. (1994). Mycelial interconnectedness. In *The Growing Fungus*, ed. N. A. R. Gow & G. M. Gadd, pp. 21–40. London: Chapman and Hall.

Rayner, A. D. M., Watkins, Z. R. & Beeching, J. R. (1999). Self-integration – an emerging concept from the fungal mycelium. In *The Fungal Colony*, ed. N. A. R. Gow, G. D. Robson & G. M. Gadd, pp. 1–24. Cambridge: Cambridge University Press.

Rees, B., Shepherd, V. A. & Ashford, A. E. (1994). Presence of a motile tubular vacuole system in different phyla of fungi. *Mycological Research* **98**, 985–92.

Smith, S. E. & Read, D. J. (1997). *Mycorrhizal Symbiosis*. London: Academic Press.

Strogatz, S. H. (2001). Exploring complex networks. *Nature* **410**, 268–76.

Tlalka, M., Watkinson, S. C., Darrah, P. R. & Fricker, M. D. (2002). Continuous imaging of amino acid translocation in intact mycelia of *Phanerochaete velutina* reveals rapid, pulsatile fluxes. *New Phytologist* **153**, 173–84.

Tlalka, M., Hensman, D., Darrah, P. R., Watkinson, S. C. & Fricker, M. D. (2003). Noncircadian oscillations in amino acid transport have complementary profiles in assimilatory and foraging hyphae of *Phanerochaete velutina*. *New Phytologist* **158**, 325–35.

Tordoff, G. M., Jones, T. H. & Boddy, L. (2006). Grazing by *Folsomia candida* (Collembola) affects the mycelial morphology of the cord-forming basidiomycetes *Hypholoma fasciculare*, *Phanerochaete velutina* and *Resinicium bicolor* differently during early outgrowth onto soil. *Mycological Research.* Available online at www.doi:10.1016/j-mycres.2005.11.012.

Uetake, Y., Kojima, T., Ezawa, T. & Saito, M. (2002). Extensive tubular vacuole system in an arbuscular mycorrhizal fungus, *Gigaspora margarita*. *New Phytologist* **4**, 761–8.

Watkinson, S. C. (1999). Metabolism and hyphal differentiation in large basidiomycete colonies. In *The Fungal Colony*, ed. N. A. R. Gow, G. D. Robson & G. M. Gadd, pp. 126–56. Cambridge: Cambridge University Press.

Watkinson, S. C., Bebber, D., Darrah, P. R., Fricker, M., Tlalka, M. & Boddy, L. (2006). The role of wood decay fungi in the carbon and nitrogen dynamics of the forest floor. In *Fungi in Biogeochemical Cycles*, ed. G. M. Gadd, pp. 151–81. Cambridge: Cambridge University Press.

2

Natural history of the fungal hypha: how Woronin bodies support a multicellular lifestyle

GREGORY JEDD

Temasek Life Sciences Laboratory, National University of Singapore

Introduction

Fungal evolution and cell biology

Fungi are one of three major clades of eukaryotic life that independently evolved multicellular organization. They have radiated into a large variety of terrestrial and aquatic niches, employing strategies ranging from symbiotic to saprobic to pathogenic, and are remarkable for their developmental diversity and ecological ubiquity, with the number of species estimated to exceed one million (Hawksworth *et al.*, 1995).

The fungi are highly varied in their mode of growth, ranging from unicellular yeasts to multicellular hyphal forms that produce complex fruiting bodies (Hawksworth *et al.*, 1995). Hyphae grow through polarized tip-extension of a tubular cell (hypha), which can be partitioned by the formation of cross-walls called septa. Phylogenetic analysis reveals four major groups of fungi: the early-diverging Chytridiomycota and Zygomycota, and the Ascomycota and Basidiomycota (Fig. 2.1) (Berbee & Taylor, 2001; Lutzoni *et al.*, 2004), which are sister clades that evolved more recently and contain the majority of fungal species (Bruns *et al.*, 1992; Hawksworth *et al.*, 1995). Hyphae are the predominant mode of vegetative cellular organization in the fungi and groups of fungi can be defined based on consistent differences in hyphal structure. The Zygomycota and Chytridiomycota can produce septa but these are infrequent in vegetative hyphae. In contrast, vegetative hyphae in the Ascomycota produce perforate septa at regular intervals and this is also found in the Basidiomycota, suggesting that this trait was present in their common ancestor (Fig. 2.1) (Berbee & Taylor, 2001).

As hyphae grow they branch and fuse, eventually forming a multicellular network of interconnected cells (Glass *et al.*, 2004). This hyphal

Fungi in the Environment, ed. G. M. Gadd, S. C. Watkinson & P. S. Dyer. Published by Cambridge University Press.

syncytium (mycelium) allows intercellular communication and trafficking of organelles and solutes within the colony, probably facilitating invasive growth and the production of multicellular reproductive structures. Indeed, protoplasmic streaming is readily observed and probably occurs to varying extents in all filamentous fungi. The eminent mycologist A. H. R. Buller described protoplasmic streaming in the Zygomycota, Basidiomycota and Ascomycota. He wrote that 'during the growth of fruit bodies, the transference of the food-materials from the vegetative mycelium into the fruit body is largely carried out by protoplasmic streaming' and concluded that 'the translocation of protoplasm from cell to cell in the mycelium by the process of streaming results in great advantage to the species in which it occurs' (Buller, 1933a).

The fungal mycelium does not readily conform to a simple cellular concept. Adjacent septa allow a convenient definition of an individual cell. However, protoplasmic streaming can rapidly replace the contents of a given hyphal compartment, including its nuclear component. Thus,

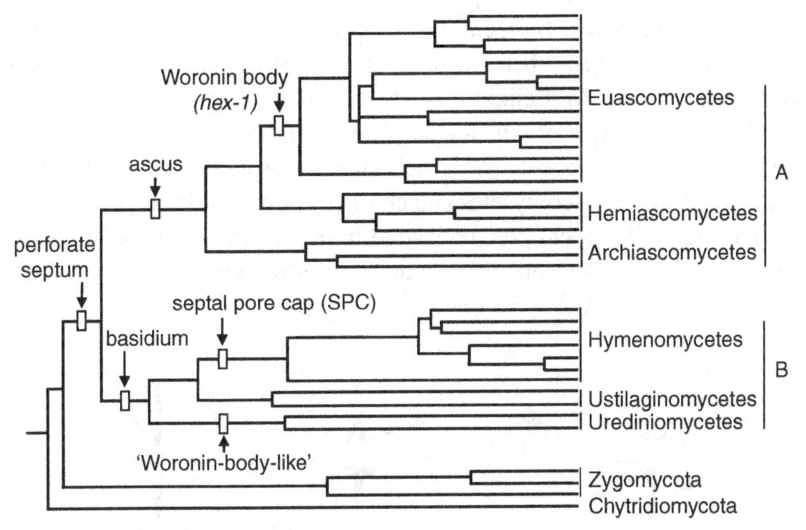

Fig. 2.1. Fungal phylogeny with selected characters indicated. The phylogeny is based on 18*S* ribosomal RNA (modified from Berbee & Taylor, 2001). A, Ascomycota; and B, Basidiomycota. The phylogenetic distribution of Woronin bodies, 'Woronin-body-like' organelles and septal pore caps is assumed to result from single evolutionary origins. The inferred phylogenetic origin of Woronin bodies in the ancestral Euascomycete is thus based on the presence of the *hex-1* gene in all Euascomycete genomes examined to date (including the earliest branching lineage[s]), whereas it is absent in other known fungal genomes (see Table 2.1). The figure is presented as a working model.

the fungal colony can be thought of as a mass of protoplasm that migrates through a growing, interconnected system of channels. The early-diverging fungi (e.g. Zygomycota) grow well in the absence of vegetative septa; what then is the benefit of the perforate septum? Septal pores confer the advantage of protoplasmic streaming and intercellular continuity but are also sufficiently small to be rapidly closed. Thus, the syncytium can 'cellularize' in response to hyphal damage, stress or old age, and during cellular differentiation. Several mechanisms exist to close the septal pore; one of these is described in detail below. Interestingly, the fungi with the most prominent and complex septal-pore-associated organelles, the Hymenomycetes and Euascomycetes (Fig. 2.1), also produce the largest and most complex multicellular fruiting bodies (Alexopolous *et al.*, 1996), suggesting that these organelles support complex multicellular organization.

Septal-pore-associated organelles

Septal pores possess a variety of associated structures that can be observed by electron microscopy. In this discussion, I focus on organelles that are circumscribed by membranes. For information on septal-pore-associated structures, their variation and phylogenetic utility, see Kimbrough (1994), McLaughlin *et al.* (1995) and Lutzoni *et al.* (2004). Septal-pore-associated organelles are found within the immediate vicinity of the septal pore, on both sides of the septum. The Woronin body (WB) is diagnostic of the Euascomycetes and characterized by a dense core, which is surrounded by a closely associated unit membrane (Markham & Collinge, 1987). Electron microscopy revealed an association with

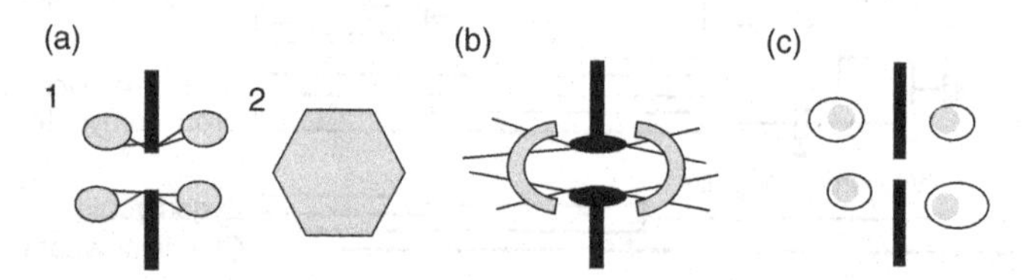

Fig. 2.2. Septal-pore-associated organelles. The structure of these organelles is shown in simplified form. The septum and its perforation are shown with a thick black line. The osmophilic core of these organelles is shown in grey and their enclosing membrane is shown as a black line. (a) The Woronin body of the Euascomycetes: (1) the oval form and (2) the hexagonal form. Tethering filaments are shown as thin black lines. (b) The septal pore cap of Hymenomycetes. For simplicity, perforations of the SPC are not shown. Associated filaments are shown as thin black lines. (c) 'Woronin-body-like' organelles found in the Urediniomycetes and the genus *Neolecta*.

microbodies (Wergin, 1973; Camp, 1977) (hereafter referred to as peroxisomes) and recent work has confirmed that Woronin bodies are peroxisome-derived (see below).

Septal-pore-associated organelles have also evolved in the Basidiomycota. Hymenomycetes produce a septum with a septal pore that is surrounded by a barrel-shaped swelling of the cell wall. This structure is associated with an organelle known as the septal pore cap (SPC) (Muller *et al.*, 1998) or parenthosome (Bracker, 1967) (Fig. 2.2). This organelle sits like a cap on the pore apparatus and consists of an electron-dense core associated with a unit membrane. Electron microscopy reveals continuity between the membrane of the SPC and the nuclear envelope (Moore, 1975; Muller *et al.*, 1998), suggesting that the SPC is a specialized domain of the endoplasmic reticulum. The SPC is also associated with cytoplasmic filaments and an osmophilic material that appears to occlude the septal pore (Moore & Marchant, 1972).

In contrast, Urediniomycetes elaborate perforate septa with associated organelles that have the appearance of peroxisomes (Bauer & Oberwinkler, 1994; Swann *et al.*, 2001). These sometimes contain electron-dense polyhedral structures in their lumen, which superficially resemble the Woronin body (Oberwinkler & Bauer, 1989, 1990) (Fig. 2.2). However, in some cases, their core shows fine striations (Bauer & Oberwinkler, 1994). This is unlike the Euascomycete Woronin body (Markham & Collinge, 1987) and suggests that the core composition of these organelles is distinct from that of Woronin bodies.

A brief history of Woronin body research

The nineteenth century

The discovery of and early work on the Woronin body is described in A. H. Reginald Buller's seminal work, *Researches on Fungi:*

> In 1886, Woronin discovered in the cells of *Ascobolus* (= *Lasiobolus*) *pulcherrimus* certain highly refractive particles a few of which can usually be seen on one side or both sides of each septum. (Buller, 1933b)

Buller further noted that similar structures had been reported in other fungi and that because of their definitive structure they needed a special name. He proposed to call them 'Woronin bodies' (Buller, 1933c). In 1900, Charlotte Ternetz further defined basic features of Woronin body cell biology (Ternetz, 1900):

> they appear in the apical cells of hyphae, irregularly dispersed in the protoplasm, and there they move every now and then here and there, and then they settle down against a longitudinal wall or a septum. (Buller, 1933b)

Today it is known that apical Woronin bodies are intermediates in biogenesis and that their formation depends on apically localized gene expression (see below).

Buller also conducted experiments on 'pore plugs and their formation under experimental conditions' and succinctly defined the problem:

> each septum in a mycelium is provided with a small, central, circular, open pore through which the protoplasm can pass, and often does pass, with the greatest ease. The acceptance of these conclusions suggest the following questions: when a cell in a living hyphae dies naturally or is killed by manipulation, does protoplasm flow through the pores of the septa from the adjacent living cells into the dead cell; and if not, what prevents it from doing so? (Buller, 1933d)

He broke hyphae using a drawn needle and found

> (1) that the two septa separating the dead cell from the two adjacent cells, which previously were plane, became convexo-concave with the convex side directed toward the lumen of the dead cell; and (2) that *the pores of the two septa became blocked by a plug* in consequence of which the *escape of living protoplasm from adjacent living cells into the dead cells is prevented.* (Buller, 1933a)

Thus, Buller first observed damage-induced septal-pore plugging but, despite the refractive nature of the plugging material, did not associate this effect with the Woronin body. Rather he concluded that

> The substance of which a plug is composed appears homogeneous, colourless, and highly refractive and may well be nothing more than a coagulum of protoplasmic origin. (Buller, 1933d)

It was roughly another 50 years before this problem was re-investigated by Collinge and colleagues, who first established the phenomenon of septal-pore plugging by Woronin bodies.

The twentieth century

Advances were mediated by the advent of the electron microscope and the careful observation and manipulation of fungal hyphae. Severing hyphae with a razor blade provided a means of inducing damage. In *Penicillium chrysogenum*, Woronin bodies rapidly plugged 90% of septal pores within the vicinity of induced damage; by contrast, only 5% were plugged in undamaged hyphae (Collinge & Markham, 1985). A similar response was observed in *Neurospora crassa*, which produces unusually large, hexagonal Woronin bodies (Trinci & Collinge, 1973). These results showed that Woronin bodies moved to septal pores adjacent to damage sites and were probably functioning to limit cellular damage. By using electron microscopy it was further shown that material resembling the septum wall is deposited over the Woronin body septal-pore complex (Markham & Collinge, 1987). This suggested that membrane re-sealing and the local deposition of new cell wall follow septal pore occlusion.

Electron microscopy showed that the Woronin body is generally oval in shape and centred on an electron-dense matrix that is circumscribed by a single membrane (Markham & Collinge, 1987). Digestion with hydrolytic enzymes showed the matrix was composed of protein (McKeen, 1971; Hoch & Maxwell, 1974). Woronin body diameter was observed to vary between 100 nm and >1 μm and was generally greater than septal pore diameter (Markham & Collinge, 1987). In most Euascomycetes, the Woronin body is oval, but in some species it presents a hexagonal appearance (Markham and Collinge, 1987). Interestingly, in a study of Woronin body development in *Fusarium oxysporum*, an early hexagonal form appeared to mature into an oval form (Wergin, 1973). Hexagonal forms have also been observed in other Euascomycetes (Markham & Collinge, 1987; Kimbrough, 1994) and together these observations suggested that hexagonal and oval Woronin bodies might share a common structural core. This hypothesis has been supported by further research (see below).

Woronin bodies were physically drawn away from the septum by using laser-induced optical light traps. Upon release they returned to their original position, suggesting the existence of a tethering mechanism (Berns *et al.*, 1992). An associated filamentous material can be seen by electron microscopy (Momany *et al.*, 2002) and this may physically attach the Woronin body to the septal pore (Fig. 2.2). The composition of this material is currently unknown.

Current research

Woronin body assembly and function

Woronin bodies have been purified from *Neurospora crassa* in independent experiments (Jedd & Chua, 2000; Tenney *et al.*, 2000). This allowed the identification of HEX-1, a key structural protein that defined a family of Euascomycete-specific proteins (Jedd & Chua, 2000; Tenney *et al.*, 2000). Antibodies to HEX-1 decorated the matrix of the Woronin body in *Neurospora crassa* (Jedd & Chua, 2000), *Aspergillus nidulans* (Momany *et al.*, 2002) and *Magnaporthe grisea* (Soundararajan *et al.*, 2004) and helped show that HEX-1 exists in an extremely large and stable protein complex (Soundararajan *et al.*, 2004). The gene *hex-1* encodes a consensus peroxisome-targeting signal (PTS-1) and its expression in yeast generates intraperoxisomal protein assemblies that are morphologically similar to the native Woronin body (Jedd & Chua, 2000). In addition, recombinant HEX-1 spontaneously crystallizes *in vitro*, suggesting that HEX-1 is a self-assembling structural protein (Jedd & Chua, 2000). A *Neurospora hex-1* mutant is devoid of visible Woronin bodies and mutant hyphae bleed protoplasm through septal pores following cellular damage. Together, these results defined HEX-1 as a key structural determinant of the Woronin body core and showed that Woronin bodies are specialized peroxisomes that function in the maintenance of cellular integrity (Jedd & Chua, 2000; Tenney *et al.*, 2000).

hex-1 Mutants have been obtained in other Euascomycetes, including *Magnaporthe grisea* (Asiegbu *et al.*, 2004; Soundararajan *et al.*, 2004), *Aspergillus oryzae* (K. Kitamoto, personal communication) and *Aspergillus nidulans* (G. Jedd, unpublished). All of these mutants are defective in septal-pore sealing, suggesting that this function is conserved. In addition, the *Magnaporthe* mutant is defective in appressorium morphogenesis and invasive growth within the plant host (Soundararajan *et al.*, 2004). Interestingly, it displays hyphal death in response to nitrogen starvation. Nitrogen starvation and growth *in planta* also appear to regulate the production of HEX-1 splice-variants, favouring the use of a particular 3′ splice acceptor sequence (Soundararajan *et al.*, 2004). These results show that the Woronin body provides *Magnaporthe* with an important defence system, and further suggest that nitrogen starvation may regulate this system through alternative splicing (Soundararajan *et al.*, 2004). The function of this regulation remains to be determined. Woronin bodies are likely to occur in all Euascomycete pathogens (Table 2.1) and thus may be attractive targets for the development of fungicides.

Crystallinity of the Woronin body core

The spontaneous crystallization of HEX-1 suggested that the crystal structure would be a good model of the native Woronin body

Table 2.1. *The hex-1 gene is specific to the Euascomycete*

Symbols + and – indicate presence or absence of the *hex-1* gene, respectively. This was determined (1) by tblastn searches of genome and EST databases by using *N. crassa* HEX-1 protein (score is given in parentheses), (2) based on authors' unpublished results from cloning, or (3) by low-stringency Southern blotting. n.d., No data. Euascomycetes that are plant and human pathogens are shown in bold.

Phylum	Group	Species	*hex-1*
Ascomycota	Euascomycetes	*Ascobolus stercorarius* (3)	+
		Aspergillus fumigatus (1)	+ (7e-44)
		Aspergillus nidulans (1)	+ (1e-49)
		Aspergillus oryzae (1)	+ (1e-51)
		Botrytis cinerea (1)	+ (4e-64)
		Capronia pilosella (2)	+
		Chaetomium globosum (1)	+ (2e-68)
		Coccidioides immitis (1)	+ (1e-54)
		Fusarium graminearum (1)	+ (1e-62)
		Magnaporthe grisea (1)	+ (5e-60)
		Morchella esculenta (3)	+
		Neurospora crassa (1)	+
		Ophiostoma floccosum (1)	+ (2e-63)
		Penicillium chrysogenum (2)	+
		Penicillium roqueforti (2)	+
		Podospora anserina (1)	+ (2e-72)
		Trichoderma reesei (1)	+ (4e-64)
		Tuber borchii (1)	+ (2e-24)
	Hemiascomycetes	*Ashbya gossypii* (1)	–
		Candida albicans (1)	–
		Saccharomyces cerevisiae (1)	–
	Archiascomycetes	*Schizosaccharomyces pombe* (1)	–
		Neolecta irregularis (3)	–?
Basidiomycota	Hymenomycetes	*Coprinus cinereus* (1)	–
		Cryptococcus neoformans (1)	–
		Phanerochaete chrysosporium (1)	–
	Urediniomycetes		n.d.
	Ustilaginomycetes	*Ustilago maydis* (1)	–
Zygomycota	Mucorales	*Rhizopus oryzae* (1)	–
Chytridiomycota	Chytridiales		n.d.

core (Jedd & Chua, 2000); the HEX-1 structure was subsequently solved at a resolution of 1.8 Å (Yuan *et al.*, 2003). This revealed a monomer with a two-domain structure consisting of mutually perpendicular antiparallel B-barrels (Yuan *et al.*, 2003). The crystal structure further revealed that HEX-1 monomers are associated through three intermolecular contacts that produce a lattice structure with the overall form of cross-linked helical filaments (Yuan *et al.*, 2003). Amino-acid residues that mediate crystal lattice formation are conserved in HEX-1 orthologs (Yuan *et al.*, 2003), suggesting that lattice formation is an evolutionarily conserved feature of these proteins.

Mutations that disrupt crystal contact residues provided a direct test of the lattice model of Woronin body structure. These abolished the ability of HEX-1 to self-assemble *in vitro* and produced aberrant Woronin bodies that possess a soluble non-crystalline core *in vivo*. The assembly-defective protein accumulated in peroxisomes and produced a spherical, osmophilic core, suggesting that the process of HEX-1 import and concentration in the Woronin body lumen is unaffected by defects in crystallization. However, these vesicles provided no Woronin-body function, as assessed by their inability to complement loss-of-function phenotypes, e.g. protoplasmic bleeding (Yuan *et al.*, 2003).

Together, these experiments showed that the HEX-1 crystal lattice is essential for Woronin body function (Yuan *et al.*, 2003). Why do these vesicles require a solid core? In a household sink, drain plugs require sufficient structural integrity to resist the pressure exerted by a column of water. Woronin bodies probably evolved a solid core for a similar reason. Returning to Buller's observation of plugged hyphal compartments, he noted that

> the two septa separating the dead cell from the two adjacent cells, which previously were plane, became convexo-concave with the convex side directed toward the lumen of the dead cell.
>
> (Buller, 1933a)

The deformation of the septum described by Buller suggests an intracellular turgor pressure, and subsequent observations further support the presence of such a force in fungal hyphae (Money & Harold, 1992; Bartnicki-Garcia *et al.*, 2000). Because this pressure needs to be resisted in the course of septal pore plugging, only a structure with a dense-core, like the Woronin body, can execute such a function.

Woronin body biogenesis in apical hyphal compartments

Woronin bodies were observed in the apical hyphal compartment (defined by the hyphal tip and first septum) by light microscopy in 1900 (Ternetz, 1900; Buller, 1933c) and later by electron microscopy in a variety of Euascomycetes, where they occupy a region of cytoplasm that underlies a cluster of tip-associated vesicles known as the *Spitzenkörper* (Brenner & Carrol, 1968; McClure *et al.*, 1968; Collinge & Markham, 1982; Momany *et al.*, 2002). Apical WBs move within the cytoplasm (Ternetz, 1900); by contrast, in sub-apical compartments (defined by two adjacent septa), the WB is immobile and localized near the septal pore or docked at the cell periphery (Markham & Collinge, 1987). In *Aspergillus* germlings that have undergone septation, fewer Woronin bodies are found at the hyphal tip, suggesting a process of retrograde transport (Momany *et al.*, 2002). Together, these observations suggest that Woronin body biogenesis is programmed to occur in the apical hyphal compartment.

Time-lapse confocal microscopy allowed the direct observation of Woronin body formation and confirmed that this process is largely tip-cell localized (Tey *et al.*, 2005). To investigate the genetic control of Woronin body formation, a fluorescent reporter protein was expressed from *hex-1* regulatory sequences. This revealed a fluorescent gradient that was maximal in apical cells. Moreover, endogenous *hex-1* transcripts were specifically enriched at the leading edge of the fungal colony, whereas other transcripts were shown to accumulate in the colony interior (Tey *et al.*, 2005). These experiments demonstrated the apical programming of gene expression and suggested that localized *hex-1* mRNA transcripts may be responsible for the apical formation of the Woronin body core. To test this model, the *hex-1* structural gene was expressed from the regulatory sequences of a transcript that normally accumulates in the interior of the fungal colony. Under these conditions, Woronin body formation was re-directed to the colony interior and these strains displayed loss-of-function phenotypes specifically in apical compartments. Thus, the local accumulation of transcripts appears to determine the site of Woronin body formation (Tey *et al.*, 2005). In this case the apical programming of gene expression ensures that mature Woronin bodies are present throughout the hypha and that apex-proximal septal pores can be efficiently sealed (Tey *et al.*, 2005).

Phylogenetic distribution and variation

The *hex-1* gene has been identified in 17 Euascomycetes, including all seven for which genome sequences are available (Table 2.1). *hex-1* has also been identified in Pezizomycetes, which are the earliest diverging Euascomycetes (Berbee & Taylor, 2001). These include an exposed sequence tag (EST) from *Tuber borchii*, and homologous sequences have been observed in *Ascobolus stercorarius* and *Morchella esculenta* by using low-stringency Southern blotting (G. Jedd & D. Hewitt, unpublished observation). In all cases where sequence is available, it encodes a consensus PTS-1 and largely conserves crystal contact residues. In contrast, *hex-1* homologues are not found in nine fully sequenced fungal genomes outside the Euascomycetes. These include yeast and filamentous Hemiascomycetes (*Saccharomyces cerevisiae* and *Ashbya gossypii*, respectively), an Archiascomycete (*Schizosaccharomyces pombe*), filamentous Hymenomyctes (*Coprinus cinereus, Phanerochaete chrysosporium*), an Ustilaginomycete (*Ustilago maydis*) and a Zygomycte (*Rhizopus oryzae*). This distribution suggests that *hex-1* arose in the ancestral Euascomycete (Fig. 2.1). An exception, however, may be found in the Archiascomycete *Neolecta.* Most Archiascomycetes grow as yeast but some are hyphal (e.g. *Taphrina* and *Neolecta*). *Neolecta* has perforate septa, fruiting bodies, forcibly discharged asci and Woronin-body-like organelles (Landvik *et al.*, 2003) and thus possesses many characters typical of the Euascomycetes. Low-stringency Southern blotting has failed to detect a *hex-1* homologue in *Neolecta* (G. Jedd & D. Hewitt, unpublished), suggesting that the Woronin-body-like organelle of *Neolecta* has a distinct evolutionary origin. However, it is possible that a highly divergent *hex-1* homologue may not be detected by this technique. Organelle purification and cloning of genes encoding resident proteins will be required to resolve the origin of the Woronin-body-like organelle in *Neolecta.* If a *hex-1* homologue is indeed present in *Neolecta,* then Woronin bodies may pre-date the ancestral Euascomycete. Alternatively, the WB-like organelle in *Neolecta* may represent a convergent, but evolutionarily independent, solution to septal-pore plugging.

Most Euascomycetes produce two forms of the HEX-1 protein through alternative splicing between the first and second exon. This results in the production of two forms of HEX-1 that differ at the amino-terminus by several kilodaltons. These two forms of HEX-1 have been documented in mycelium from *Magnaporthe grisea* (Soundararajan *et al.*, 2004), *Aspergillus oryzae* (K. Kitamoto, personal communication), *Trichoderma reesei* (Lim *et al.*, 2001; Curach *et al.*, 2004) and *Aspergillus nidulans* and *Penicillium chrysogenum* (G. Jedd, unpublished). All of these organisms produce a small, oval Woronin body, compared with the large, hexagonal form produced by a single version of HEX-1 in *Neurospora crassa* (Fig. 2.2) (Jedd & Chua, 2000). This suggests that a co-complex of HEX-1 isoforms may determine the oval morphology. In one model, the co-assembly of these isoforms may perturb monomer packing in the crystal lattice and through this mechanism alter the shape of the resultant assembly. This model is also consistent with observations showing that the oval form matures from an early hexagonal intermediate (Wergin, 1973). Alternative splicing may control other aspects of HEX-1 function (Soundararajan *et al.*, 2004). Additional research is required to determine the extent to which Woronin body structural variation can be reduced to molecular variation in HEX-1.

Septal-pore-associated organelles as adaptations

Members of the Fungi can be grouped on the basis of shared characters, some of which can be seen as adaptive. That is, they provide solutions to specific physiological problems and thereby confer a fitness advantage on the organism. For example, the Ascomycota and Basidiomycota are defined by their means of producing sexually derived spores: the Ascomycota produce meiotic spores in a sac or ascus whereas the Basidiomycota produce meiotic spores from a basidium. In both cases, these structures provide mechanisms for forcible spore discharge (Alexopolous *et al.*, 1996). Ascospores can be ejected from asci pressurized by osmosis (Fisher *et al.*, 2004) and ballistospores are discharged by an elegant 'surface-tension catapult' (Webster *et al.*, 1988; Money, 1998). These alternative means of spore dispersal probably arose in ancestral organisms upon which they conferred a significant fitness advantage and are therefore largely maintained to this day.

Organelles associated with septal pores may also be thought of as important components of a highly adapted cellular system. In one scenario the evolution of the vegetative perforate septum introduced selective pressures for the evolution of new supporting functions. One product of such selective pressures is the Woronin body, which exploited a pre-existing organelle, the peroxisome, for its formation and performs an adaptive function that is dependent on the septal pore (Jedd & Chua, 2000; Tenney *et al.*, 2000). Furthermore, a solid crystalline core of HEX-1 is essential for this function (Yuan *et al.*, 2003); this unique structural feature can be seen as advantageous in the context of the high turgor pressure characteristic of fungal hyphae (Buller, 1933b; Money & Harold, 1992; Bartnicki-Garcia *et al.*, 2000).

HEX-1 shares both sequence (Jedd & Chua, 2000) and structural (Yuan *et al.*, 2003) homology with eIF-5a proteins, suggesting that these functions may be related through gene duplication. The eIF-5a proteins are ancient (found from the Archaea through eukaryotes) and highly conserved (Kyrpides & Woese, 1998). They function in the cytoplasm and are involved in mRNA transport (Rosorius *et al.*, 1999) or stability (Zuk & Jacobson, 1998) and there is currently no evidence to suggest that eIF-5a form large protein complexes. Thus, *hex-1* may have evolved via gene duplication of an ancestral *eIF-5a* gene followed by the acquisition of new functions. Consistently, residues that are conserved between HEX-1 and eIF-5a are largely those that determine the overall protein fold, suggesting that over time, these residues were maintained in *hex-1* while surface residues evolved to acquire new functions associated with organelle targeting and self-assembly (Yuan *et al.*, 2003). This model further conforms to the hypothesis that gene duplication and the subsequent acquisition of novel function can be associated with evolutionary success (Ohno, 1970).

Molecular markers have not yet been reported for septal-pore-associated organelles of the Basidiomycota and for the Woronin-body-like organelle of *Neolecta*. Thus, for simplicity, I have presumed that these organelles arose independently (Fig. 2.1). Understanding their relationship awaits the isolation of genes encoding constituent proteins and their functional and phylogenetic characterization. Interestingly, the morphology of septal-pore-associated organelles varies considerably within their respective groups (Kimbrough, 1994; McLaughlin *et al.*, 1995), suggesting that their function may be under selection. Alternatively, the structure–function relationship of these organelles may not be highly constrained and this variation may reflect genetic drift.

Conclusions

Some aspects of Woronin body function are now understood in atomic detail (Yuan *et al.*, 2003). Some important questions remain unanswered. How is membrane re-sealing achieved and what mechanisms control association with the septal pore? What are the genetic and cellular mechanisms that regulate Woronin body formation and frequency within the hypha? Other septal-pore-associated organelles are also centred on osmophilic matrices: do these perform an analogous function or do they support a different aspect of the filamentous lifestyle? Comparing these various systems both within and between their respective taxa will help reveal the genetic, cellular and physiological basis underlying fungal diversity.

Acknowledgements

I am grateful to Swee-Peck Quek, David Hewitt and Naweed I. Naqvi for comments on the manuscript.

References

Alexopolous, C. J., Mims, C. W. & Blackwell, M. (1996). *Introductory Mycology*. New York: John Wiley & Sons.

Asiegbu, F. O., Choi, W., Jeong, J. S. & Dean, R. A. (2004). Cloning, sequencing and functional analysis of *Magnaporthe grisea* MVP1 gene, a hex-1 homolog encoding a putative 'woronin body' protein. *FEMS Microbiology Letters* **230**, 85–90.

Bartnicki-Garcia, S., Bracker, C. E., Gierz, G., Lopez-Franco, R. & Lu, H. (2000). Mapping the growth of fungal hyphae: orthogonal cell wall expansion during tip growth and role of turgor. *Biophysical Journal* **79**, 2382–90.

Bauer, R., & Oberwinkler, F. (1994). Meiosis, septal pore architecture, and systematic position of the heterobasidiomycetous fern parasite *Herpobasidium filicinum*. *Canadian Journal of Botany* **72**, 1229–42.

Berbee, M. L. & Taylor, J. W. (2001). Fungal molecular evolution: Gene trees and geological time. In *The Mycota,* vol. VII Part B, *Systematics and Evolution*, ed. D. J. McLaughlin, E. G. McLaughlin & P. A. Lemke, pp. 229–45. Berlin: Springer-Verlag.

Berns, M. W., Aist, J. R., Wright, W. H. & Liang, H. (1992). Optical trapping in animal and fungal cells using a tunable, near-infrared titanium-sapphire laser. *Experimental Cell Research* **198**, 375–8.

Bracker, C. E. (1967). Ultrastructure of fungi. *Annual Review of Phytopathology* **5**, 343–74.

Brenner, D. M. & Carrol, G. C. (1968). Fine structural correlates of growth in hyphae of *Ascodesmis sphaerospora*. *Journal of Bacteriology* **95**, 658–71.

Bruns, T. D., Vilgalys, S., Barns, D., Gonzalez, D. S., Hibbett, D. J., Lane, L., Simon, S., Stickel, T. M., Szaro, W. G., Weisburg, W. G. & Sogin, M. L. (1992). Evolutionary relationships within the fungi:analyses of nuclear small subunit RNA sequences. *Molecular Phylogenetics and Evolution* **1**, 231–241.

Buller, A. H. R. (1933a). The translocation of protoplasm through septate mycelium of certain Pyrenomycetes, Discomycetes and Hymenomycetes. In *Researches on Fungi*, vol. V, pp. 75–167. London: Longmans, Green and Co.

Buller, A. H. R. (1933b). Woronin bodies and their movements. In *Researches on Fungi*, vol. V, p. 127. London: Longmans, Green and Co.

Buller, A. H. R. (1933c). Woronin bodies and their movements. In *Researches on Fungi*, vol. V, p. 128. London: Longmans, Green and Co.

Buller, A. H. R. (1933d). Pore plugs and their formation under experimental conditions. In *Researches on Fungi*, vol. V, pp. 130–3. London: Longmans, Green and Co.

Camp, R. R. (1977). Association of microbodies, Woronin bodies, and septa in intercellular hyphae of *Cymadothea trifolii*. *Canadian Journal of Botany* **55**, 1856–9.

Collinge, A. J. & Markham, P. (1982). Hyphal tip ultrastructure of *Aspergillus nidulans* and *Aspergillus giganteus* and possible implications of Woronin bodies close to the hyphal apex of the latter species. *Protoplasma* **113**, 209–13.

Collinge, A. J. & Markham, P. (1985). Woronin bodies rapidly plug septal pores of severed *Penicillium chrysogenum* hyphae. *Experimental Mycology* **9**, 80–5.

Curach, N. C., Te'o, V. S., Gibbs, M. D., Bergquist, P. L. & Nevalainen, K. M. (2004). Isolation, characterization and expression of the hex1 gene from *Trichoderma reesei*. *Gene* **331**, 133–40.

Fisher, M., Cox, J., Davis, D. J., Wanger, A., Taylor, R., Huerta, A. J. & Money, N. P. (2004). New information on the mechanism of forcible ascospore discharge from *Ascobolus immersus*. *Fungal Genetics and Biology* **41**, 698–707.

Glass, N. L., Rasmussen, C., Roca, G. & Read, N. D. (2004). Hyphal homing, hyphal fusion and mycelial interconnectedness. *Trends in Microbiology* **12**, 135–41.

Hawksworth, D. L., Kirk, P. M., Sutton, B. C. & Pegler, D. N. (1995). *Ainsworth and Bisby's Dictionary of the Fungi*. Wallingford, UK: CAB International Publishing.

Hoch, H. C. & Maxwell, D. P. (1974). Proteinaceous hexagonal inclusion in hyphae of *Whetzelinia sclerotiorum* and *Neurospora crassa*. *Canadian Journal of Microbiology* **20**, 1029–36.

Jedd, G. & Chua, N. H. (2000). A new self-assembled peroxisomal vesicle required for efficient resealing of the plasma membrane. *Nature Cell Biology* **2**, 226–31.

Kimbrough, J. W. (1994). Septal ultrastructure and ascomycete systematics. In *Ascomycete Systematics: Problems and Perspectives in the Nineties*, ed. D. L. Hawksworth, pp. 127–41. New York: Plenum Press.

Kyrpides, N. C. & Woese, C. R. (1998). Universally conserved translation initiation factors. *Proceedings of the National Academy of Sciences of the USA* **95**, 224–8.

Landvik, S., Schumacher, T. K., Eriksson, O. E. & Moss, S. T. (2003). Morphology and ultrastructure of *Neolecta* species. *Mycological Research* **107**, 1021–31.

Lim, D. B., Hains, P., Walsh, B., Bergquist, P. & Nevalainen, H. (2001). Proteins associated with the cell envelope of *Trichoderma reesei*: a proteomic approach. *Proteomics* **1**, 899–909.

Lutzoni, F., Kauff, F., Cox, J. C., McLaughlin, D., Celio, G., Dentinger, B., Padamsee, M., Hibbett, D., James, T. Y., Baloch, E., Grube, M., Reeb, V., Hofstetter, V., Shcoch, C., Arnold, A. E., Miadlikowska, J., Spatafora, J., Johnson, D., Hambleton, S., Crockett, M., Shoemaker, R., Sung, G.-H., Lücking, R., Lumbsch, T., O'Donnell, K., Binder, M., Diederich, P., Ertz, D., Gueidan, C., Hansen, K., Harris, R. C., Hosaka, K., Lim, Y.-W., Matheny, B., Nishida, H., Pfister, D., Rogers, J., Rossman, A., Schmitt, I., Sipman, H., Stone, J., Sugiyama, J., Yahr, R. & Vilgalys, R. (2004). Assembling the fungal tree of life: progress, classification, and evolution of subcellular traits. *American Journal of Botany* **91**, 1446–80.

Markham, P. & Collinge, A. J. (1987). Woronin bodies of filamentous fungi. *FEMS Microbiology Reviews* **46**, 1–11.

McClure, W. K., Park, D. & Robinson, P. M. (1968). Apical organization in the somatic hyphae of fungi. *Journal of General Microbiology* **50**, 177–82.

McKeen, W. E. (1971). Woronin bodies in *Erysiphe graminis* DC. *Canadian Journal of Microbiology* **17**, 1557–63.

McLaughlin, D. J., Frieders, E. M. & Lu, H. (1995). A microscopist's view of heterobasidiomycete phylogeny. *Studies in Mycology* **38**, 91–109.

Momany, M., Richardson, E. A., Van Sickle, C. & Jedd, G. (2002). Mapping Woronin body position in *Aspergillus nidulans*. *Mycologia* **94**, 260–6.

Money, N. P. (1998). More g's than the space shuttle: ballistospore discharge. *Mycologia* **90**, 547–58.

Money, N. P. & Harold, F. M. (1992). Extension growth of the water mold *Achlya*: interplay of turgor and wall strength. *Proceedings of the National Academy of Sciences of the USA* **15**, 4245–59.

Moore, R. T. (1975). Early ontogenic stages in dolipore/parenthosome formation in *Polyporus biennis*. *Journal of General Microbiology* **87**, 251–9.

Moore, R. T. & Marchant, R. (1972). Ultrastructural characterization of the basidiomycete septum of *Polyporus biennis*. *Canadian Journal of Botany* **50**, 2463–9.

Muller, W. H., Montijn, R. C., Humbel, B. M., van Aelst, A. C., Boon, E. J. M. C., van der Krift, T. P. & Boekhout, T. (1998). Structural differences between two types of basidiomycete septal pore caps. *Microbiology* **144**, 1721–30.

Oberwinkler, F. & Bauer, R. (1989). The systematics of gasteroid, auricularoid Heterobasidiomycetes. *Sydowia* **41**, 224–56.

Oberwinkler, F. & Bauer, R. (1990). *Cryptomycocolax*: a new mycoparasitic heterobasidiomycete. *Mycologia* **82**, 671–92.

Ohno, S. (1970). *Evolution by Gene Duplication*. Berlin: Springer-Verlag.

Rosorius, O., Reichart, B., Kratzer, F., Heger, P., Dabauvalle, M. C. & Hauber, J. (1999). Nuclear pore localization and nucleocytoplasmic transport of eIF-5A: evidence for direct interaction with the export receptor CRM. *Journal of Cell Science* **112**, 2369–80.

Soundararajan, S., Jedd, G., Li, X., Ramos-Pamplona, M., Chua, N. H. & Naqvi, N. I. (2004). Woronin body function in *Magnaporthe grisea* is essential for efficient pathogenesis and for survival during nitrogen starvation stress. *Plant Cell* **16**, 1564–74.

Swann, E. C., Frieders, E. M. & McLaughlin, D. J. (2001). Urediniomycetes. In: *The Mycota,* vol. VII Part B, *Systematics and Evolution*, ed. D. J. McLaughlin, E. G. McLaughlin & P. A. Lemke, pp. 37–56. Berlin: Springer-Verlag.

Tenney, K., Hunt, I., Sweigard, J., Pounder, J. I., McClain, C., Bowman, E. J. & Bowman, B. J. (2000). *hex-1*, a gene unique to filamentous fungi, encodes the major protein of the Woronin body and functions as a plug for septal pores. *Fungal Genetics and Biology* **31**, 205–17.

Ternetz, C. (1900). Protoplasmabewegung und Fruchtkörperbildung bei *Ascophanus carneus. Jahrbuch für wissenschaftliche Botanik* **35**, 273–312.

Tey, W. K., North, A. J., Reyes, J. L., Lu, Y. F. & Jedd, G. (2005). Polarized gene expression determines Woronin body formation at the leading edge of the fungal colony. *Molecular Biology of the Cell* **16**, 2651–9.

Trinci, A. P. J. & Collinge, A. J. (1973). Occlusion of septal pores of damaged hyphae of *Neurospora crassa* by hexagonal crystals. *Protoplasma* **80**, 57–67.

Webster, J., Proctor, M. C. F., Davey, R. A. & Duller, G. A. (1988). Measurement of the electrical charge on some basidiospores and an assessment of two possible mechanisms of ballistospore propulsion. *Transactions of the British Mycological Society* **91**, 193–203.

Wergin, W. P. (1973). Development of Woronin bodies from microbodies in *Fusarium oxysporum* f.sp. *lycopersici. Protoplasma* **76**, 249–60.

Yuan, P., Jedd, G., Kumaran, D., Swaminathan, S., Shio, H., Hewitt, D., Chua, N. H. & Swaminathan, K. (2003). A HEX-1 crystal lattice required for Woronin body function in *Neurospora crassa. Nature Structural Biology* **10**, 264–70.

Zuk, D. & Jacobson, D. (1998). A single amino acid substitution in yeast eIF-5A results in mRNA stabilization. *EMBO Journal* **17**, 2914–25.

3

Environmental sensing and the filamentous fungal lifestyle

NICK D. READ

Institute of Cell Biology, University of Edinburgh

Introduction

The majority of fungi have a filamentous lifestyle. The evolution of the hypha has been pivotal to the success of filamentous fungi and in determining the uniqueness of their lifestyle. It has also had important consequences in determining the modes of morphogenesis of filamentous fungi, and how they operate as non-motile, heterotrophic organisms (Read, 1994). This review focuses on hyphal and colony morphogenesis and how it is influenced by environmental signals in the context of the filamentous fungal lifestyle.

The supracellular, cellular and multicellular nature of filamentous fungi

The defining cellular element of the filamentous fungi is the hypha (Figs. 3.1–3.6). Hyphae possess a unique combination of structural, behavioural and functional attributes that clearly distinguish them from uninucleate animal and plant cells. The vegetative hypha is a tip-growing cellular element (Harris *et al.*, 2005) (Figs. 3.1, 3.2) that undergoes regular branching (Trinci, 1983; Turner & Harris, 1997) (Figs. 3.3, 3.4), is typically multinucleate (Fig. 3.3) (Freitag *et al.*, 2004), and possesses incomplete cross-walls (septa) which, when open, allow movement of cytoplasm and organelles between hyphal compartments (Harris, 2001) (Fig. 3.2). In sub-peripheral regions of the colony, hyphae frequently fuse with one another (Read & Roca, 2006) (Fig. 3.3) and septal pores often become blocked (Gull, 1978) (Fig. 3.2). Vegetative hyphae thus have a supracellular nature because they are part of a network of interconnected hyphal compartments and hyphae within the colony (Fig. 3.1). The supracellular colony is a combination of being a coenocyte (in which individual mitoses

Fungi in the Environment, ed. G. M. Gadd, S. C. Watkinson & P. S. Dyer. Published by Cambridge University Press.

are not associated with individual cell divisions) and a syncytium (in which hyphae have fused together). Its supracellular nature confers novel, yet little understood, mechanisms for long-distance intracellular communication, translocation of water, and transport of nutrients (Rayner, 1996; Tlalka *et al.*, 2002, 2003). It also allows hyphae to act as a cooperative, and gives them the ability to span the typically heterogeneous environments in which they reside (Rayner, 1996; Rayner *et al.*, 1997).

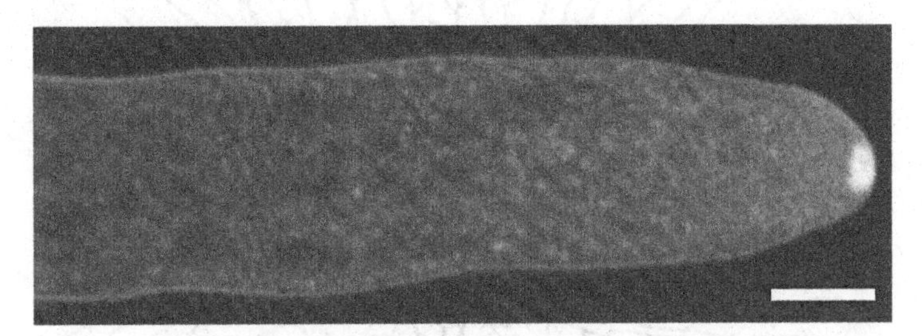

Fig. 3.1. Concentration of secretory vesicles within the Spitzenkörper of a growing hyphal tip of *Neurospora crassa* after staining with the membrane-selective dye FM4-64. Bar, 5 μm (E. R. Kalkman & N. D. Read, unpublished).

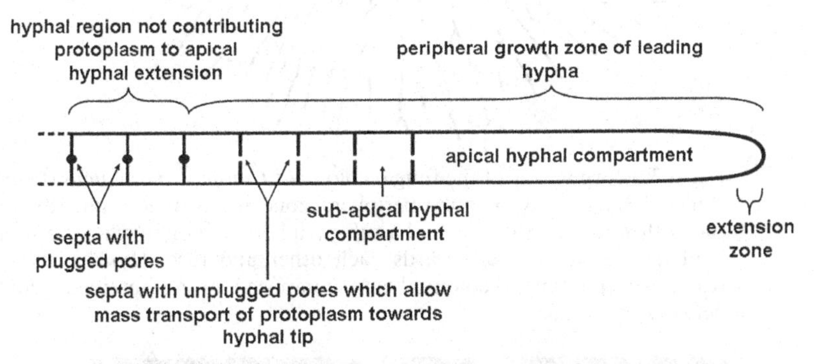

Fig. 3.2. Generalized diagram showing the structure of a leading hypha of a higher fungus such as *Neurospora crassa* (adapted from Trinci, 1971). The hypha exhibits tip growth and the extension zone occupies a few micrometres at the hyphal tip. The hypha is composed of compartments separated by septa possessing pores. In young regions of the colony these pores are open, allowing organelles and cytoplasm to pass through them. In older parts of the colony the septal pores frequently become occluded. The apical hyphal compartment is typically much longer than sub-apical compartments. The peripheral growth zone is the part of the colony that can provide materials that can support growth. This comprises hyphal compartments in the colony periphery which are in direct continuity, allowing materials to be transported to the hyphal tips.

Hyphae exhibit extraordinary developmental versatility and phenotypic plasticity, and are responsible for the enormous diversity in patterns of cellular and multicellular development within the filamentous fungi. Hyphae are also extremely versatile in terms of function and serve key

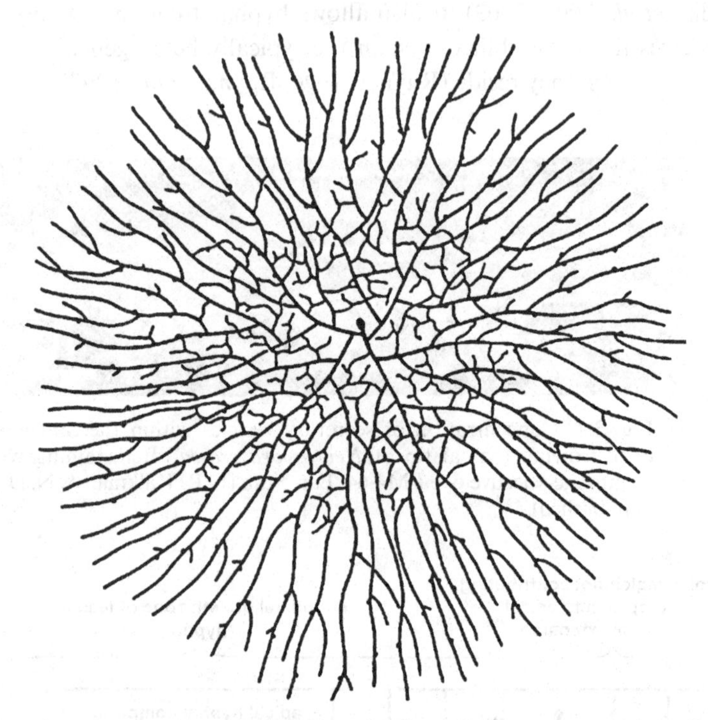

Fig. 3.3. Morphology of the fungal colony of *Coprinus sterquilinus* (from Buller 1931). It shows an outer peripheral zone in which the hyphae avoid each other and do not fuse, and a sub-peripheral region in which certain hyphal branches home towards each other and fuse. Hyphal fusion results in the fungal colony being composed of an interconnected network of hyphae.

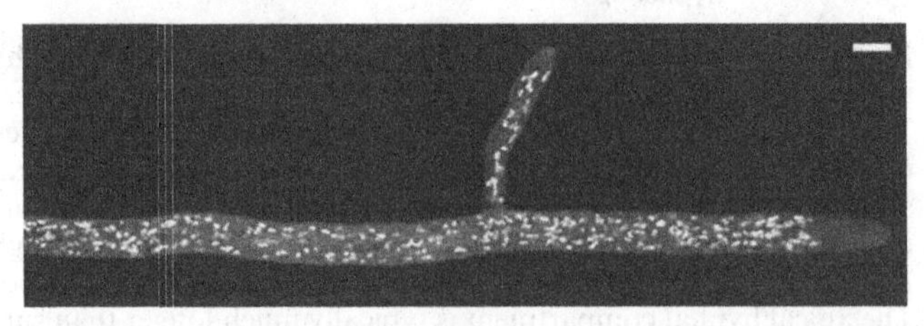

Fig. 3.4. Confocal image of *Neurospora crassa* showing the multinucleate state of its hyphae. Nuclei have been labelled with H1-GFP and membranes have been stained with FM4-64 (see Freitag *et al.*, 2004 for details). Bar, 10 μm (P. C. Hickey & N. D. Read, unpublished).

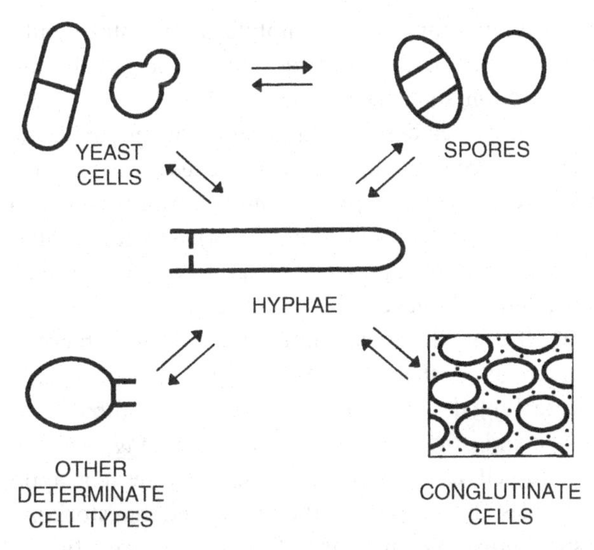

Fig. 3.5. Different cell types produced by filamentous fungi (from Read, 1994).

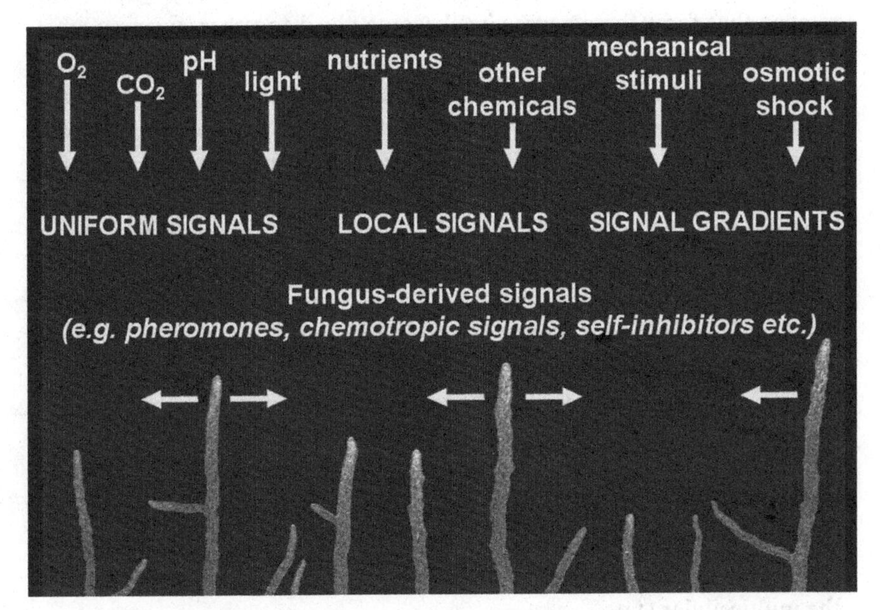

Fig. 3.6. Environmental signals influencing colony and hyphal morphogenesis.

roles in exploration, invasion, nutrient mobilization, uptake and storage, translocation of nutrients and water, defence of occupied substratum, reproduction, protection and survival (Read, 1994).

Although a hypha is often described as a cell, problems are encountered when using the term 'cell' in this context because the hypha is not a discrete, uninucleate unit of protoplasm bounded by a plasma membrane and cell wall or extracellular matrix (Read, 1994). Sometimes filamentous fungi do produce true cells (e.g. uninucleate spores or yeast cells) like those of animals and plants. However, they more commonly form what one might also term cells but which are multinucleate, although not as extensively multinucleate as most hyphae. In very general terms, uninucleate and multinucleate fungal cells can be classified into three basic types: spores, yeast cells, and other determinate cell types (Fig. 3.5). The latter class of determinate cells can include infection structures (e.g. appressoria, sub-stomatal vesicles, haustorial mother cells and haustoria), basidia, cystidia, adhesive knobs and the rings of nematophagous fungi. In filamentous fungi, all of these determinate cell types are derived from hyphae (Fig. 3.5).

More extensive multicellular development requires hyphal aggregation and the subsequent differentiation of cells within the aggregates. The hyphae that form these aggregates become glued together, forming what have been termed 'conglutinate cells' (from the Latin 'conglutinare' meaning 'to stick together') (Read, 1994). These conglutinate cells form true tissues, which are either pseudoparenchymatous (if they lose their hyphal appearance) or prosenchymatous (if the original hyphae can still be recognized). The resulting hyphal aggregates composed of different tissues can thus develop into complex multicellular systems containing numerous, discrete uninucleate and multinucleate 'cell types' as well as hyphal compartments interconnected via open septal pores. Examples of hyphal aggregates include fruit bodies (e.g. ascomata, basidiomata and conidiomata), sclerotia, rhizomorphs and mycelial cords. As an indication of how complex some of these differentiated hyphal aggregates can be, 28 morphologically distinct 'cell types' have been described in the model filamentous fungus *Neurospora crassa* (Bistis *et al.*, 2003). The mechanism by which fungi undergo multicellular development, involving the specialization of hyphal compartments originally arising from aggregated hyphae, is fundamentally different from the ways in which animals or plants achieve the multicellular state.

Physical isolation of adjacent hyphal or cell compartments is usually a prerequisite for cell specialization; septal pore occlusion probably plays a role in this during fungal multicellular development (Gull, 1978; Read, 1994). However, some septal pore plugs are very specialized, such as basidiomycete dolipore septa (Lu & McLaughlin, 2005) and those within ascogenous hyphae and between ascogenous hyphae and asci (Beckett, 1981; Read & Beckett, 1996). These complex pore occlusions may play a role in selective cell–cell communication.

In the context of this review, environmental signals play a major role in modulating the different patterns of cellular and multicellular development within the fungal colony, particularly by being important in regulating the differential expression of genes involved in these processes.

Consequences of being non-motile and heterotrophic

Animal cells are motile and heterotrophic, whereas plant cells are non-motile and autotrophic. Fungi combine features of animal and plant cells by being both non-motile and heterotrophic but, phylogenetically, fungi are more closely related to animals than plants (Taylor *et al.*, 2004). The possession of hyphae has allowed fungi to combine the attributes of non-motility and heterotrophy to great effect. Being non-motile and heterotrophic will undoubtedly have introduced selection pressures during evolution which have played a crucial role in imparting many of the characteristic features of filamentous fungi.

In common with plants, fungi are non-motile primarily because they possess walled cells. An important consequence of being non-motile is that filamentous fungi have evolved mechanisms to respond rapidly to changing environmental conditions to maximize their growth and survival in order to be successful and competitive organisms. They can rapidly adapt to, survive in, or escape from changing environments. Hyphae respond extremely sensitively and quickly to the myriad of signals within the heterogeneous microenvironments through which they invade, explore and reside. Hyphal growth direction and branching patterns can change very rapidly and hyphae can selectively grow from an unfavourable (e.g. nutrient-depleted or desiccated) to a more favourable environment (e.g. one that is nutrient-rich or contains free water). As indicated in the previous section, hyphae exhibit great developmental versatility and phenotypic plasticity. Sporulation can be initiated very quickly, resulting in spores that are adapted to remain in and resist adverse conditions until their environment becomes favourable again. Alternatively, spore production can provide a means to escape to more favourable conditions. Fungi have evolved an extraordinary array of different mechanisms for dispersing spores and propelling them into the air (Ingold, 1971). In other cases many fungi, and particularly plant pathogens, can form differentiated hyphal aggregates, which may resist adverse environments (e.g. sclerotia), allow long-distance transport of water and nutrients within adverse environments (e.g. rhizomorphs), or promote and protect spore production with the end result of the spores being able to escape from often adverse conditions (fruit bodies).

In common with animals, fungi are heterotrophic. The consequences of being heterotrophic and filamentous are manifold. First, hyphae allow the efficient exploration of their natural habitats and to rapidly capture, mobilize and translocate the heterogeneously distributed nutrients and water that they encounter. Second, filamentous fungi have evolved excellent digestive capabilities and produce a diverse range of extracellular enzymes, which are efficiently secreted into the environment. Third, soluble nutrients are efficiently absorbed by hyphae having a high surface area to volume ratio and possessing a wide range of efficient transport membrane processes to facilitate nutrient uptake.

Environmental signals influencing hyphal and colony morphogenesis

Filamentous fungi respond to a phenomenal range of environmental signals. These signals can be broadly classified into four types: (a) abiotic and global; (b) abiotic and local; (c) biotic and derived from other organisms; and (d) biotic and derived from the fungus itself.

An example of an abiotic and global environmental signal for a fungus is light, which may vary in intensity, wavelength and periodicity (especially in relation to its diurnal rhythmicity). Other signals of this type may include temperature and relative humidity. However, in their normally heterogeneous microenvironments fungi often experience external signals at the microscopic level in the form of gradients. The majority of abiotic signals experienced by fungi in their microworlds are local. They include nutrients, oxygen, carbon dioxide, pH, other chemical signals, mechanical stimuli and osmotic shock (Fig. 3.6).

Filamentous fungi respond to numerous biotic signals derived from other organisms (i.e. fungi, bacteria, plants and animals) in the natural environment. Other organisms produce a wide variety of chemical and physical signals, which can influence the growth, behaviour, metabolism and gene expression of filamentous fungi. Obvious examples include chemicals produced by other organisms that can promote or inhibit fungal growth (see, for example, French, 1985; Cooke & Whipps, 1993), compounds from other organisms that can modulate fungal secondary metabolism (Burow *et al.*, 1997), and physical and chemical signals from plant or animal hosts which can be used as cues to assist the invasion and penetration of hosts by fungi (Read *et al.*, 1992; Gow, 2004; Lucas, 2004).

Probably the least understood biotic signals that influence filamentous fungi are the extracellular chemical signals produced by the fungus itself, which have been collectively called 'autoregulators' (Ugalde, 2006). They play critical roles in regulating the filamentous organization of a colony and the various differentiated structures it produces. During germination, some act as germination self-inhibitors (Macko & Staples, 1973) whereas others influence the pattern of germ-tube emergence from spores (Robinson, 1973a, b, c). Signals that are central to regulating colony morphogenesis are those that cause hyphal avoidance (Fig. 3.1), growth away from the centre of the colony (Bottone *et al.*, 1998), and hyphal homing as a prelude to hyphal fusion (Fig. 3.1) (Hickey *et al.*, 2002; Roca *et al.*, 2005). Some autoregulatory signals function as 'quorum' (or cell density) sensors and have been shown to play roles in the regulation of yeast–hyphal dimorphism (Chen *et al.*, 2004; Hornby *et al.*, 2004; Kruppa *et al.*, 2004; Enjalbert & Whiteway, 2005), chlamydospore formation (Martin *et al.*, 2005) and conidial fusion (Roca *et al.*, 2005). Fungus-derived signals also play a role in regulating asexual and sexual reproduction. A number of self-produced compounds stimulate asexual reproduction; others have been shown to regulate the choice between sexual and asexual reproduction (Tsitsigiannis *et al.*, 2005). A significant amount has been discovered about sex pheromones and their role in sexual reproduction (Gooday, 1994). They include trisporic acid, used in the Mucorales (Gooday, 1994) and a variety of peptide pheromones used by filamentous ascomycetes (see, for example, Bobrowicz *et al.*, 2002; Kim *et al.*, 2002) and basidiomycetes (see, for example, Snetselaar *et al.*, 1996).

Intracellular signal transduction of environmental signals

Considerable progress is being made in identifying and characterizing the intracellular signalling pathways involved in perceiving and transducing environmental signals, and converting them into cellular responses. Of central importance here has been the sequencing of a variety of fungal genomes, which have provided unparalleled insights into their signal transduction machinery. To provide an indication of the complexity of the signal transduction machinery which they possess, I will briefly summarize what has been found in *Neurospora crassa*, the first filamentous fungus to have its genome sequenced (Galagan *et al.*, 2003; Borkovich *et al.*, 2004) (Fig. 3.7), and then augment this with recent information from *Magnaporthe grisea*, the first fungal plant pathogen to have its genome sequenced (Dean *et al.*, 2005).

General features

In contrast to yeast cells, filamentous fungi tend to live in more complex, heterogeneous environments, and the environmental signals they

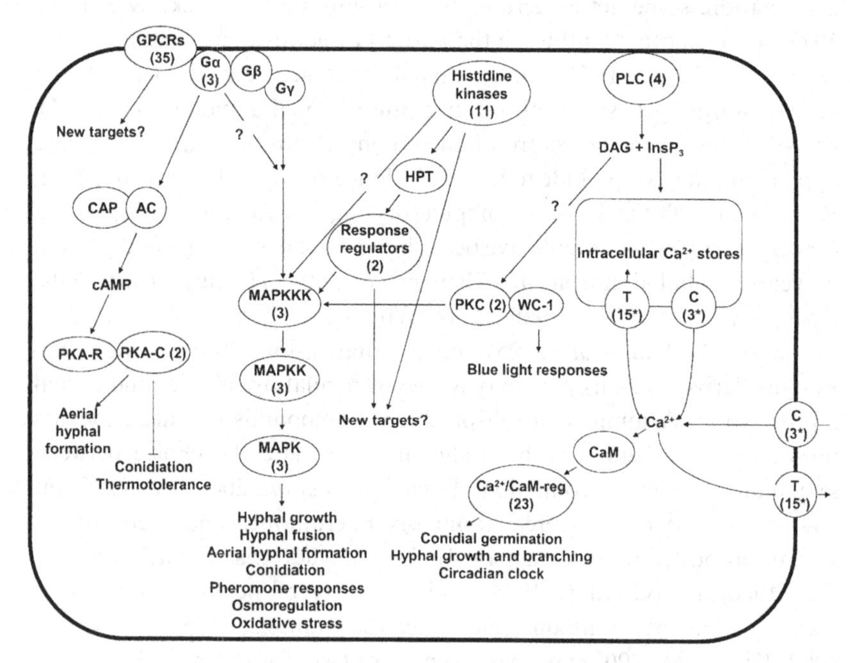

Fig. 3.7. Overview of the major intracellular signal transduction pathways in *Neurospora crassa* (incorporating data from Galagan *et al.*, 2003; Zelter *et al.*, 2004; Kulkarni *et al.*, 2005). An asterisk indicates that the location in the plasma membrane and/or organelle membranes has not been determined. Abbreviations: AC, adenylate cyclase; C, calcium channel protein; CaM, calmodulin; Ca^{2+}/CaM reg, calcium- and calmodulin-regulated; CAP, cyclase-associated protein; DAG, diacylglyerol; GPCR, G-protein coupled receptor; Gα, G protein α-subunit; Gβ, G protein β-subunit; Gγ, G protein γ-subunit; HPT, histidine-containing phosphotransfer domain protein; MAPK, MAP kinase; MAPKK, MAP kinase kinase; MAPKKK, MAP kinase kinase kinase; PKA-C, protein kinase A catalytic subunit; PKA-R, protein kinase regulatory subunit; PLC, phospholipase C; PKC, protein kinase C; T, calcium transport protein (Ca^{2+}-ATPase, H^+/Ca^{2+} exchanger, or Na^+/Ca^{2+} exchanger); WC-1, white-collar-1.

have to respond to are more varied and changing as hyphae grow through their environments. It might, therefore, be expected that filamentous fungi should possess a more extensive array of sensing and signalling capabilities. This has been borne out by the analysis of the *Neurospora* and *Magnaporthe* genomes. In some cases there are clear cases of the expansion of upstream proteins and a conserved core of downstream components (e.g. in two-component signalling systems), suggesting the presence of extensive networking and/or new signalling interactions not found in yeasts (Borkovich *et al.*, 2004). In other cases, there is an overall increase in the protein machinery, indicating greater signalling functionality, for example in G-protein coupled receptors (GPCRs) (Kulkarni *et al.*, 2005) and in calcium signalling (Borkovich *et al.*, 2004; Zelter *et al.*, 2004).

Two-component signalling systems

Two-component signalling is used extensively in prokaryotes and is also found in plants, slime moulds, yeasts and filamentous fungi but not in animals (Borkovich *et al.*, 2004). *N. crassa* possesses two-component signalling of a complex type that is uncommon in prokaryotes. Their two-component signalling systems consist of a hybrid protein, which contains both a histidine kinase and a response regulator domain. In response to an environmental stimulus, this hybrid protein signals via a phosphorelay mechanism to a histidine phosphotransferase, and then a second response regulator protein. In eukaryotes, histidine kinases are involved in the activation of MAP kinase cascades and/or transcriptional regulation (Fig. 3.7).

In *N. crassa*, only one of the eleven histidine kinases identified is predicted to be membrane-spanning. Two are involved in hyphal development (Alex *et al.*, 1996; Schumacher *et al.*, 1997) and three others appear to be involved in light sensing (Borkovich *et al.*, 2004). *N. crassa* has a larger component of histidine kinases than do yeasts (e.g. 1 in *Saccharomyces cerevisiae*; 3 in *Schizosaccharomyces pombe*) but only possesses one histidine phosphotransferase and two response regulators. The conservation of these downstream elements suggests that these proteins may serve to integrate multiple signalling inputs from many histidine kinases to evoke the proper cellular responses. Alternatively, some of the histidine kinases may not act through phosphorelays (Borkovich *et al.*, 2004).

G-protein coupled receptors and heterotrimeric G proteins

Seven-transmembrane-helix GPCRs are used by eukaryotes to sense many environmental signals (Kays & Borkovich, 2005). *N. crassa* possesses 35 predicted GPCRs and GPCR-like proteins (Fig. 3.7) whereas *S. cerevisiae* has only 6 (Kulkarni *et al.*, 2005). Ten of the *N. crassa* GPCRs fall into the five classes (A–E) in which GPCRs have been traditionally classified: microbial opsins, pheromone receptors, glucose sensors, nitrogen sensors, and a class not previously identified in fungi that shows similarity to the family of cAMP receptors in the cellular slime mould *Dictyostelium discoideum* (Galagan *et al.*, 2003; Borkovich *et al.*, 2004). Although it was proposed that *Neurospora* might use extracellular cAMP signalling in a way analogous to its role in cell chemotaxis in *D. discoideum* (Galagan *et al.*, 2003), it has recently been demonstrated that cAMP is not required for hyphal homing as a prelude to hyphal fusion (Roca *et al.*, 2005). However, extracellular cAMP may serve other, yet to be identified, functions in signalling in *N. crassa.* A novel sixth class of GPCR-like proteins (PTH11-related proteins) was recently identified in *M. grisea* and 25 of these have been found in *N. crassa* (Dean *et al.*, 2005; Kulkarni *et al.*, 2005). The targets or functions of these receptors in *Neurospora* are not known.

G-protein coupled receptors regulate the activity of second messengers through their interaction with heterotrimeric G-proteins, which consist of a Gα subunit, and a tightly associated Gβ and Gγ subunit (Neer, 1995). Binding of an extracellular ligand to the GPCR results in bound GDP being replaced by GTP on the Gα subunit, and the dissociation of the heterotrimer into Gα-GTP and Gβγ units. Gα-GTP and Gβγ can both interact with downstream effectors during signal transduction. *N. crassa* possesses more Gα subunits than does *S. cerevisiae* (three instead of two) but the same number of Gβ and Gγ subunits (one each). G-protein signalling in *N. crassa* plays roles in hyphal growth conidiation, female fertility, and stress responses (Ivey *et al.*, 1996; Baasiri *et al.*, 1997; Kays *et al.*, 2000; Yang *et al.*, 2002; Krystofova & Borkovich, 2005) (Fig. 3.7).

Ras-like GTPases

The Ras-like superfamily of monomeric small GTPases function as molecular switches that control a wide range of cellular processes, including signal transduction. Five Ras/Ras-related GTPases are present in *N. crassa*; similar proteins have been identified in the yeasts *S. cerevisiae* and *Yarrowia lipolytica* (Borkovich *et al.*, 2004). Evidence has been obtained for one of the Ras-like proteins having a role in regulating hyphal growth and conidiation (Kana-Uchi *et al.*, 1997).

cAMP signalling

cAMP is an intracellular second messenger involved in regulating aerial hyphal growth, aerial hyphal formation, conidiation and thermotolerance in *N. crassa* (Bruno *et al.*, 1996; Ivey *et al.*, 2002; Banno *et al.*, 2005). Similar to budding yeast, *N. crassa* possesses one adenylate cyclase (which synthesizes cAMP) and one adenylate cyclase associated protein. cAMP activates various downstream proteins, particularly protein kinase A. *N. crassa* possesses two catalytic subunits and one regulatory subunit of protein kinase A (*S. cerevisiae* possesses only one of each) (Borkovich *et al.*, 2004) (Fig. 3.7).

MAP kinases

Mitogen-activated protein (MAP) kinases are critical components of a central intracellular signalling switchboard that coordinates a wide variety of incoming signals in eukaryotic cells. Mitogen-activated protein (MAP) kinases consist of three types of serine/threonine protein kinase: MAP kinase kinase kinase (MAPKKK), MAP kinase kinase (MAPKK) and MAP kinase (MAPK). These three enzymes act sequentially, phosphorylating and activating each other in the aforementioned order and culminating in the phosphorylation of a target protein. Nine MAPKKK/MAPKK/MAPK protein-encoding genes have been found in the *N. crassa* genome, and these correspond to at least three MAP kinase signalling pathways (Fig. 3.7). One of these pathways is involved in hyphal growth and fusion, aerial hyphal formation, conidiation and female fertility (Pandey *et al.*, 2004; Bobrowicz *et al.*, 2005). A second pathway is involved in osmoregulation and fungicide resistance (Zhang *et al.*, 2002). A third pathway may be involved in maintaining cell integrity, as characterized in budding yeast (Levin, 2005).

Interestingly, five MAP kinase signalling pathways have been identified in *S. cerevisiae*. This finding, coupled with the diversification of GPCRs and histidine kinases, suggests greater integration of signals from multiple upstream sensory protein in *N. crassa* than in budding yeast (Borkovich *et al.*, 2004). Alternatively, the nine *N. crassa* MAP kinases may be used to create more than three MAPK modules, possibly leading to even more signalling diversity and cross-talk than observed in budding yeast. The latter hypothesis is consistent with the observation that mutations in the *N. crassa* MAP kinases have multiple phenotypic effects. Of possible relevance here is that Ste5, the scaffold protein that conditions specificity in MAP kinase signalling in *S. cerevisae*, is absent in *N. crassa* (and also *M. grisea*) (Dean *et al.*, 2005).

Calcium signalling

Calcium is an important second messenger, which is involved in regulating numerous processes in *N. crassa* including spore germination (d'Enfert *et al.*, 1999), hyphal tip growth and branching (Prokisch *et al.*, 1997), and circadian clocks (Zelter *et al.*, 2004). Transient increases in cytosolic free Ca^{2+} ($[Ca^{2+}]_c$), which are often localized in specific subcellular regions, provide intracellular signals, which can directly or indirectly regulate the activity of numerous downstream proteins. Different combinations of Ca^{2+}-channels, -pumps, -transporters and other Ca^{2+}-signalling proteins are used by cells to produce $[Ca^{2+}]_c$ signals with different dynamic, spatial and temporal characteristics (Berridge *et al.*, 2003). Three Ca^{2+}-channel proteins, nine Ca^{2+}-ATPases and eight Ca^{2+}-exchangers have been identified in the *N. crassa* genome. This 'tool kit' of Ca^{2+}-signalling proteins is substantially greater than that found in *S. cerevisiae* (Borkovich *et al.*, 2004; Zelter *et al.*, 2004).

Downstream proteins that Ca^{2+} regulates include calcineurin and calmodulin; these and numerous other Ca^{2+}/calmodulin-regulated proteins have been identified in the genome of *N. crassa* (Galagan *et al.*, 2003; Borkovich *et al.*, 2004; Zelter *et al.*, 2004). Interestingly, key proteins that are important in generating increased $[Ca^{2+}]_c$ by the release of Ca^{2+} from internal stores in animal and plants are not present in fungi (Galagan *et al.*, 2003; Zelter *et al.*, 2004). Furthermore, *Neurospora* possesses both Ca^{2+}/H^+ and Ca^{2+}/Na^+ antiporters, whereas plants only possess the former and animals the latter (Zelter *et al.*, 2004).

Differences in intracellular signalling between Magnaporthe grisea *and* Neurospora crassa

The rice blast fungus, *M. grisea*, is the most destructive pathogen of rice worldwide and the principal model organism for elucidating the molecular basis of fungal disease of plants. It was also the first fungal plant pathogen to have its genome sequenced (Dean *et al.*, 2005). Some interesting differences between the intracellular signalling machineries of *N. crassa* and *M. grisea* have been found.

M. grisea has a greatly expanded family of GPCRs and GPCR-like proteins: in comparison with 35 identified in *N. crassa* (Galagan *et al.*, 2003; Borkovich *et al.*, 2004), 76 have been found in *M. grisea* (Dean *et al.*, 2005; Kulkarni *et al.*, 2005). Sixty-one of these proteins in *M. grisea* belong to a completely new class of eukaryotic GPCRs and GPCR-like proteins (the PTH11-related proteins). Because *M. grisea* contains significantly more of these latter GPCRs than *N. crassa*, it has been suggested that

M. grisea may have greater flexibility in reacting to different extracellular signals than *N. crassa*, and this may relate to its pathogenic rather than saprotrophic lifestyle (Dean *et al.*, 2005).

Another difference in intracellular signalling between *M. grisea* and *N. crassa* relates to MAP kinase signalling. Although three MAP kinase pathways have been identified in both fungi, the processes regulated by the pathways in these two organisms are distinctly different; in particular, two of the pathways in *M. grisea* control virulence-associated appressorium development (Dean *et al.*, 2005).

The Spitzenkörper as a primary response element to environmental signals

The Spitzenkörper is a complex, multicomponent structure dominated by vesicles and located within the tips of growing hyphae (Harris *et al.*, 2005) (Fig. 3.1). There is an intimate association between the dynamic behaviour of the Spitzenkörper and hyphal morphogenesis. The Spitzenkörper is present within the tips of growing hyphae that are responding sensitively to their external environment. It moves within the apical dome (Lopez-Franco & Bracker, 1996) and its position within the dome correlates with the direction of hyphal growth, indicating that the Spitzenkörper controls the directionality of hyphal growth (Girbardt, 1957; Bracker *et al.*, 1997; Riquelme *et al.*, 1998; Dijksterhuis, 2003). In addition, new Spitzenkörper arise at sites along a hypha at locations from which new branches will emerge, giving rise to the idea that the Spitzenkörper play an important role in branch initiation (Girbardt, 1957; Reynaga-Pena & Bartnicki-Garcia, 1997; Hickey *et al.*, 2005).

A well-developed concept is that the Spitzenkörper acts as a vesicle supply centre: a moveable distribution centre for secretory vesicles involved in cell surface expansion (Bartnicki-Garcia *et al.*, 1989; Bartnicki-Garcia, 2002). Another concept that is emerging, but which has yet to be proven, is that the Spitzenkörper may be viewed as a switching station from microtubule- to actin microfilament-based vesicle transport. It is thought that microtubules are mainly responsible for the long-distance transport of secretory vesicles to the Spitzenkörper, whereas actin microfilaments primarily control vesicle organization within the Spitzenkörper and transport them to the plasma membrane (Harris *et al.*, 2005).

The Spitzenkörper is the organizing centre for hyphal and colony morphogenesis. The history of the behaviour and activity of the Spitzenkörper as a vesicle supply centre is manifest in the morphology of hyphae and the colony as a whole. This is particularly exemplified in the large range of morphogenetic mutants with altered colony morphology in *N. crassa* and *Aspergillus nidulans* (Harris *et al.*, 2005). Spitzenkörper behaviour and/or organization will have been influenced in all of these mutants. In some hyperbranching mutants, the Spitzenkörper are smaller, more mobile and less stable than in the wild type (for example, see the *spray* mutant of *N. crassa* in Hickey *et al.* (2005)). Analysis of these morphological mutants has shown that a wide range of different functions can result in altered colony morphologies and thus Spitzenkörper behaviour/organization, which indicates the genetic complexity of how these processes are regulated.

A major challenge will be to understand how local signals in the environment (e.g. touch stimuli, pheromones and other chemical signals) (Fig. 3.6) rapidly and sensitively modify Spitzenkörper behaviour and activity. The speed of these responses clearly shows that genetic regulation is not initially involved but that in many cases receptors in the apical plasma must connect through signal transduction machinery to the primary response element: the Spitzenkörper. Indeed it is known that the Spitzenkörper is tethered to the apical plasma membrane (Bracker *et al.*, 1997; Hickey *et al.*, 2005), but the identity of the molecules that provide these connections is unknown.

Conclusions

Genomic analyses of filamentous fungi are providing a new framework within which to understand how these organisms sense and respond to signals in the environment. Critical to this understanding will be the determination of the functional roles of the proteins encoded within these genomes which are relevant to signal-response coupling. Of particular importance will be the analysis of the dynamic molecular composition of the Spitzenkörper (Harris *et al.*, 2005), and the determination of how the activities of its different components sense and respond to environmental signals.

References

Alex, L. A., Borkovich, K. A. & Simon, M. I. (1996). Hyphal development in *Neurospora crassa*: involvement of a two-component histidine kinase. *Proceedings of the National Academy of Sciences of the USA* **93**, 3416–21.

Baasiri, R. A., Lu, X., Rowley, P. S., Turner, G. E. & Borkovich, K. A. (1997). Overlapping functions for two G protein α subunits in *Neurospora crassa*. *Genetics* **147**, 137–45.

Banno, S., Ochiai, N., Noguchi, R., Kimura, M., Yamaguchi, I., Kanzaki, S., Murayama, T. & Fujimura, M. (2005). A catalytic subunit of cyclic AMP-dependent protein kinase, PKAC-1, regulates asexual differentiation in *Neurospora crassa. Genes and Genetic Systems* **80**, 23–34.

Bartnicki-Garcia, S. (2002). Hyphal tip growth: outstanding questions. In *Molecular Biology of Fungal Development*, ed. H. D. Osiewacz, pp. 29–58. New York: Marcel Dekker.

Bartnicki-Garcia, S., Hergert, F. & Gierz, G. (1989). Computer simulation of fungal morphogenesis and the mathematical basis for hyphal tip growth. *Protoplasma* **153**, 46–57.

Beckett, A. (1981). The ultrastructure of septal pores and associated structures in the ascogenous hyphae and asci of *Sordaria humana. Protoplasma* **107**, 127–47.

Berridge, M. J., Bootman, M. D. & Roderick, H. L. (2003). Calcium signaling: dynamics, homeostasis and remodeling. *Nature Reviews in Molecular and Cell Biology* **4**, 517–29.

Bistis, G. N., Perkins, D. D. & Read, N. D. (2003). Different cell types in *Neurospora crassa. Fungal Genetics Newsletter* **50**, 17–19.

Bobrowicz, P., Pawlak, R., Correa, A., Bell-Pedersen, D. & Ebbole, D. J. (2002). The *Neurospora crassa* pheromone precursor genes are regulated by the mating type locus and the circadian clock. *Molecular Microbiology* **45**, 795–804.

Bobrowicz, P., Wilkinson, H. H. & Ebbole, D. J. (2005). A mitogen-activated protein kinase pathway essential for mating and contributing to vegetative growth in *Neurospora crassa. Genetics* **170**, 1091–104.

Borkovich, K. A., Alex, L. A., Yarden, O., Freitag, M., Turner, G. E., Read, N. D., Seiler, S., Bell-Pederson, D., Paietta, J., Plesofskz, N., Plamann, M., Schulte, U., Mannhaupt, G., Nargang, F., Radford, A., Selitrennikoff, C., Galagan, J. E., Dunlap, J. C., Loros, J., Catcheside, D., Inoue, H., Aramazo, R., Polzmenis, M., Selker, E. U., Sachs, M. S., Marzluf, G. A., Paulsen, I., Davis, R., Ebbole, D. J., Yelter, A., Kalkman, E., O'Rourke, R., Bowring, F., Zeadon, J., Ishii, C., Suzuki, K., Sakai, W. & Pratt, R. (2004). Lessons from the genome sequence of *Neurospora crassa*: tracing the path from genomic blueprint to multicellular organism. *Microbiological and Molecular Biology Reviews* **68**, 1–108.

Bottone, E. J., Nagarsheth, N. & Chiu, K. (1998). Evidence of self-inhibition by filamentous fungi accounts for unidirectional hyphal growth in colonies. *Canadian Journal of Microbiology* **44**, 390–3.

Bracker, C. E., Murphy, D. J. & Lopez-Franco, R. (1997). Laser beam micromanipulation of cell morphogenesis in growing fungal hyphae. In *Functional Imaging of Optical Manipulation of Living cells. Proceedings of SPIE,* vol. 2983, ed. D. L. Farkas & B. J. Tromberg, pp. 67–80. Bellingham, Washington, USA: International Society of Optical Engineering.

Bruno, K. S., Aramayo, R., Minke, P. F., Metzenberg, R. L. & Plamann, M. (1996). Loss of growth polarity and mislocalization of septa in a *Neurospora* mutant altered in the regulatory subunit of cAMP-dependent protein kinase. *EMBO Journal* **15**, 5772–82.

Buller, A. H. R. (1931). *Researches on Fungi*, vol. 4. London: Longman.

Burow, G. B., Nesbitt, J., Dunlap, J. & Keller, N. P. (1997). Seed lipoxygenase products modulate *Aspergillus* mycotoxin biosynthesis. *Molecular Plant-Microbe Interactions* **10**, 380–7.

Chen, H., Fujita, M., Feng, Q., Clardy, J. & Fink, G. R. (2004). Tyrosol is a quorum-sensing molecule in *Candida albicans. Proceedings of the National Academy of Sciences of the USA* **101**, 5048–52.

Cooke, R. C. & Whipps, J. M. (1993). *Ecophysiology of Fungi.* London: Blackwell Scientific Publications.

Dean, R. A., Talbot, N. J., Ebbole, D. J., Farman, M. L., Mitchell, T. K., Orbach, M. J., Thon, M., Kulkarni, R., Xu, J. R., Pan, H., Read, N. D., Lee, Y. H., Carbone, I., Brown, D., Oh, Y. Y., Donofrio, N., Jeong, J. S., Soanes, D. M., Djonovic, S., Kolomiets, E., Rehmeyer, C., Li, W., Harding, M., Kim, S., Lebrun, M. H.,

Bohnert, H., Coughlan, S., Butler, J., Calvo, S., Ma, L. J., Nicol, R., Purcell, S., Nusbaum, C., Galagan, J. E. & Birren, B. W. (2005). The genome sequence of the rice blast fungus *Magnaporthe grisea*. *Nature* **434**, 980–6.

Dijksterhuis, J. (2003). Confocal microscopy of Spitzenkörper dynamics during growth and differentiation of rust fungi. *Protoplasma* **222**, 53–9.

d'Enfert, C., Bonini, B. M., Zapella, P. D., Fontaine, T., da Silva, A. M. & Terenzi, H. F. (1999). Neutral trehalases catalyse intracellular trehalose breakdown in the filamentous fungi *Aspergillus nidulans* and *Neurospora crassa*. *Molecular Microbiology* **32**, 471–83.

Enjalbert, B. & Whiteway, M. (2005). Release from quorum-sensing molecules triggers hyphal formation during *Candida albicans* resumption of growth. *Eukaryotic Cell* **4**, 1203–10.

Freitag, M., Hickey, P. C., Raju, N. B., Selker, E. U. & Read, N. D. (2004). GFP as a tool to analyze the organization, dynamics and function of nuclei and microtubules in *Neurospora crassa*. *Fungal Genetics and Biology* **41**, 897–910.

French, F. C. (1985). The bioregulatory action of flavor compounds on fungal spores and other propagules. *Annual Review of Phytopathology* **23**, 173–99.

Galagan, J., Calvo, S., Borkovich, K., Selker, E., Read, N. D., FitzHugh, W., Ma, L. P.-J., Smirnov, S., Purcell, S., Rehman, B., Elkins, T., Engels, R., Wang. S., Nielsen, C. B., Butler, J., Jaffe, D., Endrizzi, M., Qui, D., Planakiev, P., Bell-Pedersen, D., Nelson, M. A., Werner-Washburne, M., Selitrennikoff, C. P., Kinsey, J. A., Braun, E. L., Zelter, A., Schulte, U., Kothe, G. O., Jedd, G., Mewes. W., Staben, C., Marcotte, E., Greenberg, D., Roy, A., Foley, K., Naylor, J., Stange-Thomann, N., Barrett, R., Gnerre, S., Kamal, M., Kamvysselis, M., Bielke, C., Rudd, S., Frishman, D., Krystofova, S., Rasmussen, C., Metzenberg, R. L., Perkins, D. D., Kroken, S., Catcheside, D., Li, W., Pratt, R. J., Osmani, S. A., DeSouza, C. P. C., Glass, L., Orbach, M. J., Berglund, J. A., Voelker, R., Yarden, O., Plamann, M., Seiler, S., Dunlap, J., Radford, A., Amraayo, R., Natvig, D. O., Alex, L. A., Mannhaupt, G., Ebbole, D. J., Freitag, M., Paulsen, I., Sachs, M. S., Lander, E. S., Nusbaum, C., Birren, B. (2003). The genome sequence of the filamentous fungus *Neurospora crassa*. *Nature* **422**, 859–68.

Girbardt, M. (1957). Der Spitzenkörper von *Polystictus versicolor*. *Planta* **50**, 47–59.

Gooday, G. W. (1994). Hormones in mycelial fungi. In *The Mycota*, vol. 1, *Growth, Differentiation and Sexuality*, 1st edn, ed. J. G. H. Wessels & F. Meinhardt, pp. 401–11. Berlin: Springer-Verlag.

Gow, N. A. R. (2004). New angles in mycology: studies in directional growth and directional motility. *Mycological Research* **108**, 5–13.

Gull, K. (1978). Form and function of septa in filamentous fungi. In *The Filamentous Fungi*, vol. 3, *Developmental Mycology*, ed. J. E. Smith & D. R. Berry, pp. 78–93. London: Edward Arnold.

Harris, S. D. (2001). Septum formation in *Aspergillus nidulans*. *Current Opinion in Microbiology* **4**, 736–9.

Harris, S. D., Read, N. D., Roberson, R. W.., Shaw, B., Seiler, S., Plamann, M. & Momany, M. (2005). Polarisome meets Spitzenkörper: microscopy, genetics, and genomics converge. *Eukaryotic Cell* **4**, 225–9.

Hickey, P. C., Jacobson, D., Read, N. D. & Glass, N. L. (2002). Live-cell imaging of vegetative hyphal fusion in *Neurospora crassa*. *Fungal Genetics and Biology* **37**, 109–19.

Hickey, P. C., Swift, S. R., Roca, M. G. & Read, N. D. (2005). Live-cell imaging of filamentous fungi using vital fluorescent dyes and confocal microscopy. In *Methods in Microbiology*, vol. 34, *Microbial Imaging*, ed. T. Savidge & C. Pothoulakis, pp. 63–87. Amsterdam: Elsevier.

Hornby, J. M., Jensen, E. C., Lisec, A. D., Tasto, J. J., Jahnke, B., Shoemaker, R., Dussault, P. & Nickerson, K. W. (2004). Quorum sensing in the dimorphic fungus *Candida albicans* is mediated by farnesol. *Applied and Environmental Microbiology* **67**, 2982–92.

Ingold, C. T. (1971). *Fungal Spores, their Liberation and Dispersal.* Oxford: Clarendon Press.

Ivey, F. D., Hodge, P. N., Tunrer, G. E. & Borkovich, K. A. (1996). The Gα_i homologue gna-1 controls multiple differentiation pathways in *Neurospora crassa. Molecular Biology of the Cell* **7**, 1283–97.

Ivey, F. D. Kays, A. M. & Borkovich, K. A. (2002). Shared and independent roles for a Gα protein and adenylyl cyclase in regulating development and stress responses in *Neurospora crassa. Eukaryotic Cell* **1**, 634–42.

Kana-Uchi, A., Yamashiro, C. T., Tanabe, S. & Murayama, T. (1997). A *ras* homologue of *Neurospora crassa* regulates morphology. *Molecular and General Genetics* **254**, 427–32.

Kays, A. M. & Borkovich, K. A. (2005). Signal transduction pathways mediated by heterotrimeric G proteins. In *The Mycota*, vol. 3, *Biochemistry and Molecular Biology*, 2nd edn., ed. R. Brambl & G. A. Marzluf, pp. 175–207. Berlin: Springer-Verlag.

Kays, A. M., Rowley, P. S., Baasiri, R. A. & Borkovich, K. A. (2000). Regulation of conidiation and adenylyl cyclase levels by the protein GNA-3 in *Neurospora crassa. Molecular Cell Biology* **20**, 7693–705.

Kim, H., Metzenberg, R. L. & Nelson, M. A. (2002). Multiple functions of *mfa-1*, a putative pheromone precursor gene of *Neurospora crassa. Eukaryotic Cell* **1**, 987–99.

Kruppa, M., Krom, B. P., Chauhan, N., Bambach, A. V., Cihlar, R. L. & Calderone, R. A. (2004). The two-component signal transduction protein Chk1p regulates quorum sensing in *Candida albicans. Eukaryotic Cell* **3**, 1062–5.

Krystofova, S. & Borkovich, K. A. (2005). The heterotrimeric G-protein subunits GNG-1 and GNB-1 form a G$\beta\gamma$ dimer required for normal female fertility, asexual development, and Gα protein levels in *Neurospora crassa. Eukaryotic Cell* **4**, 365–78.

Kulkarni, R. D., Thon, M. R., Pan, H. & Dean, R. A. (2005). Novel G-protein-coupled receptor-like proteins in the plant pathogenic fungus *Magnaporthe grisea. Genome Biology* **6**, R24.

Levin, D. E. (2005). Cell wall integrity signaling in *Saccharomyces cerevisiae. Microbiological and Molecular Biology Reviews* **69**, 292–305.

Lopez-Franco, R. & Bracker, C. E. (1996). Diversity and dynamics of the Spitzenkörper in growing hyphal tips of higher fungi. *Protoplasma* **195**, 90–111.

Lu, H. & McLaughlin, D. J. (2005). Ultrastructure of the septal pore apparatus and early septum initiation in *Auricularia auricula-judae. Mycologia* **83**, 322–34.

Lucas, J. A. (2004). Survival, surfaces and susceptibility – the sensory biology of pathogens. *Plant Pathology* **53**, 679–91.

Macko, V. & Staples, R. C. (1973). Regulation of uredospore germination and germ tube development. *Bulletin of the Torrey Botanical Club* **100**, 223–9.

Martin, S. W., Douglas, L. M. & Konopka, J. B. (2005). Cell cycle dynamics and quorum sensing in *Candida albicans* chlamydospores are distinct from budding and hyphal growth. *Eukaryotic Cell* **4**, 1191–202.

Neer, E. J. (1995). Heterotrimeric G proteins: organizers of transmembrane signals. *Cell* **80**, 249–57.

Pandey, A., Roca, G., Read, N. D. & Glass, N. L. (2004). Role of a MAP kinase pathway during conidial germination and hyphal fusion in *Neurospora crassa. Eukaryotic Cell* **3**, 348–58.

Prokisch, H., Yarden, O., Dieminger, M., Tropschug, M. & Barthelmess, I. B. (1997). Impairment of calcineurin function in *Neurospora crassa* reveals its essential role in hyphal growth, morphology and maintenance of the apical Ca^{2+} gradient. *Molecular and General Genetics* **256**, 104–14.

Rayner, A. D. M. (1996). Interconnectedness and individualism in fungal mycelia. In *A Century of Mycology*, ed. B. C. Sutton, pp. 193–232. Cambridge: Cambridge University Press.

Rayner, A. D. M., Watkins, Z. R. & Beeching, J. R. (1997). Self integration – an emerging concept from the fungal mycelium. In *The Fungal Colony*, ed. N. A. R. Gow, G. D. Robson & G. M. Gadd, pp. 1–24. Cambridge: Cambridge University Press.

Read, N. D. (1994). Cellular nature and multicellular morphogenesis of higher fungi. In *Shape and Form in Plants and Fungi*, ed. D. S. Ingram & A. Hudson, pp. 251–69. London: Academic Press.

Read, N. D. & Beckett, A. (1996). Ascus and ascospore morphogenesis. *Mycological Research* **100**, 1281–314.

Read, N. D. & Roca, M. G. (2006). Vegetative hyphal fusion in filamentous fungi. In *Cell–Cell Channels*, ed. F. Baluska, D. Volkmann & P. W. Barlow, pp. 87–98. Georgetown, Texas: Landes Bioscience.

Read, N. D., Kellock, L. J., Knight, H. & Trewavas, A. J. (1992). Contact sensing during infection by fungal pathogens. In *Perspectives in Plant Cell Recognition*, ed. J. A. Callow & J. R. Green, pp. 137–172. Cambridge: Cambridge University Press.

Reynaga-Pena, C. G. & Bartnicki-Garcia, S. (1997). Apical branching in a temperature sensitive mutant of *Aspergillus niger*. *Fungal Genetics and Biology* **22**, 153–67.

Riquelme, M., Reynaga-Pena, C. G., Gierz, G. & Bartnicki-Garcia, S. (1998). What determines growth direction in fungal hyphae? *Fungal Genetics and Biology* **24**, 101–9.

Robinson, P. M. (1973a). Chemotropism in fungi. *Transactions of the British Mycological Society* **61**, 303–13.

Robinson, P. M. (1973b). Autotropism in fungal spores and hyphae. *Botanical Review* **39**, 367–84.

Robinson, P. M. (1973c). Oxygen-positive chemotropic factor for fungi? *New Phytologist* **72**, 1349–56.

Roca, M. G., Arlt, J., Jeffree, C. E. & Read, N. D. (2005). Cell biology of conidial anastomosis tubes in *Neurospora crassa*. *Eukaryotic Cell* **4**, 911–19.

Schumacher, M. M., Enderlin, C. S. & Selitrennikoff, C. P. (1997). The osmotic-1 locus of *Neurospora crassa* encodes a putative histidine kinase similar to osmosensors of bacteria and yeast. *Current Microbiology* **34**, 340–7.

Snetselaar, K. M., Bolker, M. & Kahmann, R. (1996). *Ustilago maydis* mating hyphae orient their growth toward pheromone sources. *Fungal Genetics and Biology* **20**, 299–312.

Taylor, J. W., Spatafora, J., O'Donnell, K., Lutzoni, F., Hibbet, D. S., Geiser, D., Bruns, T. D. & Blackwell, M. (2004). The Fungi. In *Assembling the Tree of Life*, ed. J. Cracraft & M. J. Donoghue, pp. 171–94. Oxford: Oxford University Press.

Tlalka, M., Watkinson, S. C., Darrah, P. R. & Fricker, M. D. (2002). Continuous imaging of amino acid translocation in intact mycelia of *Phanerochaete velutina* reveals rapid, pulsatile fluxes. *New Phytologist* **153**, 173–84.

Tlalka, M., Hensman, D., Darrah, P. R., Watkinson, S. C. & Fricker, M. D. (2003). Noncircadian oscillations in amino acid transport have complementary profiles in assimilatory and foraging hyphae of *Phanerochaete velutina*. *New Phytologist* **158**, 325–35.

Trinci, A. P. J. (1971). Influence of the width of the peripheral growth zone on the radial growth rate of fungal colonies on solid medium. *Journal of General Microbiology* **67**, 325–44.

Trinci, A. P. J. (1983). Regulation of hyphal branching and hyphal orientation. In *The Ecology and Physiology of the Fungal Mycelium*, ed. D. H. Jennings & A. D. M. Rayner, pp. 243–8. Cambridge, Cambridge University Press.

Tsitsigiannis, D. I., Kowieski, T. M., Zarnowski, R. & Keller, N. P. (2005). Three putative oxylipin biosynthetic genes integrate sexual and asexual development in *Aspergillus nidulans*. *Microbiology* **151**, 1809–21.

Turner, G. & Harris, S. D. (1997). Genetic control of polarized growth and branching in filamentous fungi. In *The Fungal Colony*, ed. N. A. R. Gow, G. D. Robson & G. M. Gadd, pp. 229–60. Cambridge: Cambridge University Press.

Ugalde, U. (2006). Autoregulatory signals in mycelial fungi. In *The Mycota*, vol. 1, *Growth, Differentiation and Sexuality*, 2nd edn, ed. U. R. Kües & R. Fischer, pp. 203–213. Berlin: Springer-Verlag.

Yang, Q., Pool, S. I. & Borkovich, K. A. (2002). A G-protein β subunit required for sexual and vegetative development and mainteneance of normal Gα protein levels in *Neurospora crassa*. *Eukaryotic Cell* **1**, 378–90.

Zelter, A., Bencina, M., Bowman, B. J., Yarden, O. & Read, N. D. (2004). A comparative genomic analysis of the calcium signaling machinery in *Neurospora crassa*, *Magnaporthe grisea*, and *Saccharomyces cerevisiae*. *Fungal Genetics and Biology* **41**, 827–41.

Zhang, Y., Lamm, R., Pillonel, C., Lam, S. & Xu, J. R. (2002). Osmoregulation and fungicide resistance: the *Neurospora crassa os-2* gene encodes a HOG1 mitogen-activated protein kinase homologue. *Applied and Environmental Microbiology* **68**, 532–8.

4

Berkeley Award Lecture: Mathematical modelling of the form and function of fungal mycelia

FORDYCE A. DAVIDSON

Division of Mathematics, University of Dundee

Introduction

In most environments, the spatial distribution of nutrient resources is not uniform. Such *heterogeneity* is particularly evident in mineral soils, where an additional level of spatial complexity prevails owing to the complex pore network in the solid phases of the soil. Mycelial fungi are well adapted to growth in such spatially complex environments, since the filamentous hyphae can grow with ease across surfaces and also bridge air gaps between such surfaces. This ability is significantly enhanced by the propensity of many species to translocate materials through hyphae between different regions of the mycelium. Thus, it has been suggested that hyphae growing through nutritionally impoverished zones of soil, or deleterious regions (e.g. localized deposits of organic pollutants, toxic metals, dry or waterlogged zones), can be supplemented by resources imported from distal regions of the mycelium (Morley *et al.*, 1996). This has profound implications for the growth and functioning of mycelia and attendant effects upon the environment in which they live. Thus, the fungal mycelium represents an extremely efficient system for spatial exploration and exploitation.

The study of filamentous fungi through experimental means alone can be very difficult owing to the complexity of their natural growth habitat and the range of scales over which they grow and function. Mathematical modelling provides a complementary, powerful and efficient method of investigation and can provide new insight into the complex interaction between the developing mycelium and its environment.

In his seminal paper on morphogenesis, Turing (1952) wrote that a mathematical model is 'a simplification and an idealisation'. This captures the aim of mathematical modelling. It is *not* the goal of the mathematical

Fungi in the Environment, ed. G. M. Gadd, S. C. Watkinson & P. S. Dyer. Published by Cambridge University Press.

modeller to form an extremely complex system of equations in an attempt to mirror reality. All that achieves is the replacement of one form of impenetrable complexity with another. Instead, the aim is to reduce a complex (biological) system to a simpler (mathematical) system where the rigorous, logical structure of the latter can be used to isolate, identify and investigate key properties. However, as Einstein is famously quoted, 'everything should be made as simple as possible, but no simpler'. Hence, mathematical modelling is not about what to include, but instead, about what can be omitted, where the art is in achieving a meaningful balance between the two.

Mathematical modelling of the growth and function of fungi has been conducted for several decades. For an extensive review of the work conducted up to the mid-1990s, see Prosser (1995). One of the main problems that faces modellers is the choice of scale. Clearly, it must first be decided what specific biological questions are to be addressed. It must then be decided at what scale these are most likely to be expressed. Of course the end goal would be to construct models that operate at a range of scales and which transfer information across scale boundaries from the action of individual genes through to the growth and function of large-scale mycelia. Indeed, this problem of *multi-scale* modelling of biological systems is currently the subject of intense interest and we discuss later how our new modelling approach fits into this framework. Consequently, until recently, attempts at mathematical modelling of fungal growth have in general either focused on the mycelium, using variables such as biomass yield (see, for example, Paustian & Schnürer, 1987; Lamour *et al.*, 2000), or focused on growth at the hyphal level, such as hyphal tip growth, branching and anastomosis (see, for example, Prosser & Trinci, 1979; Heath, 1990; Regalado *et al.*, 1997). In the former, spatial properties are generally ignored; in the latter, temporal effects are often neglected. Large-scale, spatio-temporal properties of fungal mycelia have been less extensively addressed, but progress has been made, in particular by Edelstein (1982), Edelstein & Segel (1983), Edelstein-Keshet & Ermentrout (1989), Davidson *et al.* (1996a, b, 1997a, b), Davidson (1998), Davidson & Park (1998), Davidson & Olsson (2000) and most recently by Boswell *et al.* (2002, 2003a, b). In these studies, the approach has been to derive systems of equations (nonlinear partial differential equations) that represent the interaction of fungal biomass and a growth-limiting substrate (e.g. a carbon source). Such an approach is ideal when modelling dense mycelia, for example growth in Petri dishes or on the surfaces of solid substrates such as foodstuffs, plant surfaces and building materials. This modelling strategy has, for example, allowed the study of biomass distribution within the mycelium in homogeneous and heterogeneous conditions and the study of translocation in a variety of habitat configurations as well as the study of certain functional consequences of fungal growth, such as acid production. However, when growth is sparse, as in nutrient-poor conditions or in structurally heterogeneous environments such as soils, then this continuum approach is less relevant. In this case, a discrete modelling approach is more appropriate, where individual hyphae are identified. Such discrete models usually take the form of computer simulations (Hutchinson *et al.*, 1980; Bell, 1986; Kotov & Reshetnikov, 1990; Ermentrout & Edelstein-Keshet, 1993; Soddell *et al.*, 1995; Regalado *et al.*, 1996; Meskauskas *et al.*, 2004a, b) and are often derived from the statistical properties of the experimental system under investigation. Although these models can yield images that are almost indistinguishable from real fungi grown in uniform conditions, they often employ non-mechanistic rules to generate hyphal tip extension and hyphal branching and thus must be re-formulated (or at the very least completely re-calibrated) to consider the growth of the same species in different environments or to consider the growth of a different species. Certainly, there are significant advances that can be made by using these models (see in particular the work conducted by Meskauskas *et al.*, 2004a, b). However, such advances may lie more in the testing of hypotheses concerning basic growth architecture. Difficulties arise in attempting to make and test hypotheses concerning growth dynamics and function when the underlying mechanisms for growth are not modelled appropriately. Furthermore, because of the overwhelming computational difficulties, discrete models have until now neglected two key properties of fungi, anastomosis and translocation, processes crucial in the extension of hyphal tips with particular relevance to growth in heterogeneous environments.

In the following account, the recent development of a new mathematical model is detailed. The model connects physiology at the hyphal level (e.g. tip growth and branching) to growth and function at the mycelial level (see Boswell *et al.*, 2002, 2003a, b). First a continuum formulation is outlined that details the number and location of hyphae, hyphal tips, and concentration of a growth-limiting substrate and which makes useful predictions concerning the roles of nutrient translocation within mycelia. Moreover, it also allows the study of the functional consequences of fungal growth in a variety of habitat configurations. It will then be described how this continuum approach can be used to construct a model in which the mycelium is represented as a discrete structure. In this formulation, anastomosis and translocation are explicitly included. However, to facilitate the latter, the substrates are, as is most appropriate, treated as continuous variables. Therefore the model is of the *hybrid cellular automaton* type.

A new model for mycelial growth

The mycelium is modelled as a distribution consisting of three components: *active hyphae* (corresponding to those hyphae involved in the translocation of internal metabolites), *inactive hyphae* (denoting those hyphae not involved in translocation or growth, e.g. moribund hyphae) and *hyphal tips*. An important distinction is made between nutrients located within the fungus (*internal*) and those free in the outside environment (*external*). Internally located material is used for metabolism and biosynthesis, e.g. in the extension of hyphal tips (creating new hyphae), branching (creating new hyphal tips), maintenance and the uptake of external nutrient resources. In most environments, a combination of nutrients is necessary for growth (carbon, nitrogen, oxygen, etc.) but, for simplicity, in the model system it is assumed that a single generic element is limiting for growth. This element is assumed to be carbon, given its central role in growth and function and also since nitrogen and oxygen were abundant in the model calibration experiment (Boswell *et al.*, 2002). This model is based on the physiology and growth characteristics of the ubiquitous soil-borne saprophyte *Rhizoctonia solani* and it is to the results of growth experiments using this fungus that the model has been initially compared. However, many aspects of the model (and results thereby obtained) are applicable to a large class of fungi growing in a variety of habitats.

In terms of the five variables outlined above (active hyphae, inactive hyphae, hyphal tips, internal substrate and external substrate), the model has the following structure:

change in active hyphae
in a given area = *new hyphae (laid down by moving tips)*
+ *reactivation of inactive hyphae*
− *inactivation of active hyphae*, (1)

change in inactive hyphae
in a given area = *inactivation of active hyphae*
− *reactivation of inactive hyphae*
− *degradation of inactive hyphae*, (2)

change in hyphal tips
in a given area = *tip movement out of/into area*
+ *branching from active hyphae*
− *anastomosis of tips with hyphae*, (3)

change in internal substrate
in a given area = *translocation (active and passive mechanisms)*
+ *uptake into the fungus from external sources*
− *maintenance costs of hyphae*
− *growth costs of hyphal tips*
− *active translocation costs*, (4)

change in external substrate in a
given area = *diffusion of external substrate out of/ into area*
− *uptake by fungus*. (5)

It is commonly observed that hyphal tips have a tendency to move in a straight line but with small random fluctuations in the direction of growth (due to the manner in which new wall material is incorporated at the tip) and that the rate of tip growth depends on the status of internally located material. The model includes these important growth characteristics. In the model, the 'trails' left behind moving hyphal tips correspond to the creation of hyphae. It has been widely reported that hyphal branching in mycelial fungi is related to the status of internally located material: turgor pressure and the build-up of tip vesicles have been implicated (Webster, 1980; Gow & Gadd, 1995). Thus, the branching process is modelled as being proportional to the internal substrate concentration. In mycelial fungi, the uptake of nutrients occurs by active transport across the plasma membrane. Hence, in the model system, the uptake process depends not only on the concentration of the external substrate, but also on the concentration of the internal substrate (i.e. the energy available for the active uptake) and on the amount of hyphae (i.e. membrane surface area). It is known that many species of fungi possess both active (i.e. metabolically driven) and passive (i.e. diffusive) translocation mechanisms for carbon (Olsson, 1995). Active substrate translocation, unlike diffusion, depletes the energy reserves within the mycelium and is modelled as a process that moves internal substrate towards hyphal tips since they represent the major component of mycelial growth and are therefore likely to be the largest net energy sinks.

A continuum approach

As a first step, it is assumed that the variables in the model system are *continuous* (i.e. can be viewed as densities) and as such a system of partial differential equations is formed. This treatment results in a system that is best suited to modelling dense mycelial growth of the type often observed in laboratory experiments. However, the true, branched (fractal) nature of the mycelial network is not disregarded entirely in the formulation of our model: this is taken account of by carefully modelling translocation so as to best represent movement inside a branching (fractal) structure (essentially it is assumed that the *transit time* of vesicles transported around the network is less than if they were diffusing in free space; see Boswell *et al.* (2003a)).

Although the core of the model is formed from a consideration of the general growth characteristics of mycelial fungi, as mentioned above, for direct comparison with experimental observations, the results presented here were obtained in conjunction with experiments using *Rhizoctonia solani* Kühn anastomosis group 4 (R3) (IMI 385768) cultured on mineral salts medium (MSM) containing 2% (w/v) glucose. The model was calibrated by using simple growth experiments, and approximate tip velocities and branching and anastomosis rates were estimated by inspecting enlarged images of a mycelium grown over a 15 h time period. Other parameters were taken from the literature, namely the diffusion of internal and external substrate and the uptake rate of the substrate (see Boswell *et al.*, 2002).

The model equations were solved on a computer using a finite-difference approximation, which involves dividing time and space into discrete units. A square grid is superimposed on the (continuous) growth domain so that each square (or 'cell') in the grid contains a quantity of active and inactive biomass, hyphal tips, and internal and external substrate. Thus the densities and concentrations of the model system are stored on the computer in a series of two-dimensional arrays. These quantities change in subsequent time steps, as determined by the model equations, according to the status of each 'cell' and that of its neighbouring 'cells'. In this way, both local concentrations and gradients of concentrations of the five model variables can be considered. By repeatedly applying the above process using finer grids and smaller time steps, the numerical approximation obtained progressively resembles the true solution of the model equations.

A simple quantitative test of the model's predictive power is given by comparing the colony radial expansion, measured experimentally, to the biomass expansion obtained from the solution of the model equations. The colony radial expansion of *R. solani* was obtained by inoculating MSM with 0.25 cm radius plugs of the fungus previously grown on tap water agar. The plates were incubated at 15 °C and 30 °C and at regular time intervals the colony radii were measured in perpendicular directions. The radius of growth is defined as the mean of these distances once the radius of the initial inoculation has been subtracted. The mean radial growth from five replicates was determined (Fig. 4.1).

The calibrated model equations were solved with initial data representing the experimental protocol described above for growth at 30 °C. The radius of growth in the model system was determined in a consistent manner to that in the experimental system (Fig. 4.1). The total hyphal density (i.e. active and inactive hyphae) is shown in Fig. 4.2 (a)–(d) and there is good quantitative agreement between the experimental and model biomass values obtained (Boswell *et al.*, 2003a). By assuming that tip

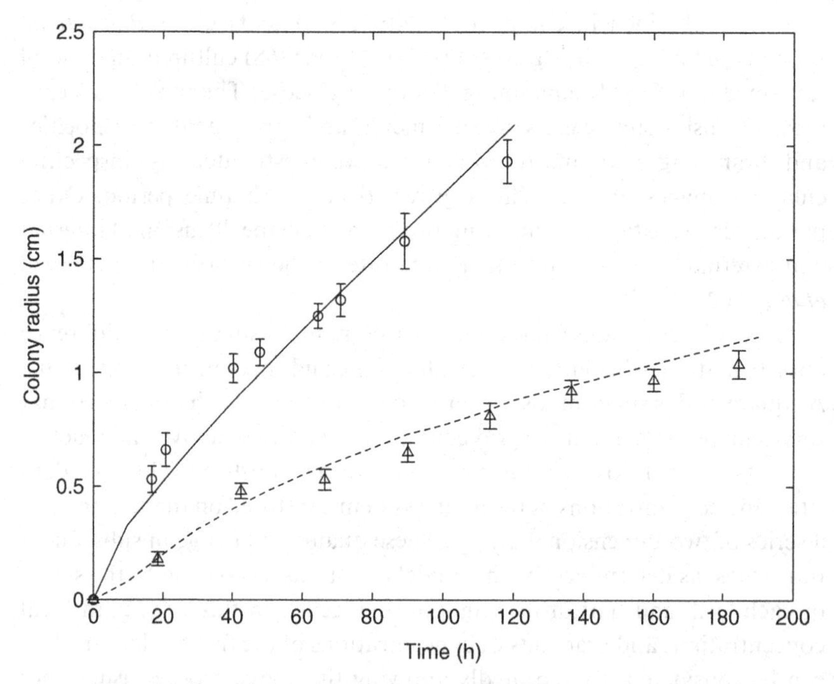

Fig. 4.1. Experimental and model colony radial growth plotted over time. Experimental data for growth measurements at 30 °C are denoted by circles; experimental data for growth measured at 15 °C are denoted by triangles; both are augmented with standard error bars. The solid line corresponds to the model biomass radius growth for the 30 °C calibration and the dashed line corresponds to the model biomass radius growth at 15 °C using the reduction of tip-velocity corresponding to the Q_{10}-rule.

velocity captures the cumulative effects of temperature on a cascade of metabolic processes within the mycelium, a simple reduction of tip velocity by an amount consistent with the Q_{10}-rule generates a predictive radial growth rate at the lower temperature of 15 °C that is surprisingly consistent with experimental data (Fig. 4.1) (the Q_{10}-rule is a rule of thumb that states a metabolic reaction approximately halves with a 10 K decrease in temperature). Thus it appears that the apparently complex effects of temperature on the growth of *R. solani* may be easily accounted for in the mathematical model by varying a *single* parameter.

A remarkable result arises when the active translocation term in the model is turned off (i.e. the parameter associated with active translocation is set to zero): the radial growth rate and biomass distributions are largely unaffected. A conclusion from this observation is that modelling translocation by diffusion of internal substrate is sufficient to accurately replicate the experimental growth behaviour of *R. solani* in uniform, substrate-rich conditions. Therefore, the model predicts that *R. solani* does not use active translocation in nutrient-rich, uniform habitats and instead relies on the 'energy-free' process of diffusion for the re-distribution of internal metabolites.

In addition to modelling fungal growth and the subsequent nutrient depletion, the production of acidity can be modelled, generating further qualitative and quantitative data. It has been shown that acidity (which can arise from, for example, proton efflux and organic acid excretion) is produced by *R. solani* only in the presence of a utilizable carbon source (Jacobs *et al.*, 2002a, b). Because the internal substrate in the model system represents such a carbon source, the production of acidity can be modelled

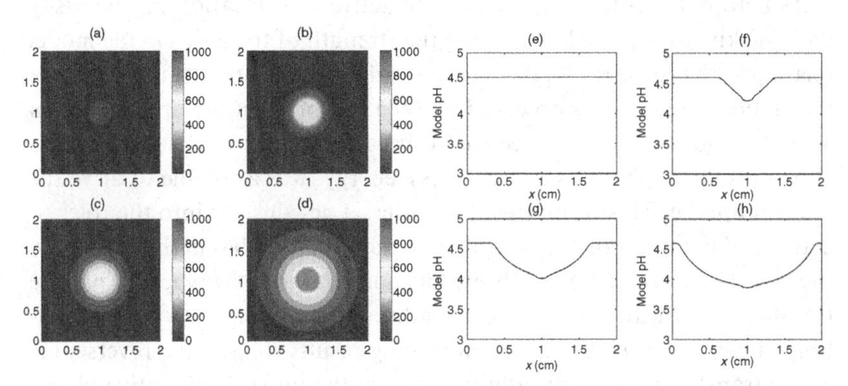

Fig. 4.2. (a)–(d) Biomass densities (cm hyphae cm^{-2}) from the time of 'inoculation' up to a time representing 2 days. (e)–(h) Cross-section of the corresponding model acidification of the growth medium.

as being proportional to the concentration of internal substrate (Boswell *et al.*, 2003a). This assumption provides model pH profiles shown in Fig. 4.2 (e)–(h), which accurately replicate (and extend) results obtained when pH gradients are measured experimentally (see, for example, Sayer & Gadd, 1997).

The cases considered above are all concerned with mycelial growth in initially uniform conditions. However, the model can easily be adapted to consider nutritionally heterogeneous environments, for example the tessellated agar droplet system discussed by Jacobs *et al.* (2002b). In that system, molten MSM agar was pipetted onto the bases of 9 cm diameter Petri dishes, forming a hexagonal array comprising 19 circular droplets each of diameter 10 mm and separated at their closest point by a nutrient-free gap of 2 mm. In total, 16 combinations of tessellations were considered by using MSM, MSM amended with glucose, MSM amended with insoluble calcium phosphate, and MSM amended with both glucose and calcium phosphate, to form the (seven) interior and (twelve) exterior droplets.

Recall that, in the model system, fungal growth depends on a single generic element, which is assumed to be the carbon source. Hence, the model can be applied without alteration to a sub-set of the tessellations considered in Jacobs *et al.* (2002b) corresponding to those four configurations constructed by using standard MSM and glucose-amended MSM (Fig. 4.3a, d, g). The model predicts general growth characteristics that are similar to those observed experimentally (Fig. 4.3c, f, i). In fact, the model extends the experimental results by, for example, explicitly mapping internal substrate concentrations. The production of acidity can thus be modelled, allowing for further extension of experimental results.

As before, the roles and functions of active translocation can be easily and quickly investigated by altering the strength of the terms in the model relating to that process. Upon decreasing the rate of active translocation in the model system, it was observed that the rate of substrate uptake rate on a newly colonized droplet decreased, because less internal substrate was carried at the hyphal tips and thus less ‘energy’ was available to drive the (active) uptake. This result therefore offers a new insight into the roles of active and passive translocation. Active translocation has in the main been thought to be associated with exploration (outgrowth) whereas passive translocation (diffusion) has been traditionally associated with exploitation (substrate utilization). The modelling results suggest the reverse, i.e. active translocation is crucially involved in the initial exploitative phase, whereas diffusive translocation is in the main used as a short-range explorative mechanism.

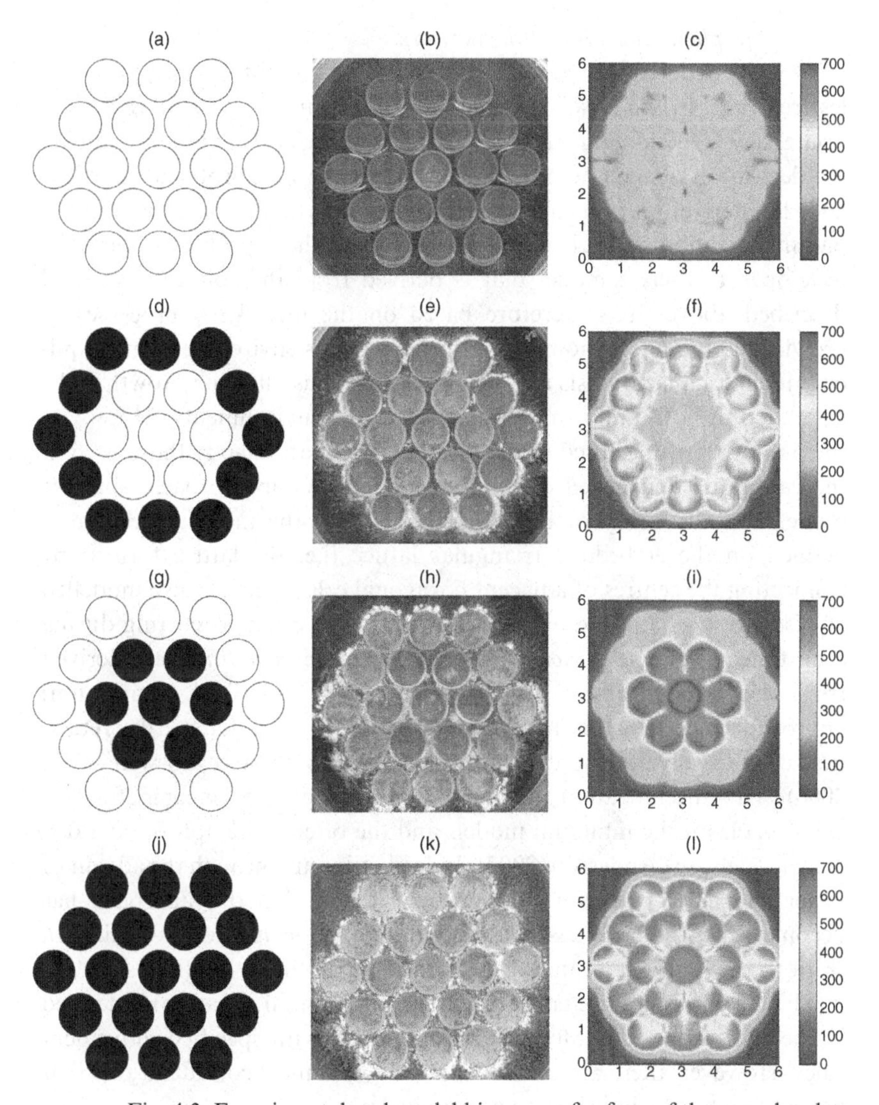

Fig. 4.3. Experimental and model biomasses for four of the agar droplet tessellations described by Jacobs *et al.* (2002b). (a), (d), (g) and (j): the tessellations where solid discs denote agar droplets formed by using glucose-amended MSM and open discs denote agar droplets formed by using unamended MSM. The images (b), (e), (h), (k) show the fungus 7 days after inoculation of the central droplet (corresponding, respectively, to tessellations (a), (d), (g), (j)) and are obtained by scanning the underside of a 9 cm diameter Petri dish. The images (c), (f), (i), (l) show the corresponding model biomass densities after a time representing 7 days (cm hyphae cm^{-2}).

Explicit modelling of the network

As detailed above, when growth is sparse a continuum approach is less relevant. In this case, a discrete modelling approach is more appropriate, in which individual hyphae are identified. As also discussed above, the derivation of such discrete models is fraught with problems, particularly the derivation of meaningful rules for growth and function and the parameterization of such rules. To overcome these problems, we have developed a discrete model that is derived from the continuum model described above. It is therefore based on the underlying processes of growth and the interaction of the fungus with its environment and explicitly includes anastomosis and translocation, thus allowing growth to be simulated in both uniform and heterogeneous environments. Moreover, the parameter values used in the discrete model are exactly those used in the calibrated and tested continuum model. In our approach, space is modelled as an array of hexagonal 'cells' and the model mycelium is defined on the embedded triangular lattice (i.e. the lattice formed by connecting the centres of adjacent hexagonal cells). Time is also modelled as discrete steps and the probabilities of certain events occurring during each time interval (the *movement* or *transition probabilities*) are derived from the assumptions used in the previously described (continuum) approach: essentially they are derived from the finite-difference discretization used to solve the continuum model numerically (see Boswell *et al.*, 2006). This discretization method offers a simple and meaningful link between classical continuum models and the often more appropriate discrete models (see Anderson, 2003). Indeed, it is only through the advent of faster and more powerful computer processors that this approach has become practicable. In essence, this discretization method is similar in spirit to the reinforced random walk theory of Othmer & Stevens (1997), in which a continuous equation (a partial differential equation) is derived in the limit from a (reinforced) random walk with specified movement rules. However, there are certain important technical considerations that mean these two techniques are not equivalent.

In the context of the continuum fungal model, the discretization procedure allows certain key processes, including hyphal inactivation and reactivation, branching and anastomosis to be treated in a more detailed manner. To the author's knowledge, a novel property of the model system detailed here is the simultaneous use of a combination of 'cell' models (which are used for modelling internal/external substrate and hyphal tips) and 'bond' models (which are used for modelling active/inactive hyphae). 'Cell' models are ideal for modelling the movement of individual particles because each cell (or site) can take a value corresponding to the current state of that cell (for example, the presence or absence of a tip). However, this cellular approach is not suitable for modelling the development of a network, because adjacent cells need not be connected. For example, consider two adjacent hyphae that run parallel to one another. If a cell-based approach was used, the two hyphae would automatically be connected to one another and thus would allow the transfer of internally located material. In reality, this is unlikely to be the case: such connections will only occur at anastomosis points. A bond-based approach explicitly models any such connections and thus can accurately model the formation of, and internal substrate re-distribution inside, parallel hyphae.

This discrete model replicates many of the important qualitative features associated with mycelial growth in uniform conditions (Fig. 4.4). Moreover, quantitative features such as *fractal dimension* (how well the network fills space) are also consistent with experimental observations and we have been able to obtain relations between substrate concentration and fractal dimension (see Boswell *et al.* 2006).

Growth and function in soils

Soils exhibit spatiotemporal, nutritional and structural heterogeneity. The structural heterogeneity in soils is determined by the relative location of soil particles and the resulting pore space. Nutritional heterogeneity is strongly modulated by the ground-water distribution, which

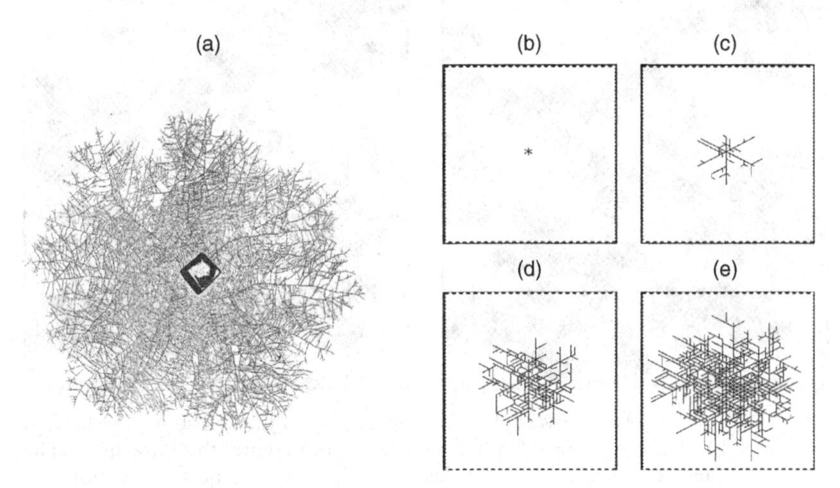

Fig. 4.4. (a) A typical mycelium of *Rhizoctonia solani*. (b)–(e) The development over time of a model mycelial network generated by the hybrid model.

itself depends on the architecture of the pore space. In non-saturated soils, water films prevail around pore walls and larger pores are air-filled. Nutrients (with the exception of oxygen) are in general confined to such water films. The effects of water surface tension mean that these nutrients diffuse within the film but not across its outer surface. Experimental studies of mycelial growth in soils typically consist of examining thin slices of soils. These soil slices are in essence a two-dimensional object and hence

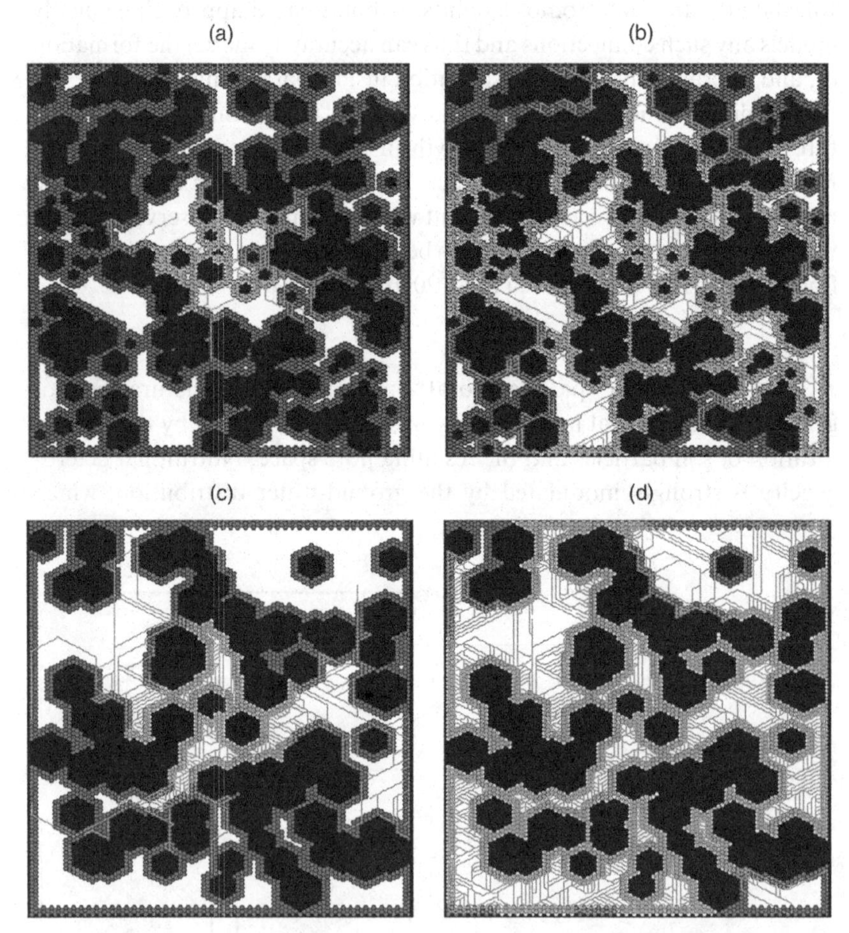

Fig. 4.5. Model mycelial growth generated by the hybrid model, (a) and (b). The developing hyphal network is represented by dark lines. The black cells denote soil particles; the lighter cells respectively denote the water film and the white regions correspond to air-filled pore spaces. Typical biomass growth in a similar environment but with reduced water surface tension is shown in (c) and (d).

various properties, such as the fractal dimension of the growth habitat and the location and abundance of biomass within the growth habitat, can be easily compared between the model and experimental systems. We have constructed artificial structures that emulate heterogeneous porous media, such as soils, by randomly 'removing' (possibly overlapping) hexagonal blocks of cells from the growth domain. The distribution of the remaining habitat, which corresponds to the pore space, may be connected or fragmented. Fundamental properties of the model pore space, such as its fractal dimension, can be computed, enabling qualitative and quantitative comparisons with real soil systems as indicated above.

By constructing a model soil as detailed above, the spatiotemporal development of the model mycelium and the predicted acidification of the surrounding environment can be studied in 'real time'. From the model, it is seen that early biomass growth is confined to the region representing the water film; the external substrate in this region is taken up and used for tip extension (Fig. 4.5a, b). A small number of tips emerge from the region representing the water film and extend rapidly across the model pore space and by doing so, locate new substrate resources, which are subsequently colonized and exploited. Surface tension of the water film was observed to play a significant role in determining biomass and acidity distribution. Reducing surface tension in the model results in a greater biomass distribution in the pore space and a faster overall biomass expansion (Fig. 4.5c, d).

Conclusions

Our model captures key experimentally observed qualitative and quantitative features, and makes predictions regarding colony radial expansion rate, biomass distribution, and acidification of the growth environment. The hybrid approach discussed here overcomes certain intrinsic difficulties in deriving meaningful cellular automata by basing the rules for growth and function and associated parameter values on those derived for the calibrated and tested continuum model. This approach enables growth and function to be modelled in a far more precise manner, which explicitly details the development of the hyphal network. This detail is essential for the study of growth and function in structured media, for example soils. We conclude that our modelling approach provides a powerful tool to augment experimental studies of growth, function and morphogenesis of mycelial fungi in uniform and heterogeneous environments.

Certain fungi have significant uses as biological control agents against pests and diseases of plants; others have the ability to transform toxic metals and are important factors in bioremediation. As nutrient recyclers, biocontrol agents and bioremediation agents, fungi are growing in environments exhibiting spatio-temporal nutritional and structural heterogeneity. Appropriate mathematical models can yield considerable insight into the functional consequences of such fungi and the roles they play. By combining modelling with experimental data, more detailed qualitative *and* quantitative results can be obtained. Consequently, we predict that mathematical modelling can play a central role in the successful application of fungi to biotechnological areas such as biocontrol and bioremediation as well as providing a better understanding of their significant roles in environmental biogeochemistry.

Acknowledgements

I am deeply honoured to have received the Berkeley Award and thank the British Mycological Society for their continued support. I am also indebted to my colleagues Professor Geoff Gadd, Dr Graeme Boswell, Professor Karl Ritz and Dr Helen Jacobs and acknowledge the financial support from the Biotechnology and Biological Sciences Research Council (BBSRC 94/MAF 12243) as part of the Mathematics and Agriculture and Food Systems Initiative.

References

Anderson, A. R. A. (2003). A hybrid discrete-continuum technique for individual based migration models. In *Polymer and Cell Dynamics*, ed. W. Alt, M. Chaplain, M. Griebel & J. Lenz, pp. 251–9. Zurich: Birkhauser.

Bell, A. D. (1986). The simulation of branching patterns in modular organisms. *Philosophical Transactions of the Royal Society of London*, **B313**, 143–60.

Boswell, G. P., Jacobs, H., Davidson, F. A., Gadd, G. M. & Ritz, K. (2002). Functional consequences of nutrient translocation in mycelial fungi. *Journal of Theoretical Biology* **217**, 459–77.

Boswell, G. P., Jacobs, H., Davidson, F. A., Gadd, G. M. & Ritz, K. (2003a). Growth and function of mycelial fungi in heterogeneous environments. *Bulletin of Mathematical Biology* **65**, 447–77.

Boswell, G. P., Jacobs, H., Gadd, G. M., Ritz, K. & Davidson, F. A. (2003b). A mathematical approach to studying fungal mycelia. *Mycologist* **17**, 165–71.

Boswell, G. P., Jacobs, H., Ritz, K., Gadd, G. M. & Davidson, F. A. (2006). The development of fungal networks in complex environments. *Bulletin of Mathematical Biology* (in press).

Davidson, F. A. (1998). Modelling the qualitative response of fungal mycelia to heterogeneous environments. *Journal of Theoretical Biology* **195**, 281–91.

Davidson, F. A. & Olsson, S. (2000). Translocation induced outgrowth of fungi in nutrient-free environments. *Journal of Theoretical Biology* **205**, 73–84.

Davidson, F. A. & Park, A. W. (1998). A mathematical model for fungal development in heterogeneous environments. *Applied Mathematics Letters* **11**, 51–6.

Davidson, F. A., Sleeman B. D., Rayner, A. D. M., Crawford, J. W. & Ritz, K. (1996a). Context-dependent macroscopic patterns in growing and interacting mycelial networks. *Proceedings of the Royal Society of London* **B263**, 873–80.

Davidson, F. A., Sleeman B. D., Rayner, A. D. M., Crawford, J. W. & Ritz, K. (1996b). Large-scale behaviour of fungal mycelia. *Mathematical and Computer Modelling* **24**, 81–7.

Davidson, F. A., Sleeman, B. D., Rayner, A. D. M., Crawford, J. W. & Ritz, K. (1997a). Travelling waves and pattern formation in a model for fungal development. *Journal of Mathematical Biology* **35**, 589–608.

Davidson, F. A., Sleeman, B. D. & Crawford, J. W. (1997b). Travelling waves in a reaction-diffusion system modelling fungal mycelia. *IMA Journal of Applied Mathematics* **58**, 237–57.

Edelstein, L. (1982). The propagation of fungal colonies – a model for tissue growth. *Journal of Theoretical Biology* **98**, 679–701.

Edelstein, L. & Segel, L. (1983). Growth and metabolism in mycelial fungi. *Journal of Theoretical Biology* **104**, 187–210.

Edelstein-Keshet, L. & Ermentrout, B. (1989). Models for branching networks in two dimensions. *SIAM Journal of Applied Mathematics* **49**, 1136–57.

Ermentrout, B. & Edelstein-Keshet, L. (1993). Cellular automata approaches to biological modeling. *Journal of Theoretical Biology* **160**, 97–133.

Gow, N. A. R. & Gadd, G. M. (eds.) (1995). *The Growing Fungus*. London: Chapman and Hall.

Heath, I. B. (1990). *Tip Growth in Plant and Fungal Cells*. London: Academic Press.

Hutchinson, S. A., Sharma, P., Clark, K. R. & MacDonald, I. (1980). Control of hyphal orientation in colonies of *Mucor hiemalis*. *Transactions of the British Mycological Society* **75**, 177–91.

Kotov, V. & Reshetnikov, S. V. (1990). A stochastic model for early mycelial growth. *Mycological Research* **94**, 577–86.

Jacobs, H., Boswell, G. P., Harper, F. A., Ritz, K., Davidson, F. A. & Gadd, G. M. (2002a). Solubilization of metal phosphates by *Rhizoctonia solani*. *Mycological Research* **106**, 1468–79.

Jacobs, H., Boswell, G. P., Ritz, K., Davidson, F. A. & Gadd, G. M. (2002b). Solubilization of calcium phosphate as a consequence of carbon translocation by *Rhizoctonia solani*. *FEMS Microbiology Ecology* **40**, 64–71.

Lamour, A., van den Bosch, F., Termorshuizen, A. J. & Jeger, M. J. (2000). Modelling the growth of soil-borne fungi in response to carbon and nitrogen. *IMA Journal of Mathematics Applied to Medicine and Biology* **17**, 329–46.

Meskauskas, A., McNulty, A. J. & Moore, D. (2004a). Concerted regulation of all hyphal tips generates fungal fruit body structures: experiments with computer visualizations produced by a new mathematical model of hyphal growth. *Mycological Research* **108**, 341–53.

Meskauskas, A., Fricker, M. D. & Moore, D. (2004b). Simulating colonial growth of fungi with the Neighbour-Sensing model of hyphal growth. *Mycological Research* **108**, 1241–56.

Morley, G., Sayer, J., Wilkinson, S., Gharieb, M. & Gadd, G. M. (1996). Fungal sequestration, solubilization and transformation of toxic metals. In *Fungi and Environmental Change*, ed. J. C. Frankland, N. Magan & G. M. Gadd, pp. 235–56. Cambridge: Cambridge University Press.

Olsson, S. (1995). Mycelial density profiles of fungi on heterogeneous media and their interpretation in terms of nutrient reallocation patterns. *Mycological Research* **99**, 143–53.

Othmer, H. G. & Stevens, A. (1997). Aggregation, blowup, and collapse: the ABCs of taxis in reinforced random walks. *SIAM Journal of Applied Mathematics* **57**, 1044–81.

Paustian, K. & Schnürer, J. (1987). Fungal growth response to carbon and nitrogen limitation: a theoretical model. *Soil Biology and Biochemistry* **19**, 613–20.

Prosser, J. I. (1995). Mathematical modelling of fungal growth. In: *The Growing Fungus*, ed. N. A. R. Gow & G. M. Gadd, pp. 319–35. London: Chapman and Hall.

Prosser, J. I. & Trinci, A. P. J. (1979). A model for hyphal growth and branching. *Journal of General Microbiology* **111**, 153–64.

Regalado, C. M., Crawford, J. W., Ritz, K. & Sleeman B. D. (1996). The origins of spatial heterogeneity in vegetative mycelia: a reaction-diffusion model. *Mycological Research* **100**, 1473–80.

Regalado, C. M., Sleeman, B. D. & Ritz, K. (1997). Aggregation and collapse of fungal wall vesicles in hyphal tips: a model for the origin of the Spitzenkörper. *Philosophical Transactions of the Royal Society of London* B**352**, 1963–74.

Sayer, J. A. & Gadd, G. M. (1997). Solubilization and transformation of insoluble metal compounds to insoluble metal oxalates by *Aspergillus niger*. *Mycological Research* **101**, 653–61.

Soddell, F., Seviour, R. & Soddell, J. (1995). Using Lindenmayer systems to investigate how filamentous fungi may produce round colonies. *Complexity International*, **2**. Available online: http://www.csu.edu.au/ci/vol2/f_soddel/f_soddel.html

Turing, A. (1952). The chemical basis for morphogenesis. *Philosophical Transactions of the Royal Society of London* B**237**, 37–72.

Webster, J. (1980). *Introduction to Fungi*. Cambridge: Cambridge University Press.

II

Functional ecology of saprotrophic fungi

5
Mineral transformations and biogeochemical cycles: a geomycological perspective

GEOFFREY M. GADD, EUAN P. BURFORD, MARINA FOMINA AND KARRIE MELVILLE
School of Life Sciences, University of Dundee

Introduction

Rocks and minerals represent a vast reservoir of elements, many of them are essential to life. Bulk biological metals, such as Na, K, Mg and Ca, are among the eight most abundant elements in the Earth's crust and together make up 11.06% of crustal rock (Fraústo da Silva & Williams, 1993; Gadd, 2004). Rocks and minerals also include essential metals (e.g. Mn, Mo, Fe, Co, Ni, Cu, Zn) and, crucial for microbial and plant growth, phosphorus. Many elements have essential functional potential for the synthesis of biological macromolecules and energy capture (e.g. C, N, H, O, P, S), for the transmission of information (e.g. Na, K, Ca), for catalysis (e.g. Fe, Cu, Zn, Mo), for transfer of electrons (e.g. Fe), and for building solid structures (e.g. Ca, P, Si) (Fraústo da Silva & Williams, 1993). All these elements must be released into bioavailable forms that can be assimilated by the biota. Their release occurs via weathering of rock substrates and their mineral constituents through physical (mechanical), chemical and biological processes (Burford *et al.*, 2003). Near-surface weathering of rocks and minerals (sub-aerial and sub-soil environments) often involves an interaction between all three types (White *et al.*, 1992). In addition to mobilization of essential nutrients during lithospheric weathering, non-essential toxic metals (e.g. Cs, Al, Cd, Hg, Pb) may also be mobilized (Gadd, 1993, 2001a, b). Metals can exert toxic effects in many ways: they can block functional groups of enzymes, displace and substitute for essential metal ions, disrupt cellular and organellar membranes, and interact with systems that normally protect against the harmful effects of free radicals generated during normal metabolism (Gadd, 1993; Howlett & Avery, 1997). However, fungi employ a variety of mechanisms, both active and incidental, which contribute to tolerance and survival (Gadd, 1993).

Fungi in the Environment, ed. G. M. Gadd, S. C. Watkinson & P. S. Dyer. Published by Cambridge University Press.

Bioweathering can be defined as the erosion, decay and decomposition of rocks and minerals mediated by living organisms (Goudie, 1996). One of the most important processes of bioweathering is weathering mediated by microorganisms, including fungi (Burford *et al.*, 2003; Gadd, 2006; Gadd *et al.*, 2006). Because bacteria exhibit extensive metabolic versatility, growing under both aerobic and anaerobic conditions for example, and play highly significant geochemical roles in cycling of elements, most geomicrobiological studies have very much focused on prokaryotes (Sterflinger, 2000; Burford *et al.*, 2003). However, there is increasing evidence that fungi can dissolve minerals and mobilize metals at higher pH values, and over a wider redox range, more efficiently than bacteria (Gu *et al.*, 1998; Castro *et al.*, 2000; Burford *et al.*, 2003). Fungi are well suited as weathering agents because they can be highly resistant to extreme environmental conditions such as metal toxicity, UV radiation, and desiccation; they can adopt a variety of growth, metabolic and morphological strategies; they can exude protons and metal-complexing metabolites, and form mutualistic symbiotic associations with plants, algae and cyanobacteria (Burford *et al.*, 2003; Braissant *et al.*, 2004; Gadd, 2004, 2006; Gadd *et al.*, 2006). Most fungi exhibit a filamentous growth habit, which gives an ability to increase or decrease their surface area, to adopt either exploration or exploitation strategies, and also to form linear organs of aggregated hyphae for protected fungal translocation (Fomina *et al.*, 2003). Some fungi are polymorphic, occurring as filamentous mycelium and unicellular yeasts or yeast-like cells, e.g. the black meristematic or microcolonial rock-dwelling fungi (Gorbushina & Krumbein, 2000; Sterflinger, 2000; Gorbushina *et al.*, 2003). The ability of fungi to translocate nutrients within the mycelial network is another very important feature for exploring and exploiting heterogeneous environments (Lindahl & Olsson, 2004).

Fungal involvement in biogeochemical processes in the environment

Fungi have been components of microbial communities of many terrestrial environments, including deserts, metal-rich and hypersaline habitats, since the Ordovician period (480–460 MYBP) (Heckman *et al.*, 2001; Burford *et al.*, 2003). The fungal filamentous explorative growth habit, and high surface-area-to-mass ratio, together with the ability to translocate nutrients and form hydrated mucilaginous sheaths surrounding hyphae (which provides an interconnected microenvironment within the fungal network) ensures that fungal processes are an integral component of biogeochemical change (Gadd, 2004, 2006). Fungal communities in soil are very diverse, with mycorrhizal fungi being a particularly important group in the cycling of elements (Hoffland *et al.*, 2002; Fomina *et al.*, 2004; Gadd, 2006; Gadd *et al.*, 2006).

Many oligotrophic fungi can scavenge nutrients from the air and rainwater, and this enables them to grow on rock surfaces (Wainwright *et al.*, 1993). In the sub-aerial rock environment, they can also use organic and inorganic residues on mineral surfaces or within cracks and fissures, waste products of other microorganisms, decaying plants and insects, dust particles, aerosols and animal faeces as nutrient sources (Sterflinger, 2000). Rock-inhabiting communities must also be able to deal with varying extremes in microclimatic conditions, e.g. light, salinity, pH, and water potential, in order to thrive over considerable periods of time. Fungi may achieve protection by the presence of melanin pigments and mycosporines in their in cell wall, and by embedding colonies into mucilaginous polysaccharide slime that may entrap clay particles, providing extra protection (Gorbushina *et al.*, 2003; Volkmann *et al.*, 2003). The symbiotic lichen association with algae and/or cyanobacteria, where photosynthetic symbionts provide a source of carbon and surface protection from light and irradiation, is one of the most successful means for fungi to survive in extreme sub-aerial environments (Gorbushina *et al.*, 1993; Sterflinger, 2000) and is an extremely biogeochemically active fungal growth form (Banfield *et al.*, 1999).

Rock- and stone-dwelling microorganisms can be classified as (a) epilithic, (b) hypolithic, (c) endolithic, (d) chasmolithic, (e) cryptoendolithic and (f) euendolithic organisms (May, 2003). Epiliths occur on the surface of rocks and building stone, whereas hypolithic microorganisms are associated with pebbles. Endoliths inhabit the rock sub-surface and may form distinct masses or brightly coloured layers. Endolithic microorganisms can occur as chasmoliths, cryptoendoliths and euendoliths. Chasmoliths grow in pre-existing cracks and fissures within rock, and are often visible from the rock surface. Conversely, cryptoendoliths grow inside cavities and among crystal grains and cannot be observed from the rock surface. Euendolithic microbes are a specialized group of cryptoendoliths that are capable of actively penetrating (boring) into submerged rock (Ehrlich, 2002). Epiliths are often represented by microcolonial black-coloured fungi that occur on the surface as spherical clusters of tightly packed cells with thick pigmented walls or as moniliform thick-walled hyphae (Gorbushina *et al.*, 1993; Gorbushina & Krumbein, 2000) (Fig. 5.1B). Filamentous fungi, including zygomycetes, ascomycetes and basidiomycetes, can also occur on rock surfaces (epiliths) and in cracks, fissures and pores of the rock sub-surface (endoliths), as well as apparently burrowing into rock substrates (cryptoendoliths) (Fig. 5.1A, C) (Kumar & Kumar, 1999; Sterflinger, 2000; Verrecchia, 2000). Fungi have been found in many rock types (e.g. limestone, soapstone, marble, granite, sandstone, andesite, basalt, gneiss, dolerite, amphibolite and quartz) (Fig. 5.1) (Staley *et al.*, 1982; Gorbushina *et al.*, 1993; Sterflinger, 2000; Verrecchia, 2000; Burford *et al.*, 2003). It is likely that fungi are ubiquitous components of the microflora of all rocks and building stone, throughout a wide range of geographical and climatic zones.

Interactions of fungi with rocks and minerals can directly and indirectly lead to bioweathering (Burford *et al.*, 2003; Gadd *et al.*, 2006). In general, natural interactions between minerals, metals and non-metallic species in an aqueous fluid nearly always involve the presence of microbes or their metabolites (Banfield *et al.*, 1999). Fungal bioweathering of basaltic outcrops in sub-polar areas is believed to be chronologically the first process of weathering, followed by subsequent cryogenic processes (Etienne &

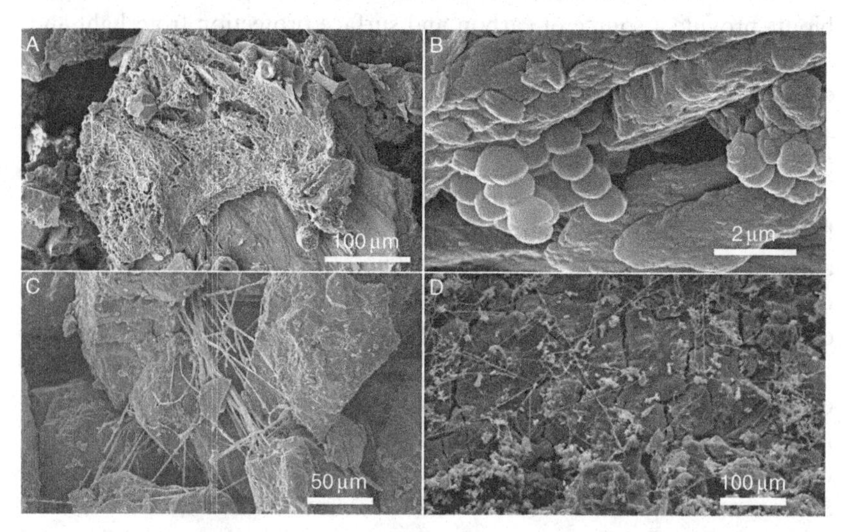

Fig. 5.1. Electron microscopic images of fungal communities on naturally weathered samples of (A) pegmatitic granite (Ireland), (B, C) Arctic sandstone (Eurica) and (D) experimentally deteriorated cracked surface of concrete block exposed to fungal (*Aspergillus niger*) weathering for 2 years. Specimens were air-dried and coated with 30 nm Au/Pd with a Cressington 208 HR sputter coater, with subsequent examination under a Philips XL30 environmental scanning electron microscope. Images A, C, D illustrate widespread hyphal networks and B shows microcolonial fungi adhering to the rock surface. Scale bars are (A, D) 100 µm, (B) 2 µm, (C) 50 µm (K. Melville, M. Fomina & G. M. Gadd, unpublished).

Dupont, 2002). Alkaline (basic) rocks seem more susceptible to fungal attack than acidic rocks (Eckhardt, 1985; Kumar & Kumar, 1999). Along with other organisms, fungi are believed to contribute to the weathering of silicate-bearing rocks (mica and orthoclase), and iron- and manganese-bearing minerals (biotite, olivine and pyroxene) (Kumar & Kumar, 1999). Fungi can also degrade olivine, natural glass, and artificial antique and medieval glass (Callot *et al.*, 1987; Krumbein *et al.*, 1991). *Aspergillus niger* and *Penicillium expansum* can degrade olivine, dunite, serpentine, muscovite, feldspar, spodumene, kaolin, nepheline and basalt; *Penicillium simplicissimum* and *Scopulariopsis brevicaulis* both released aluminium from aluminosilicates (Mehta *et al.*, 1979; Rossi, 1979; Sterflinger, 2000). Fungal weathering of limestone, sandstone and marble has been also been described (Gomez-Alarcon *et al.*, 1994; Hirsch *et al.*, 1995; Kumar & Kumar, 1999; Sterflinger, 2000; Ehrlich, 2002). Clump-like colonies of epi- and endolithic darkly pigmented microcolonial fungi are common inhabitants of limestone, sandstone, marble and granite as well as other rock types in hot and cold deserts and semi-arid regions, and are associated with pitting and etching (Fig. 4.1B) (Staley *et al.*, 1982; Gorbushina *et al.*, 1993; Gorbushina & Krumbein, 2000; Sterflinger, 2000). Microcolonial fungi are also common inhabitants of biogenic oxalate crusts on granitic rocks (Blazquez *et al.*, 1997).

Mechanisms of weathering

Microorganisms can influence mineral replacement reactions and dissolution–re-precipitation processes in rocks by mineral dissolution, precipitation of secondary minerals, and alteration of mineral surface chemistry and reactivity (Hochella, 2002; Putnis, 2002). Two synergistic actions by which fungi degrade mineral substrates are biomechanical and biochemical weathering (Burgstaller & Schinner, 1993; Banfield *et al.*, 1999; Burford *et al.*, 2003; Fomina *et al.*, 2005a) (Fig. 5.2).

Biomechanical weathering of minerals by fungi can be direct or indirect. Direct biomechanical degradation of minerals can occur through extensive penetration by fungal hyphae into decayed rocks and by tunnelling into otherwise intact mineral matter, which can occur along crystal planes, cleavage, cracks and grain boundaries in sandstone, calcitic and dolomitic rocks (Banfield *et al.*, 1999; Kumar & Kumar, 1999; Sterflinger, 2000; Burford *et al.*, 2003; Money, 2004). Fungal hyphae can exert considerable mechanical force, which derives from the osmotically generated turgor pressure within hyphae (Money, 2004). Biomechanical penetration into mineral matter is facilitated by thigmotropic reactions and lubrication with mucilaginous slime, which may contain acidic and metal-chelating metabolites (Burford *et al.*, 2003). Thigmotropism or contact guidance is a directed mode of fungal growth towards grooves, ridges and pores in solid material and may explain how fungal hyphae explore and exploit weakened sites in mineral surfaces (Watts *et al.*, 1998). Indirect biomechanical weathering can also occur because of shrinking and swelling effects of the hydrated mucilage produced by many fungi (Warscheid & Krumbein, 1994).

All the processes involved in biomechanical weathering of rocks by fungi are strongly connected with biochemical processes, which are believed to be much more important than mechanical mineral degradation (Kumar & Kumar, 1999; Burford *et al.*, 2003). The main mechanisms of solubilization of minerals and insoluble metal compounds by fungi are acidolysis, complexolysis and redoxolysis, all of which may be enhanced by metal accumulation in and/or around the fungal biomass (Burgstaller & Schinner, 1993; Burford *et al.*, 2003; Gadd, 2004). The primary fungal impact on biogeochemical cycling appears to result from acidolysis and complexolysis. Acidolysis (or proton-promoted dissolution) occurs when fungi acidify their microenvironment as a result of the excretion of protons and organic acids and the formation of carbonic acid resulting from respiratory CO_2 (Burgstaller & Schinner, 1993). Many fungi

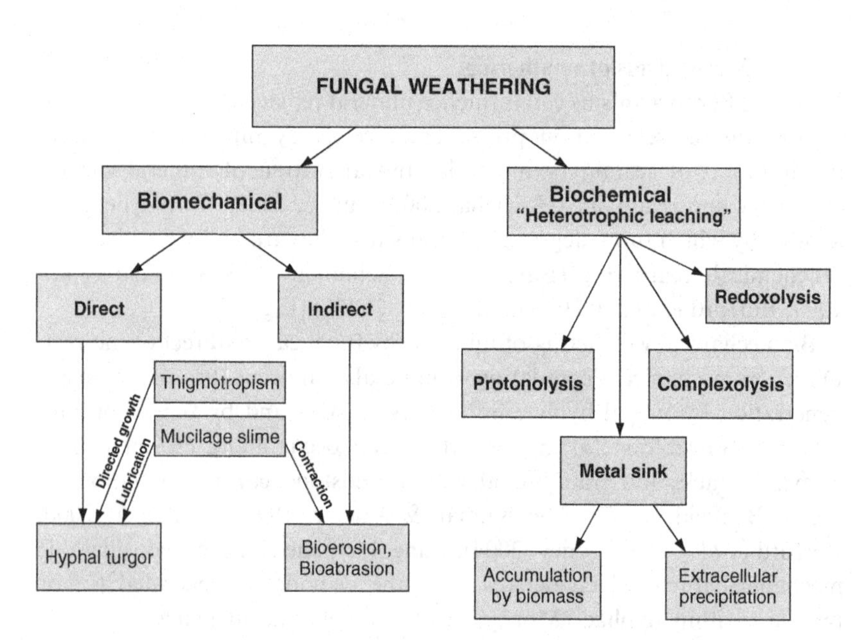

Fig. 5.2. Mechanisms of fungal weathering.

are able to excrete metal-complexing metabolites, which are associated with complexolysis or ligand-promoted dissolution (Burford *et al.*, 2003). These include carboxylic acids, amino acids, siderophores and phenolic compounds (Manley & Evans, 1986; Muller *et al.*, 1995; Gadd, 1999). Organic acid excretion by fungi is inter- and intraspecific and can be strongly influenced by the presence of toxic metal minerals (Fomina *et al.*, 2004). Fungal-derived carboxylic acids with strong chelating properties (e.g. oxalic and citric acid) perform aggressive attack on mineral surfaces (Gadd, 1999; Fomina *et al.*, 2005a). Moreover, carboxylic acids provide a source of protons for solubilization and chelating anions, which can complex metal cations (Devevre *et al.*, 1996). Acidolysis was the main mechanism of fungal dissolution of toxic metal phosphates grown in the presence of ammonium as a nitrogen source (Fomina *et al.*, 2004, 2005b). If the fungal culture is capable of excreting large amounts of a strong chelator (e.g. oxalate), the solubilization mechanism shifts to ligand-promoted dissolution (Fomina *et al.*, 2004, 2005a).

Metal immobilization mechanisms reduce the external free metal activity and may shift the equilibrium to release more metal into aqueous solution (Gadd, 1993, 2000; Sterflinger, 2000). Mobile metal species can be bound, accumulated or precipitated by fungal biomass via biosorption to biomass (cell walls, pigments and extracellular polysaccharides), transport and intracellular accumulation, and extracellular precipitation (Fomina *et al.*, 2005a). Fungi can be highly efficient bioaccumulators of soluble and particulate forms of metals (e.g. Ni, Zn, Ag, Cu, Cd and Pb), especially from dilute external concentrations (Gadd, 1993, 2000, 2001a, b; Baldrian, 2003). Metal binding by fungi can be an important passive process in both living and dead fungal biomass (Gadd, 1990, 1993).

Precipitation of secondary mycogenic minerals

The formation of secondary organic and inorganic minerals by fungi can occur through metabolism-independent and -dependent processes. Precipitation, nucleation and deposition of crystalline material on and within cell walls are influenced by such factors as environmental pH and the composition of cell walls (Gadd, 1990, 1993; Ferris *et al.*, 1994; Arnott, 1995; Urzi *et al.*, 1999; Sterflinger, 2000; Verrecchia, 2000; Gorbushina *et al.*, 2003). This process may be important in soil as precipitation of carbonates, phosphates and hydroxides increases soil aggregation. Cations such as Si^{4+}, Fe^{3+}, Al^{3+} and Ca^{2+} (which may be released through dissolution mechanisms) stimulate precipitation of compounds that may act as bonding agents for soil particles. Hyphae can enmesh soil particles, alter alignment and also release organic metabolites that enhance aggregate stability (Bronick & Lal, 2005).

Carbonates

Microbial carbonate precipitation coupled with silicate weathering could provide an important sink for CO_2 in terrestrial environments (Verrecchia *et al.* 1990; Rivadeneyra *et al.* 1993; Bruand & Duval, 1999; Knorre & Krumbein, 2000; Riding, 2000; Warren *et al.* 2001; Hammes & Verstraete, 2002). In limestone, fungi and lichens are considered to be important agents of mineral deterioration. Many near-surface limestones (calcretes), calcic and petrocalcic horizons in soils are often secondarily cemented with calcite ($CaCO_3$) and whewellite (calcium oxalate monohydrate, $CaC_2O_4.H_2O$) (Verrecchia, 2000). The presence of fungal filaments mineralized with calcite ($CaCO_3$), together with whewellite (calcium oxalate monohydrate, $CaC_2O_4.H_2O$), has been reported in limestone and calcareous soils from a range of localities (Kahle, 1977; Callot *et al.*, 1985; Monger & Adams, 1996; Bruand & Duval, 1999; Verrecchia, 2000). Calcium oxalate can also be degraded to calcium carbonate, e.g. in semi-arid environments, where such a process may again act to cement pre-existing limestones (Verrecchia *et al.*, 1990; Verrecchia, 2000). During decomposition of fungal hyphae, calcite crystals can act as sites of further secondary calcite precipitation (Verrecchia, 2000). Manoli *et al.* (1997) demonstrated that chitin, the major component of fungal cell walls, was a substrate on which calcite will readily nucleate. Other experimental work has demonstrated fungal precipitation of secondary calcite, whewellite and glushkinskite ($MgC_2O_4.2H_2O$) (Burford *et al.*, 2003; E. Burford & G. M. Gadd, unpublished data).

Oxalates

Fungi can produce metal oxalates with a variety of different metals and metal-bearing minerals (Ca, Cd, Co, Cu, Mn, Sr, Zn, Ni and Pb) (Arnott, 1995; Gadd, 1999, 2000; Sayer *et al.*, 1999; Magyarosy *et al.*, 2002; Fomina *et al.*, 2005a) (Fig. 5.3). Calcium oxalate dihydrate (weddelite) and the more stable calcium oxalate monohydrate (whewellite) are the most common forms of oxalate in nature and are associated with various ecophysiological groups of fungi (Arnott, 1995; Burford *et al.*, 2003). Depending on physicochemical conditions, biotic fungal calcium oxalate can exhibit a variety of crystalline forms (tetragonal, bipyramidal, plate-like, rhombohedral or needles) (Fig. 5.3). Precipitation of calcium oxalate can act as a reservoir for calcium in the ecosystem and also influences phosphate availability (Gadd, 1993, 1999). The formation of toxic metal oxalates may provide a mechanism whereby fungi can tolerate environments containing potentially high concentrations of toxic metals. For example, oxalate over-excreting *Beauveria caledonica* was able to transform cadmium, copper, lead and zinc from a variety of toxic metal minerals into oxalates and was tolerant to all tested minerals (Fomina *et al.*,

Fig. 5.3. ESEM images illustrating mineral transformations into metal oxalate crystals by *B. caledonica* grown on agar media. (A) A crust of calcium oxalate (weddelite and whewellite) crystals and tubular crystalline sheath around hyphae formed on medium containing $CaCO_3$ and $Cu_3(PO_4)_2$ (ESEM dry mode). (B) Zinc oxalate hydrate formed by the mycelium grown on zinc phosphate and covered with a mucilaginous hyphal net (ESEM wet mode). (C) Moolooite (copper oxalate) crystals precipitated on hyphae grown on copper phosphate medium (ESEM wet mode). Scale bars are (A) 20 μm, (B) 10 μm, (C) 50 μm. (Fomina *et al.*, 2005a; M. Fomina & G. M. Gadd, unpublished).

2005a) (Fig. 5.3). It has been reported that oxalate excretion by fungi is enhanced with NO_3^- as a nitrogen source in contrast to NH_4^+, and also by the presence of HCO_3^-, Ca^{2+} and some toxic metals (e.g. Cu, Al) or minerals (e.g. pyromorphite, zinc phosphate) (Lapeyrie *et al.*, 1991; Ahonen-Jonnarth *et al.*, 2000; Whitelaw, 2000; Van Leerdam *et al.*, 2001; Arvieu *et al.*, 2003; Casarin *et al.*, 2003; Clausen & Green, 2003; Fomina *et al.*, 2004).

Reductive and oxidative precipitation

Reduced forms of metals and metalloids (e.g. elemental silver, selenium, tellurium) within and around fungal cells can be precipitated by many fungi (Gadd, 1993, 2000, 2004). The reductive ability of fungi is manifest by black colouration of fungal colonies precipitating elemental Ag or Te, or red colouration for those precipitating elemental Se (Gharieb *et al.*, 1999). An oxidized metal layer (patina) a few millimetres thick found on rocks and in soils of arid and semi-arid regions, called desert varnish, is also believed to be of microbial origin with some proposed fungal involvement. Fungi can oxidize manganese and iron in metal-bearing minerals such as siderite ($FeCO_3$) and rhodochrosite ($MnCO_3$) and precipitate them as oxides, and also form dark Fe(II)- and Mn(II)- patinas on glass surfaces (Eckhardt, 1985; Grote & Krumbein, 1992).

Other mycogenic minerals

A specific combination of biotic and abiotic factors can lead to the deposition of a variety of other secondary minerals associated with fungi, e.g. birnessite, MnO and FeO, ferrihydrite, iron gluconate, calcium formate, forsterite, goethite, halloysite, hydroserussite, todorokite, moolooite and montmorillonite (Grote & Krumbein, 1992; Hirsch *et al.*, 1995; Verrecchia, 2000; Gorbushina *et al.*, 2001; Burford *et al.*, 2003). Precipitation immobilizes metals in the soil environment and therefore limits bioavailability (Gadd, 2000).

Fungal symbioses in mineral transformations

One of the most remarkable adaptations of fungi for exploitation of soil and rock environments is their ability to form mutualistic partnerships with land plants (mycorrhizas) and algae or cyanobacteria (lichens). Symbiotic fungi are provided with carbon by the photosynthetic partners while the fungi protect the symbiosis from harsh environmental conditions (e.g. desiccation, metal toxicity), increase the absorptive area of the symbiotic associations and provide increased access to mineral nutrients

(Smith & Read, 1997; Meharg & Cairney, 2000; Adriaensen *et al.* 2003; Meharg, 2003; Colpaert *et al.*, 2004).

Lichens

Lichens, fungi that exist in facultative or obligate symbioses with one or more photosynthesizing partners, play an important role in many biogeochemical processes. They are commonly thought of as pioneer colonizers of fresh rock outcrops, and were possibly one of the earliest life forms to occupy Earth's land surfaces. The lichen symbiosis formed between the fungal partner (mycobiont) and the photosynthesizing partner (algal or cyanobacterial photobiont) enables lichens to grow in practically all surface terrestrial environments: an estimated 6% of the Earth's land surface is covered by lichen-dominated vegetation (Haas & Purvis, 2006). Globally, lichens play an important biogeochemical role in the retention and distribution of nutrient (e.g. C, N) and trace elements, in soil formation processes, and in rock weathering (Barker *et al.*, 1997; Banfield *et al.*, 1999). Lichens can accumulate metals such as lead (Pb), copper (Cu) and others of environmental concern, including radionuclides (Purvis, 1996), and also form a variety of metal–organic biominerals, especially during growth on metal-rich substrates (Purvis & Halls, 1996). A detailed account of lichen biogeochemistry is outwith the scope and length of this chapter, but referral to the cited references is recommended, especially Haas & Purvis (2006), which provides the most recent overview.

Mycorrhizas

Nearly all land plants depend on symbiotic mycorrhizal fungi (Smith & Read, 1997). Two main types of mycorrhiza are the endomycorrhizas, in which the fungus colonizes the interior of host plant root cells (e.g. ericoid and arbuscular mycorrhizas) and ectomycorrhizas, in which the fungus is located outside the root cells of the host plant (Fig. 5.4). It has been demonstrated that mycorrhizal fungi are involved in proton- and ligand-promoted metal mobilization from mineral sources, metal immobilization via biosorption and accumulation within biomass, and extracellular precipitation of mycogenic toxic metal oxalates (Fomina *et al.*, 2004, 2005b).

Biogeochemical activities of mycorrhizal fungi lead to changes in the physicochemical characteristics of the root environment and enhanced weathering of soil minerals, resulting in metal cation release (Olsson & Wallander, 1998; Lundstrom *et al.*, 2000; Whitelaw, 2000). Dissolution of soil weatherable calcium-bearing minerals by ectomycorrhizal fungi has been well documented (Callot *et al.*, 1985; Lapeyrie *et al.*, 1990, 1991). It has been shown that ectomycorrhizal mycelia may respond to the presence of different soil silicate and phosphate minerals (apatite, quartz, potassium feldspar) by regulating their growth and activity, e.g. colonization, carbon allocation and substrate acidification (Rosling *et al.*, 2004a, b). Carbon allocation within the mycelium was significantly greater in *Hebeloma crustuliniforme*/*Pinus sylvestris* ectomycorrhizas colonizing potassium feldspar patches than in those colonizing quartz patches (Rosling *et al.*, 2004b).

During their growth, mycorrhizal fungi often excrete low-molecular-mass carboxylic acids (e.g. malic, succinic, gluconic, oxalic), contributing to the process of 'heterotrophic leaching' (Ahonen-Jonnarth *et al.*, 2000; Martino *et al.*, 2003; Fomina *et al.*, 2004). In podzol E horizons under European coniferous forests, the weathering of hornblendes, feldspars and granitic bedrock has been attributed to oxalic, citric, succinic, formic and malic acid excretion by ectomycorrhizal hyphae. Ectomycorrrhizal hyphal tips could produce micro- to millimolar concentrations of these organic acids and were associated with micropores (3–10 μm) in weatherable soil minerals (van Breemen *et al.*, 2000; Smits *et al.*, 2005). The ectomycorrhizal fungus *Piloderma* was able to extract K and/or Mg from biotite, microcline and chlorite to satisfy nutritional requirements, and precipitated mycogenic calcium oxalate crystals on the hyphae (Glowa *et al.*,

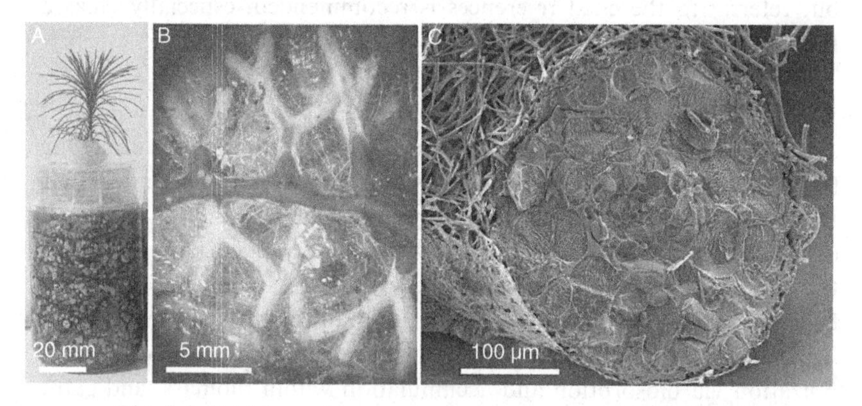

Fig. 5.4. (A) Mason jar mesocosm system containing sterile cellophane bags with zinc phosphate in the matrix with Scots pine seedlings. (B) Light microscopic image and (C) typical Cryo-FESEM image of cryopreserved and cryofractured ectomycorrhizal roots of *Paxillus involutus*/*Pinus sylvestris* association showing (B) extramatrical mycelium and (C) hyphal mantle surrounding root and well-developed Hartig net formed by fungus between cortical cells of pine root. Scale bars are (A) 20 mm, (B) 5 mm, (C) 100 μm. (M. Fomina & G. M. Gadd, unpublished).

2003). The ectomycorrhizal fungi *Suillus granulatus* and *Paxillus involutus* were able to release elements from apatite and wood ash (K, Ca, Ti, Mn, Pb) and accumulate them in the mycelia, especially rhizomorphs (Wallander *et al.*, 2003).

Ericoid mycorrhizal and ectomycorrhizal fungi can dissolve a variety of cadmium-, copper-, zinc- and lead-bearing minerals, including metal phosphates (Leyval & Joner, 2001; Martino *et al.*, 2003; Fomina *et al.*, 2004, 2005b). Mobilization of phosphorus is generally regarded as one of the most important functions of mycorrhizal fungi (Lapeyrie *et al.*, 1991; Wallander *et al.*, 1997; Whitelaw, 2000). An experimental study of zinc phosphate dissolution by the ectomycorrhizal association of *Paxillus involutus* with Scots pine (*Pinus sylvestris*) demonstrated that phosphate mineral dissolution, phosphorus acquisition and zinc accumulation by the plant depended on the mycorrhizal status of the pines, the zinc tolerance of the fungal strain and the phosphorus status of the environment (Fig. 5.4) (M. Fomina & G. M. Gadd, unpublished data). Under both phosphorus-replete and -depleted conditions, ectomycorrhizal roots accumulated significantly higher amount of phosphorus than did non-mycorrhizal roots. Under phosphorus-replete conditions, total zinc mobilization from zinc phosphate and zinc accumulation in pines was the least for the ectomycorrhizal association with a zinc-tolerant strain of *P. involutus*. It was possible that this strain employed a toxic metal avoidance strategy aimed at restricting metal entry inside cells, and thereby protecting plant tissues from excessive zinc accumulation (Fomina *et al.*, 2004). The highest zinc accumulation was observed for non-mycorrhizal plants. In contrast, under phosphorus deficiency, zinc-tolerant ectomycorrhiza mobilized and accumulated the highest amount of zinc. It was concluded that, although both non-mycorrhizal and ectomycorrhizal plants were able to dissolve zinc phosphate and acquire phosphorus, zinc phosphate dissolution and host plant protection against the zinc was strongly dependent on the phosphorus status of the matrix (M. Fomina & G. M. Gadd, unpublished data).

Environmental significance of mineral mycotransformations

At local and global scales, fungal involvement in the biogeochemical cycling of elements has important implications for living organisms, plant production and human health (Gadd, 2006). The ability of fungi to transform toxic metals and metalloids has potential for treatment of contaminated land and substrates (Gadd, 2000, 2002). Dissolution of toxic metal minerals and release of mobile and bioavailable metal cations may increase toxicity of the local microenvironment, but conversely could be used in cleaning up soils and solid industrial wastes and by-products, low-grade ores and metal-bearing minerals (Gadd, 2000, 2002; Brandl, 2001). The ability of fungi to immobilize metals via biosorption, accumulation within biomass and extracellular organic and inorganic precipitation turns metals into chemically more inert forms and therefore results in detoxification (Gadd, 2000; Fomina & Gadd, 2002). Fungi with Cr(VI)-reducing activity may be useful for treatment of Cr-polluted soils (Cervantes *et al.*, 2001).

In metal-contaminated soils, the mineral-solubilizing activity of fungi may affect some *in situ* methods of chemical remediation. For example, the formation of metal phosphates in metal-contaminated soils (so-called phosphate-induced metal stabilization) has been proposed as a cost-effective remediation technology, which leads to increased deposition of toxic metal phosphates (e.g. hopeite, pyromorphite) in treated soils (Chen *et al.*, 1997; Conca, 1997; Brown *et al.*, 2004). However, fungi may transform the final mineral products of such remediation (Sayer *et al.*, 1995, 1999; Fomina *et al.*, 2004, 2005a). Assessment of long-term environmental consequences of chemical remediation technologies should therefore take into account the biogeochemical activity of soil fungal (and bacterial) communities.

Interactions between mycorrhizas and metal pollution have raised increased interest because of the potential of mycorrhizas to enhance plant growth on contaminated soils and to remediate and remove pollutants (Jentschke & Godbold, 2000; Meharg & Cairney, 2000; Van Tichelen *et al.*, 2001; Adriaensen *et al.*, 2003; Meharg, 2003; Colpaert *et al.*, 2004). Ectomycorrhizal mycobionts can filter toxic metals in the hyphal sheath or Hartig net by sorption, restrict metal mobility due to hydrophobicity of the fungal sheath, and also complex metals by released organic acids and other substances (Jentschke & Godbold, 2000; Krupa & Kozdroj, 2004). It has been reported that arbuscular mycorrhizas also reduce toxic metal (e.g. Zn) uptake by plants (Christie *et al.*, 2004). Most studies on ectomycorrhizal associations that have demonstrated amelioration of metal toxicity to the host seem to be most efficient when the mycobiont was metal resistant (Colpaert & Van Assche, 1992; Hartley-Whitaker *et al.*, 2000; Jentschke & Godbold, 2000; Van Tichelen *et al.*, 2001; Adriaensen *et al.*, 2003). Such associations could possibly be used in re-vegetation and re-forestation programmes as well as in phyto- and rhizo-remediation strategies. However, these mycorrhizal interactions are variable and context-dependent and their outcome may vary in response to biotic and abiotic factors (Setala *et al.*, 1997). When mycorrhizas enhance accumulation of toxic metals in host plants, this may be relevant for phytoextraction techniques.

Another negative effect of fungal mineral transformations is their significance for deterioration of mineral-based building materials. Any type of building or ceramic material, concrete and cement can be degraded by fungi (Diercks *et al.*, 1991; Gaylarde & Morton, 1999; Kikuchi & Sreekumari, 2002; Roberts *et al.*, 2002) (Fig. 5.5). Under certain conditions, deterioration of concrete by fungi may be more efficient than that caused by bacteria (Perfettini *et al.*, 1991; Gu *et al.*, 1998; Nica *et al.*, 2000). Fungal attack on concrete can arise from mechanisms similar to those described earlier (Sand & Bock, 1991). Microbial ability to corrode concrete barrier materials raises concern with respect to both existing and future nuclear waste storage. An experimental study of the effect of microfungi on barrier concrete showed that fungi were able to colonize and corrode the surface of concrete, avoiding the areas filled with granite (Figs. 5.1D, 5.5) (Fomina *et al.*, 2005c). Fungi can also adapt to severe radioactive contamination: studies of the walls of the 'Shelter' over Reactor No. 4 of the Chernobyl nuclear power plant (radiation $\alpha = 500\ \mathrm{Bq\ cm^{-2}}$, $\beta = 20\,000\ \mathrm{Bq\ cm^{-2}}$, $\gamma = 700\ \mathrm{mR\ h^{-1}}$) demonstrated that

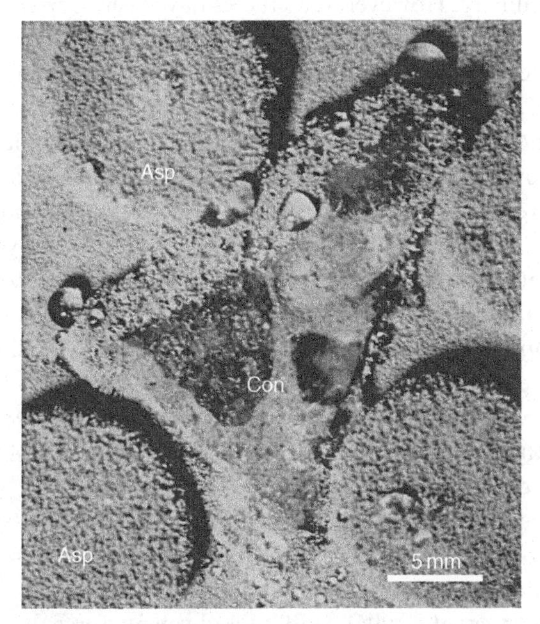

Fig. 5.5. Six-month Petri dish microcosm: *Aspergillus flavipes* (Asp) colonization of a piece of barrier concrete (Con) inserted into Czapek–Dox agar. Scale bar, 5 mm. Adapted from Fomina *et al.* (2005c).

microfungi, especially melanized strains of *Alternaria*, *Cladosporium* and *Aureobasidium*, were able to survive and colonize highly radioactive concrete (Mironenko *et al.*, 2000; Zhdanova *et al.*, 2000). Myco-corrosion may therefore reduce the theoretically calculated service life of concrete, causing a potential risk from the release of radionuclides into the environment.

Assessment of geoactive fungal communities

It is apparent that lithic fungal communities are ubiquitous in the geological environment and play many roles important in nutrient, metal and carbon cycling (Gadd, 2000; Nannipieri *et al.*, 2003). Such organisms are essential components of the microbial communities found associated with a wide variety of mineral types (Burford *et al.*, 2003; Hughes & Lawley, 2003; Sigler *et al.*, 2003). Diverse microbial communities have been found living in the pore spaces of exposed rock from the McMurdo Dry Valleys of Antarctica (Torre *et al.*, 2003), extreme hot and cold deserts (Staley *et al.*, 1982; Ehrlich, 1998; Sterflinger, 2000) and in rock substrates buried several kilometres below the Earth's surface (Kerr, 1997; Banfield *et al.*, 1999). Defining microbial communities to ascertain their diversity and aid elucidation of their impact on the environment is an important aspect of biogeochemistry. However, few studies have been carried out on fungal communities living in association with rocks and minerals.

Determining the composition of microbial communities remains a feasible, yet difficult, goal for microbiologists to accomplish. Traditional culture-based methods for assessing the diversity of intricate microbial communities are now generally recognized as insufficient. However, several studies have proven to be very useful for characterizing fungal communities living in association with rocks. Microscopical analyses to characterize phototrophic epilithic microbial communities from the Roman Catacombs found that the most frequently encountered fungal isolates were *Sporotrichum* and *Aspergillus* spp. (Albertano & Urzi, 1999). Cultivation-dependent studies have also been carried out on sandstone, where fungal isolates were identified as members of the genera *Pleospora*, *Acremonium*, *Engyodontium*, *Stemphylium* and *Penicillium* (K. Melville, S. Zhu & G. M. Gadd, unpublished data).

However, it is now evident that the majority of environmental species are refractory to laboratory cultivation, e.g. some biotrophs (rust or smut fungi), many basidiomycetes, arbuscular endomycorrhizas, in particular the Glomales (Thorn *et al.*, 1996), and rock-dwelling fungi with low metabolic activity (Sterflinger *et al.*, 1998; Gomes *et al.*, 2003). This has led to a tendency to underestimate the complexity and diversity of fungal communities, owing to the bias associated with traditional microbiological methods (Smit *et al.*, 1999; Nikolcheva *et al.*, 2003). For example, it has been well documented that, at best, perhaps only 5% of environmental fungi can be cultivated (Hawksworth, 2001). Consequently, reliance on morphological and physiological identification of cultivable species will never elucidate the inherent complexities of community dynamics; this is especially true for complex *in situ* ecosystems. Novel methodologies are therefore necessary to enhance conventional microbiological methods.

The advent of DNA-based characterization techniques is becoming increasingly important for ecological studies (Morris *et al.*, 2002). Such culture-independent approaches have significantly advanced bacterial community studies (see Fig. 5.6). Molecular characterization of fungal communities is becoming increasingly common, although to a much lesser extent when compared with published articles on bacterial diversity. A large proportion of fungal studies have concentrated on soil, principally investigations that involve the rhizosphere (see, for example, Buscot *et al.*, 2000; Hodge, 2000; Daniell *et al.*, 2001; Burke *et al.*, 2005). Studies that address the molecular analysis of fungal communities in rocks and minerals are extremely rare.

Molecular methods for fungal community analysis

The majority of culture-independent studies rely on analysis of rRNA genes amplified from DNA extracted directly from environmental samples. Soil DNA extraction kits are available, but are often unsuccessful for rock samples (D. B. Gleeson, personal communication). Extracting DNA from most environmental sources, including DNA extraction from a mineral substrate, requires extraction of high-molecular-mass DNA, free from inhibitors to allow downstream molecular manipulations, that represents the full microbial consortium present. In most cases, DNA extraction methods include various combinations of bead beating, detergents, enzymatic lysis and solvent extraction to obtain a crude nucleic acid preparation (Nannipieri *et al.*, 2003). However, it should be noted that bias is still a problem with nucleic acid extraction yields, because lysis efficiency may vary between different species and between spores and mycelia (Prosser, 2002). Gabor *et al.* (2003) investigated the efficiency of four DNA extraction methods for a range of sediments and soils: soft lysis, hard lysis, blending and cation exchange. Variations in DNA recovery were estimated by denaturing gradient gel electrophoresis (DGGE) profiles and blunt-end cloning efficiencies. This study emphasized that exploration into the most effective and appropriate nucleic acid extraction technique should be executed prior to any molecular community analysis.

Employing PCR-based methodologies to amplify specific regions of the ribosomal RNA gene of the microbial populations of interest has been the main methodology to date for both bacteria and fungi. Amplification using primers specific to the target microbial group will result in the production of a population of gene fragments, which is considered to represent the entire community of a particular sample. Characterization of the fungal members of a microbial community usually involves PCR amplification of either the small subunit SSU (18*S*) or internal transcribed spacer (ITS) rDNA regions. Once amplified, the pool of community gene fragments representing all the species amplifiable by the specific primers can be investigated by using a number of molecular approaches. Variation within the bacterial 16*S* rRNA gene has been exploited with great effect to distinguish between species. In fungi, the 18*S* rRNA gene has also been the subject of a high volume of study. However, there is a marked reduction in sequence resolution due to the relative lack of evolutionary variation between closely related fungal species (Hugenholtz & Pace, 1996). It should also be noted that there are problems associated with PCR bias of cycling conditions, the formation of chimeric sequences, and the introduction of sequence error and primer choice (Amann *et al.*, 1995; von Wintzingerode *et al.*, 1997; Prosser, 2002; Anderson *et al.*, 2003). Increasingly, the ITS region, located between the 18*S* rRNA and 28*S* rRNA genes and incorporating the 5.8*S* rRNA gene, has been used in fungal ecological studies owing to a higher degree of heterogeneity between species, thus allowing an improvement in community resolution. Although rRNA genes give quality phylogenetic information, there is a great limitation when it comes to rRNA gene quantification. Problems arise because the number of copies of the rRNA operon is known to vary between different fungal species (Hibbet, 1992). Copy numbers are still largely unknown for most species, making quantification of fungal species from a mixed community very difficult (Crosby & Criddle, 2003; Anderson & Cairney, 2004).

Phylogenetic structure of fungal communities

Molecular ecology can be split into two main groups of analysis, and the choice of method will depend upon the questions that are being asked about a particular community (Fig. 5.6). If the question was 'What species are present?' the best approach would be to use a phylogenetic method to identify the species present in the system. Most fine-scale information can be obtained by using a cloning approach of whole environmental DNA. By cloning the amplified rRNA gene pool, followed by sequencing and database sequence comparison, a list of species present can provide a detailed insight into a community. Classically, prokaryotic bioinformatic databases are more extensive compared with fungal data, but nevertheless fungal databases are growing in size. Characterization of

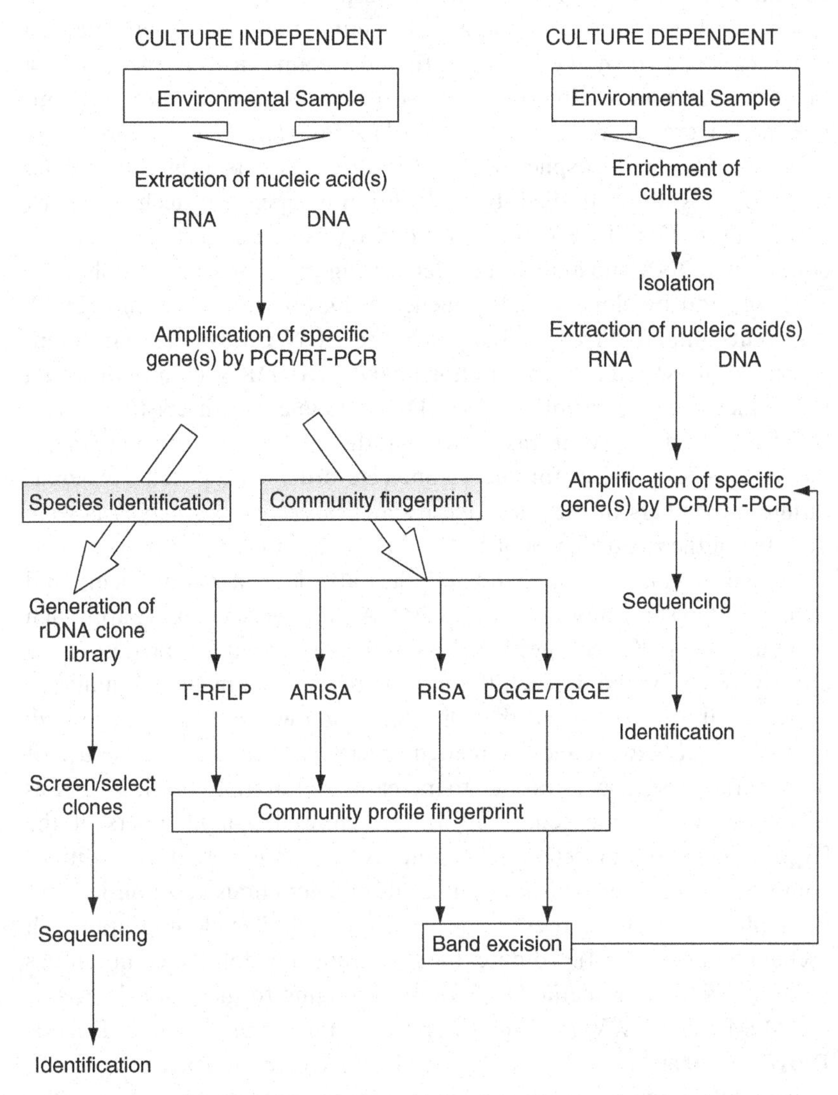

Fig. 5.6. Flow chart showing culture-dependent and culture-independent procedures for microbial diversity analysis of environmental samples.

fungi based on DNA extraction from environmental samples, followed by PCR amplification, cloning, sequencing and ultimately identification, has been the main approach to study a variety of problems (Morris *et al.*, 2002). Ectomycorrhizal fungal communities are extremely diverse and have been studied to a great degree by molecular techniques (see Horton & Bruns, 2001; Nara *et al.*, 2003). Filion *et al.* (2004) studied microbial communities associated with the rhizosphere of both healthy and unhealthy black spruce seedlings and found that there were marked differences caused by the health of the host plant. Healthy plants had a higher proportion of Homobasidiomycete clones, whereas diseased samples had a greater proportion of Sordariomycetes. Fungal population dynamics in the rhizospheres of two maize cultivars originating from tropical soils were studied by cultivation-independent techniques by Gomes *et al.* (2003). 18*S* ribosomal DNA was amplified from the total community DNA and analysed by denaturing gradient gel electrophoresis (DGGE) and by cloning and sequencing. Direct cloning of 18*S* rDNA fragments amplified from soil or rhizosphere DNA provided 39 different amplified ribosomal DNA restriction analysis (ARDRA) clones. Based on the sequence similarity of the 18*S* rDNA fragment with existing fungal isolates in the database, it was shown that the rhizospheres of young maize plants seemed to select for the ascomycete order Pleosporales, whereas different members of the Ascomycetes and basidiomycetous yeasts were detected in the rhizospheres of senescent maize plants.

Fungal diversity in grassland soils has also been a recent theme and point of interest for molecular ecologists. An 18*S* rDNA cloning approach by Hunt *et al.* (2004) followed by ARDRA showed that this process could effectively screen and detect differences between soil fungal communities. However, it was also found that the 18*S* region was not variable enough to distinguish between closely related species, suggesting that the use of more variable genes were needed to increase resolution. In total, 28 species of fungus were sequenced; identification showed that members of the Zygomycetes, Ascomycetes and Basidiomycetes were present. In addition, comparisons between fungal populations of leguminous and non-leguminous plant roots have been made by using a similar cloning approach (Scheublin *et al.*, 2004). To date, the best studied endolithic communities by non-DNA-based methods are those belonging to the polar deserts of the McMurdo Dry Valleys in Antarctica (Maurice *et al.*, 2002; De Los Ríos *et al.*, 2004; Wierzchos *et al.*, 2004, 2005). However, Torre *et al.* (2003) carried out a large investigation using a cloning approach to determine the microbial diversity of cryptoendolithic communities from the McMurdo

Dry Valleys. Target populations for cloning included bacteria, archaea, cyanobacteria and eukaryotes, including fungi. Their results showed that there were several distinct cryptoendolithic communities present. Putative identification of several abundant fungal clones revealed most homology to *Texosporium sancti-jacobi*, *Bullera unica* and *Geomyces pannorum*. It is evident that by using a similar approach fungal communities inhabiting a variety of rock types and minerals could provide an excellent baseline for further geomycological studies. The growth of molecular phylogenetics has, however, created several persistent problems relating to the underlying statistical and computational support. Tree-rooting problems, data set conflicts and computational management of data sets are an ever-increasing and important consideration (Sanderson & Shaffer, 2002).

Fingerprinting techniques for analysis of amplified genes

Fingerprinting methods can allow multiple sample analysis to study spatial and temporal variation in the composition of environmental samples. If the research to be undertaken relates to community shifts caused by, for example, a particular treatment, then the most appropriate approach would be to use a fingerprinting method such as denaturing gradient gel electrophoresis (DGGE), temperature gradient gel electrophoresis (TGGE) (Muyzer *et al.*, 1993), ribosomal intergenic spacer analysis (RISA) (Borneman & Triplett, 1997), automated ribosomal intergenic spacer analysis (ARISA) (Fisher & Triplett, 1999) or terminal restriction fragment length polymorphism (T-RFLP) (Liu *et al.*, 1997). These fingerprinting techniques have somewhat superseded cloning-based approaches when large sample numbers are involved, owing to the laborious and costly nature of gene fragment cloning. These monitoring techniques provide band profiles, as with DGGE, TGGE and RISA, or a high-resolution automated capillary electrophoresis profile, e.g. ARISA and T-RFLP, that is representative of the genetic structure of a community as defined by selected primers.

Gradient gel electrophoresis

Denaturing gradient gel electrophoresis (DGGE) and temperature gradient gel electrophoresis (TGGE) are electrophoretic fingerprinting techniques that exploit differences in sequence composition of the 18*S* rRNA gene. Usually, a portion of the 18*S* rRNA gene is amplified (with all fragments having the same size of amplicon, usually less than 500 bp) from environmental DNA. Members of the community are separated by differences in the melting behaviour caused by variation within the 18*S* sequence arrangement for each species. Rapid fingerprints can be generated and banding patterns can be visualized to compare communities. Subsequent identification of bands of interest can be carried out by gel excision and sequencing. DGGE and TGGE have been used to study fungal communities from a variety of environments including decomposing leaves (Nikolcheva *et al.*, 2003, 2005), coniferous forests (Cullings *et al.*, 2005; Hernesmaa *et al.*, 2005) and the wheat rhizosphere (Smit *et al.*, 1999). One molecular characterization paper focused on endolithic cyanobacterial communities associated with an exposed dolomite in central Switzerland (Sigler *et al.*, 2003). This study did not include fungal communities but highlights the applicability of DGGE to determine any population of microorganisms within rocks. In our own work (K. Melville and G. M. Gadd, unpublished data), lithic fungal DGGE profiles were generated by using DNA extracted directly from sandstone. Profiles demonstrated considerable variations in the banding patterns between samples. Such large dissimilarities between rocks originating from the same geographical origin suggested that rock-dwelling fungal populations are diverse and the selection of indigenous populations is a highly complex process.

Ribosomal intergenic spacer analysis (RISA and ARISA)

RISA is a valuable tool for monitoring the structure and dynamics of complex communities where excising, cloning and sequencing of RISA bands can be used to identify the populations involved in community adaptation (Ranjard *et al.*, 2001). The basis for this technique is gel separation of gene fragments by length differentiation of the ITS region of the rDNA. Automated ribosomal intergenic spacer analysis (ARISA), developed by Fisher and Triplett (1999), is a high-throughput, automated version of RISA. ARISA is a highly sensitive, culture-independent, reproducible technique suitable for environmental community analysis. This PCR-based technology uses a fluorescently labelled primer to exploit the length variations within the ITS region of the rRNA gene. Following amplification of the DNA of interest, the labelled products are analysed by using an automated capillary electrophoresis system, resulting in an electrophorogram where each peak corresponds to a DNA fragment, called a ribotype or operational taxonomic unit (OTU). Community complexity has been reported to range from ~30 to over 200 fungal peaks per profile, highlighting the sensitivity of this approach. The recent development of high-throughput automated methods to fingerprint soil fungal structure has provided fuller characterization of soil fungal populations

(Ranjard *et al.*, 2001; Hansgate *et al.*, 2005; Kennedy *et al.*, 2005). There are, however, several disadvantages. The main drawback is that this technique is unable to generate direct sequence information and so does not provide species identification: only a community fingerprint is produced, therefore making it necessary to combine other phylogenetic methodologies.

A study utilizing RISA and ARISA to determine the phylogeny and fungal community profile of weathered pegmatitic granite was carried out to explore the extent of mineralogical influences on indigenous fungal communities (Gleeson *et al.*, 2005). This study used molecular methods in combination with geochemical analyses and multivariate statistics to exploit the contrasting elemental chemistry of pegmatitic granite. Briefly, large intact crystals of variably weathered muscovite, plagioclase, K-feldspar, quartz and whole granite were sampled from a single outcrop from the Wicklow Mountains, Ireland. The major elemental chemistry was determined along with fungal community profiles for each mineral type. Average sample ribotype numbers ranged from 25 on K-feldspar, 23 on granite and muscovite, 20 on plagioclase, to 4 on quartz. A randomization test was then undertaken, which indicated that fungal community structure differed significantly between each mineral sample. Canonical correspondence analysis (CCA) revealed broad-scale mineralogical influences on community make-up of the 16 most abundant ribotypes. Strong relations were found between certain ribotypes and particular chemical elements, as individual ribotypes were almost exclusively restricted to single mineral types (Fig. 5.7). This study was able to demonstrate that distinct fungal populations inhabited rocks of a particular mineralogical type, indicating that individual chemical elements were able to exert a selective pressure on fungal community structure (Gleeson *et al.*, 2005).

Terminal restriction fragment length polymorphism (T-RFLP)

T-RFLP is another technique that can analyse whole microbial community nucleic acids to provide information about the most abundant microbial amplicons in a community. This approach has been used to describe bacterial and fungal communities in soil. Like ARISA, it is capable of higher resolution when compared with commonly used gel electrophoresis systems. In brief, fluorescently labelled PCR products are amplified from total environmental DNA. Products are digested by restriction enzymes and analysed by a gene fragment sequencer to produce an electrophorogram that will allow rapid comparison of samples. Buchan *et al.* (2003) successfully used T-RFLP to investigate the dynamics of fungal and bacterial communities of a salt marsh in the southeastern USA. Assignment of distinct fungal terminal restriction fragments (T-RFs) to a clone or isolate was achieved by comparing peak position with peaks gained from empirical evidence generated from a previous study (Buchan *et al.*, 2002). Of the five samples collected over one year, *Phaeosphaeria spartinicola*, *P. halima* and *Mycosphaerella* sp. were present in all samples. This study found that there was considerable temporal variability, with little to moderate spatial variability. Again, the main disadvantage is lack of instant phylogenetic information unless other empirical evidence is available. Spurious peaks have also been reported where pseudo-terminal restriction fragments can be generated by incomplete digestion of PCR products in T-RFLP profiles of environmental samples (Osborn *et al.*, 2000). Incompletely digested PCR products from a complex microbial community may result in additional peaks and, consequently, an overestimation of diversity (Egert & Friedrich, 2003).

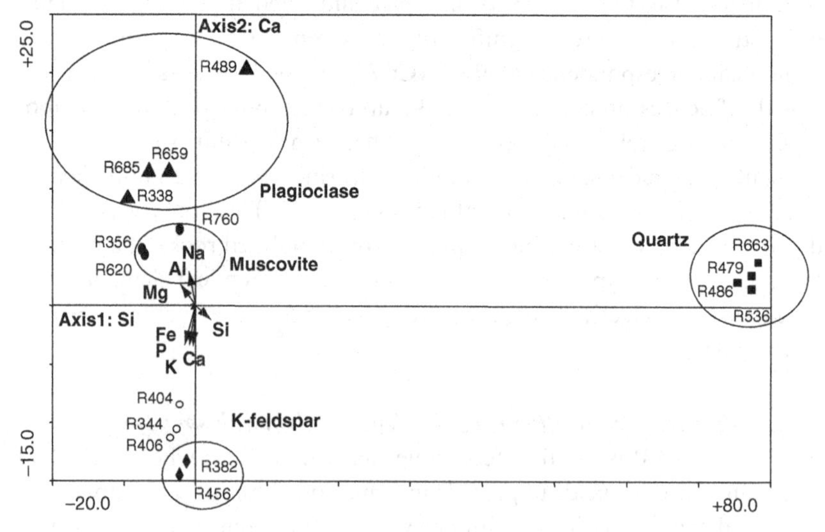

Fig. 5.7. Canonical correspondence analysis (CCA) ordination diagram of fungal ARISA data, with chemical variables represented as arrows and ribotypes (labelled according to size in bp) represented by the following symbols: filled circles, muscovite; triangles, plagioclase; open circles, granite; squares, quartz; diamonds, K-feldspar. Canonical intra-set correlations for the chemical factors for each axis indicated that Axis 1 was most strongly correlated with Si and Al and Axis 2 with Ca, although K and Na were also important factors (adapted from Gleeson *et al.*, 2005).

Conclusions

Rock and mineral surfaces support complex fungal communities that are only beginning to be studied by culture-independent means. Application of PCR technology has permitted a view of soil microbial diversity that is not distorted by culturing bias and revealed that the *in situ*, uncultured majority is highly diverse. In general, these techniques are still in their relative infancy and many obstructions need to be overcome. However, molecular methodologies are among the most important tools currently available and are capable of revealing important aspects of fungal ecology.

One of the most significant challenges in ecological studies is to understand how microbial communities are influenced by spatial and temporal environmental heterogeneity (Torsvik *et al.*, 2002). Overcoming this challenge has been aided by combining environmental molecular studies with a variety of descriptive statistics such as principal components analysis (PCA) and canonical correspondence analysis (CCA), which can provide comprehensive representations of the similarities and differences between genetic fingerprints (Hartmann *et al.*, 2005). Applying statistical analysis can give an accurate estimate of microbial diversity from assorted environments as well as delivering a quantitative means of relating microbial community composition to environmental factors and conditions (Haack *et al.*, 2004; Gleeson *et al.*, 2005, 2006). The rRNA approach, in combination with a range of molecular techniques, has the potential to analyse any fungal community where total environmental DNA extraction is successful, including rock and mineral sources. Community profiling techniques have only been applied to rocks and minerals in a select few examples but it is likely that such DNA-based approaches to study geomicrobial communities will not only enhance our understanding but also generate a wealth of further fundamental questions.

Acknowledgements

G.M.G. and M.F. gratefully acknowledge research support from the Biotechnology and Biological Sciences Research Council, the Natural Environment Research Council, and British Nuclear Fuels plc. E.P.B. gratefully acknowledges receipt of a NERC postgraduate research studentship. K.M. gratefully acknowledges receipt of a BBSRC postgraduate research studentship. Thanks are also due to Mr Martin Kierans (Centre for High Resolution Imaging and Processing (CHIPs), School of Life Sciences, University of Dundee) for assistance with scanning electron microscopy, and to Drs Deirdre Gleeson and Nicholas Clipson (University College Dublin) for advice and assistance regarding molecular analysis of fungal communities.

References

Adriaensen, K., Van der Lelie, D., Van Laere, A., Vangrosveld, J. & Colpaert, J. V. (2003). A zinc-adapted fungus protects pines from zinc stress. *New Phytologist* **161**, 549–55.

Ahonen-Jonnarth, U., Van Hees, P. A. W., Lundstrom, U. S. & Finlay, R. D. (2000). Organic acids produced by mycorrhizal *Pinus sylvestris* exposed to elevated aluminium and heavy metal concentrations. *New Phytologist* **146**, 557–67.

Albertano, P. & Urzi, C. (1999). Structural interactions among epilithic cyanobacteria and heterotrophic microorganisms in Roman Hypogea. *Microbial Ecology* **38**, 244–52.

Amann, R. I., Wolfgang, L. & Schleifer, K. H. (1995). Phylogenetic identification and *in situ* detection of individual microbial cells without cultivation. *Microbiological Reviews* **59**, 143–69.

Anderson, I. C. & Cairney, J. W. G. (2004). Diversity and ecology of soil fungal communities: increased understanding through the application of molecular techniques. *Environmental Microbiology* **6**, 769–79.

Anderson, I. C., Campbell, C. D. & Prosser, J. I. (2003). Potential bias of fungal 18S rDNA and internal transcribed spacer polymerase chain reaction primers for estimating fungal biodiversity in soil. *Environmental Microbiology* **5**, 36–47.

Arnott, H. J. (1995). Calcium oxalate in fungi. In *Calcium Oxalate in Biological Systems*, ed. S. R. Khan, pp. 73–111. Boca Raton, FL: CRC Press.

Arvieu, J. C., Leprince, F. & Plassard, C. (2003). Release of oxalate and protons by ectomycorrhizal fungi in response to P-deficiency and calcium carbonate in nutrient solution. *Annals of Forest Science* **60**, 815–21.

Baldrian, P. (2003). Interaction of heavy metals with white-rot fungi. *Enzyme and Microbial Technology* **32**, 78–91.

Banfield, J. P., Barker, W. W., Welch, S. A. & Taunton, A. (1999). Biological impact on mineral dissolution: application of the lichen model to understanding mineral weathering in the rhizosphere. *Proceedings of the National Academy of Sciences of the USA* **96**, 3404–11.

Barker, W. W., Welch, S. A. & Banfield, J. F. (1997). Biogeochemical weathering of silicate minerals. In *Geomicrobiology: interactions between Microbes and Minerals, Reviews in Mineralogy*, vol. 35, ed. J. F. Banfield & K. H. Nealson, pp. 391–428. Chelsea, MI:. Mineralogical Society of America.

Blazquez, F., Garcia-Vallez, M., Krumbein, W. E., Sterflinger, K. & Vendrell-Saz, M. (1997). Microstromatolithic deposits on granitic monuments: development and decay. *European Journal of Mineralogy* **9**, 899–901.

Borneman, J. & Triplett, E. W. (1997). Molecular microbial diversity in soils from eastern Amazonia: evidence for unusual microorganisms and microbial population shifts associated with deforestation. *Applied and Environmental Microbiology* **63**, 2647–53.

Braissant, O., Cailleau, G., Aragno, M. & Verrecchia, E. P. (2004). Biologically induced mineralization in the tree *Milicia excelsa* (Moraceae): its causes and consequences to the environment. *Geobiology* **2**, 59–66.

Brandl, H. (2001). Heterotrophic leaching. In *Fungi in Bioremediation*, ed. G. M. Gadd, pp. 383–423. Cambridge: Cambridge University Press.

van Breemen, N., Lundström, U. S. & Jongmans, A. G. (2000). Do plants drive podzolization via rock-eating mycorrhizal fungi? *Geoderma* **94**, 163–71.

Bronick, C. J. & Lal, R. (2005). Soil structure and management: a review. *Geoderma* **124**, 3–22.

Brown, S., Chaney, R., Hallfrisch, J., Ryan, J. A. & Berti, W. R. (2004). *In situ* soil treatments to reduce the phyto- and bioavailability of lead, zinc, and cadmium. *Journal of Environmental Quality* **33**, 522–31.

Bruand, A. & Duval, O. (1999). Calcified fungal filaments in the petrocalcic horizon of Eutrochrepts in Beauce, France. *Soil Science Society of America Journal* **63**, 164–9.

Buchan, A., Newell, S. Y., Moreta, J. I. & Moran, M. A. (2002). Molecular characterization of bacterial and fungal decomposer communities in a Southeastern U. S. salt marsh. *Microbial Ecology* **43**, 329–40.

Buchan, A., Newell, S. Y., Butler, M., Biers, E. J., Hollibaugh, J. T. & Moran, M. A. (2003). Dynamics of bacterial and fungal communities on decaying salt marsh grass. *Applied and Environmental Microbiology* **69**, 6676–87.

Burford, E. P., Fomina, M. & Gadd, G. M. (2003). Fungal involvement in bioweathering and biotransformation of rocks and minerals. *Mineralogical Magazine* **67**, 1127–55.

Burgstaller, W. & Schinner, F. (1993). Leaching of metals with fungi. *Journal of Biotechnology* **27**, 91–116.

Burke, D. J., Martin, K. J., Rygiewicz, P. T. & Topa, M. A. (2005). Ectomycorrhizal fungi identification in single and pooled root samples: terminal restriction fragment length polymorphism (TRFLP) and morphotyping compared. *Soil Biology and Biochemistry* **37**, 1683–94.

Buscot, F., Munch, J. C., Charcosset, J. Y., Gardes, M., Nehls, U. & Hampp, R. (2000). Recent advances in exploring physiology and biodiversity of ectomycorrhizas highlight the functioning of these symbioses in ecosystems. *FEMS Microbiology Reviews* **24**, 601–14.

Callot, G., Maurette, M., Pottier, L. & Dubois, A. (1987). Biogenic etching of microfractures in amorphous and crystalline silicates. *Nature* **328**, 147–9.

Callot, G., Mousain, D. & Plassard, C. (1985). Concentrations de carbonate de calcium sur les parois des hyphes mycéliens. *Agronomie* **5**, 143–50.

Casarin, V., Plassard, C., Souche, G. & Arvieu, J.-C. (2003). Quantification of oxalate ions and protons released by ectomycorrhizal fungi in rhizosphere soil. *Agronomie* **23**, 461–9.

Castro, I. M., Fietto, J. L. R., Vieira, R. X., Gutiérrez-Corona, F. G., Loza-Tavera, H., Carlos, J., Torres-Guzmán, J. C. & Moreno-Sánchez, R. (2000). Bioleaching of zinc and nickel from silicates using *Aspergillus niger* cultures. *Hydrometallurgy* **57**, 39–49.

Cervantes, C., Campos-Garcia, J., Devars, S. Gutiérrez-Corona, F. G., Loza-Tavera, H., Carlos, J., Torres-Guzmán, J. C. & Moreno-Sánchez, R. (2001). Interactions of chromium with microorganisms and plants. *FEMS Microbiology Reviews* **25**, 335–47.

Chen, X.-B., Wright, J. V., Conca, J. L. & Peurrung, L. M. (1997). Evaluation of heavy metal remediation using mineral apatite. *Water, Air and Soil Pollution* **98**, 57–78.

Christie, P., Li, X. L. & Chen, B. D. (2004). Arbuscular mycorrhiza can depress translocation of zinc to shoots of host plants in soils moderately polluted with zinc. *Plant and Soil* **261**, 209–17.

Clausen, C. A. & Green, F. (2003). Oxalic acid overproduction by copper-tolerant brown-rot basidiomycetes on southern yellow pine treated with copper-based preservatives. *International Biodeterioration and Biodegradation* **51**, 139–44.

Colpaert, J. V. & Van Assche, J. A. (1992). Zinc toxicity in ectomycorrhizal *Pinus sylvestris*. *Plant and Soil* **143**, 201–11.

Colpaert, J. V., Muller, L. A. H., Lambaerts, M., Adriaensen, K. & Vangronsveld, J. (2004). Evolutionary adaptation to Zn toxicity in populations of Suilloid fungi. *New Phytologist* **162**, 549–59.

Conca, J. L. (1997). *Phosphate-induced metal stabilization (PIMS)*. Final report to the U.S. Environmental Protection Agency 68D60023, Res. Triangle Park, NC.

Crosby, L. D. & Criddle, C. S. (2003). Understanding bias in microbial community analysis techniques due to *rrn* operon copy number heterogeneity. *BioTechniques* **34**, 790–802.

Cullings, K., Raleigh, C., New, M. H. & Henson, J. (2005). Effects of artificial defoliation of pines on the structure and physiology of the soil fungal community of a mixed pine-spruce forest. *Applied and Environmental Microbiology* **71**, 1996–2000.

Daniell, T. J., Husband, R., Fitter, A. H. & Young, J. P. W. (2001). Molecular diversity of arbuscular mycorrhizal fungi colonizing arable crops. *FEMS Microbiology Ecology* **36**, 203–9.

De Los Ríos, A., Wierzchos, J., Sancho, L. G. & Ascaso, C. (2004). Exploring the physiological state of continental Antarctic endolithic microorganisms by microscopy. *FEMS Microbiology Ecology* **50**, 143–52.

Devevre, O., Garbaye, J. & Botton, B. (1996). Release of complexing organic acids by rhizosphere fungi as a factor in Norway spruce yellowing in acidic soils. *Mycological Research* **100**, 1367–74.

Diercks, M., Sand, W. & Bock, E. (1991). Microbial corrosion of concrete. *Experientia* **47**, 514–16.

Eckhardt, F. E. W. (1985). Solubilisation, transport, and deposition of mineral cations by microorganisms-efficient rock-weathering agents. In *The Chemistry of Weathering*, ed. J. Drever, pp. 161–73. Dordrecht: D. Reidel Publishing Company.

Egert, M. & Friedrich, M. W. (2003). Formation of pseudo-terminal restriction fragments, a PCR-related bias affecting terminal restriction fragment length polymorphism analysis of microbial community structure. *Applied and Environmental Microbiology* **69**, 2555–62.

Ehrlich, H. L. (1998). Geomicrobiology: its significance for geology. *Earth-Science Reviews* **45**, 45–60.

Ehrlich, H. L. (2002). *Geomicrobiology*. New York: Marcel Dekker.

Etienne, S. & Dupont, J. (2002). Fungal weathering of basaltic rocks in a cold oceanic environment (Iceland): comparison between experimental and field observations. *Earth Surface Processes and Landforms* **27**, 737–48.

Ferris, F. G., Wiese, R. G. & Fyfe, W. S. (1994). Precipitation of carbonate minerals by microorganisms: implications for silicate weathering and the global carbon dioxide budget. *Geomicrobiology Journal* **12**, 1–13.

Filion, M., Hamelin, R. C., Bernier, L. & St-Arnaud, M. (2004). Molecular profiling of rhizosphere microbial communities associated with healthy and diseased black spruce (*Picea mariana*) seedlings grown in a nursery. *Applied and Environmental Microbiology* **70**, 3541–51.

Fisher, M. M. & Triplett, E. W. (1999). Automated approach for ribosomal intergenic spacer analysis of microbial diversity and its application to freshwater bacterial communities. *Applied and Environmental Microbiology* **65**, 4630–6.

Fomina, M. & Gadd, G. M. (2002). Metal sorption by biomass of melanin-producing fungi grown in clay-containing medium. *Journal of Chemical Technology and Biotechnology* **78**, 23–34.

Fomina, M., Ritz, K. & Gadd, G. M. (2003). Nutritional influence on the ability of fungal mycelia to penetrate toxic metal-containing domains. *Mycological Research* **107**, 861–71.

Fomina, M. A., Alexander, I. J., Hillier, S. & Gadd, G. M. (2004). Zinc phosphate and pyromorphite solubilization by soil plant-symbiotic fungi. *Geomicrobiology Journal* **21**, 351–66.

Fomina, M., Hillier, S., Charnock, J. M., Melville, K., Alexander, I. J. & Gadd, G. M. (2005a). Role of oxalic acid over-excretion in toxic metal mineral transformations by *Beauveria caledonica*. *Applied and Environmental Microbiology* **71**, 371–81.

Fomina, M. A., Alexander, I. J., Colpaert, J. V. & Gadd, G. M. (2005b). Solubilization of toxic metal minerals and metal tolerance of mycorrhizal fungi. *Soil Biology and Biochemistry* **37**, 851–66.

Fomina, M. A., Olishevska, S. V., Kadoshnikov, V. M., Zlobenko, B. P. & Podgorsky, V. S. (2005c). Concrete colonization and destruction by mitosporic fungi in model experiment, *Mikrobiologichny Zhurnal* **67**, 97–106 (in Russian with English summary).

Fraústo da Silva, J. J. R. & Williams, R. J. P. (1993). *The Biological Chemistry of the Elements.* Oxford: Oxford University Press.

Gabor, E. M., de Vries, E. J. & Janssen, D. B. (2003). Efficient recovery of environmental DNA for expression cloning by indirect extraction methods. *FEMS Microbiology Ecology* **44**, 153–63.

Gadd, G. M. (1990). Fungi and yeasts for metal accumulation. In *Microbial Mineral Recovery*, ed. H. L. Ehrlich & C. Brierley, pp. 249–75. New York: McGraw-Hill.

Gadd, G. M. (1993). Interactions of fungi with toxic metals. *New Phytologist* **124**, 25–60.

Gadd, G. M. (1999). Fungal production of citric and oxalic acid: importance in metal speciation, physiology and biogeochemical processes. *Advances in Microbial Physiology* **41**, 47–92.

Gadd, G. M. (2000). Bioremedial potential of microbial mechanisms of metal mobilization and immobilization. *Current Opinion in Biotechnology* **11**, 271–9.

Gadd, G. M. (ed.) (2001a). *Fungi in Bioremediation.* Cambridge: Cambridge University Press.

Gadd, G. M. (2001b). Accumulation and transformation of metals by microorganisms. In *Biotechnology, a Multi-volume Comprehensive Treatise*, vol. 10, *Special Processes*, ed. H.-J. Rehm, G. Reed, A. Puhler & P. Stadler, pp. 225–64. Weinheim, Germany: Wiley-VCH Verlag GmbH.

Gadd, G. M. (2002). Interactions between microorganisms and metals/radionuclides: the basis of bioremediation. In *Interactions of Microorganisms with Radionuclides*, ed. M. J. Keith-Roach & F. R. Livens, pp. 179–203. Amsterdam: Elsevier.

Gadd, G. M. (2004). Mycotransformation of organic and inorganic substrates. *Mycologist* **18**, 60–70.

Gadd, G. M. (ed.) (2006). *Fungi in Biogeochemical Cycles.* Cambridge: Cambridge University Press.

Gadd, G. M., Fomina, M. & Burford, E. P. (2006). Fungal roles in rock, mineral and soil transformations. In *Micro-organisms and Earth Systems – Advances in Geomicrobiology*, ed. G. M. Gadd, K. T. Semple & H. M. Lappin-Scott, pp. 201–31. Cambridge: Cambridge University Press.

Gaylarde, C. C. & Morton, L. H. G. (1999). Deteriogenic biofilms on buildings and their control: a review. *Biofouling* **14**, 59–74.

Gharieb, M. M., Kierans, M. & Gadd, G. M. (1999). Transformation and tolerance of tellurite by filamentous fungi: accumulation, reduction and volatilization. *Mycological Research* **103**, 299–305.

Gleeson, D. B., Clipson, N., Melville, K., Gadd, G. M. & McDermott, F. P. (2005). Characterization of fungal community structure on a weathered pegmatitic granite. *Microbial Ecology* **50**, 360–8.

Gleeson, D. B., Kennedy, N. M., Clipson, N., Melville, K., Gadd, G. M. & McDermott, F. P. (2006). Characterization of bacterial community structure on a weathered pegmatitic granite. *Microbial Ecology* **51**, 526–34.

Glowa, K. R., Arocena, J. M. & Massicotte, H. B. (2003). Extraction of potassium and/or magnesium from selected soil minerals by *Piloderma*. *Geomicrobiology Journal* **20**, 99–111.

Gomes, N. C. M., Fagbola, O., Costa, R., Rumjanek, N. G., Buchner, A., Mendona-Hagler, L. & Smalla, K. (2003). Dynamics of fungal communities in bulk and maize rhizosphere soil in the tropics. *Applied and Environmental Microbiology* **69**, 3758–66.

Gomez-Alarcon, G., Munoz, M. L. & Flores, M. (1994). Excretion of organic acids by fungal strains isolated from decayed limestone. *International Biodeterioration and Biodegradation* **34**, 169–80.

Gorbushina, A. A. & Krumbein, W. E. (2000). Subaerial microbial mats and their effects on soil and rock. In *Microbial Sediments*, ed. R. E. Riding & S. M. Awramik, pp. 161–9. Berlin: Springer-Verlag.

Gorbushina, A. A., Krumbein, W. E., Hamann, R., Panina, L., Soucharjevsky, S. & Wollenzien, U. (1993). On the role of black fungi in colour change and biodeterioration of antique marbles. *Geomicrobiology Journal* **11**, 205–21.

Gorbushina, A. A., Boettcher, M., Brumsack, H. J., Krumbein, W. E. & Vendrell-Saz, M. (2001). Biogenic forsterite and opal as a product of biodeterioration and lichen stromatolite formation in table mountain systems (tepuis) of Venezuela. *Geomicrobiology Journal* **18**, 117–32.

Gorbushina, A. A., Whitehead, K., Dornieden, T., Niesse, A., Schulte, A. & Hedges, J. I. (2003). Black fungal colonies as units of survival: hyphal mycosporines synthesized by rock-dwelling microcolonial fungi. *Canadian Journal of Botany* **81**, 131–8.

Goudie, A. S. (1996). Organic agency in calcrete development. *Journal of Arid Environments* **32**, 103–10.

Grote, G. & Krumbein, W. E. (1992). Microbial precipitation of manganese by bacteria and fungi from desert rock and rock varnish. *Geomicrobiology Journal* **10**, 49–57.

Gu, J. D., Ford, T. E., Berke, N. S. & Mitchell, R. (1998). Biodeterioration of concrete by the fungus *Fusarium*. *International Biodeterioration and Biodegradation* **41**, 101–9.

Haack, S. K., Fogarty, L. R., West, T. G., Alm, E. W., McGuire, J. T., Long, D. T., Hyndman, D. W. & Forney, L. J. (2004). Spatial and temporal changes in microbial community structure associated with recharge-influenced chemical gradients in a contaminated aquifer. *Environmental Microbiology* **6**, 438–48.

Haas, J. R. & Purvis, O. W. (2006). Lichen biogeochemistry. In *Fungi in Biogeochemical Cycles*, ed. G. M. Gadd, pp. 344–76. Cambridge: Cambridge University Press.

Hammes, F. & Verstraete, W. (2002). Key roles of pH, and calcium metabolism in microbial carbonate precipitation. *Reviews in Environmental Science and Biotechnology* **1**, 3–7.

Hansgate, A. M., Schloss, P. D., Hay, A. G. & Walker, L. P. (2005). Molecular characterization of fungal community dynamics in the initial stages of composting. *FEMS Microbiology Ecology* **51**, 209–14.

Hartley-Whitaker, J, Cairney, J. W. G. & Maharg, A. A. (2000). Sensitivity to Cd and Zn of host and symbiont of ectomicorrhizal *Pinus sylvestris* L. (Scots pine) seedlings. *Plant and Soil* **218**, 31–42.

Hartmann, M., Frey, R., Kölliker, R. & Widmer, F. (2005). Semi-automated genetic analyses of soil microbial communities: comparison of T-RFLP and RISA based on descriptive and discriminative statistical approaches. *Journal of Microbiological Methods* **61**, 349–60.

Hawksworth, D. L. (2001). The magnitude of fungal diversity: the 1.5 million species estimate revisited. *Mycological Research* **105**, 1422–32.

Heckman, D. S., Geiser, D. M., Eidell, B. R., Stauffer, R. L., Kardos, N. L. & Hedges, S. B. (2001). Molecular evidence for the early colonisation of land by fungi and plants. *Science* **293**, 1129–33.

Hernesmaa, A., Björklöf, K., Kiikkilä, O., Fritze, H., Haahtela, K. & Romantschuk, M. (2005). Structure and function of microbial communities in the rhizosphere of Scots pine after tree-felling. *Soil Biology and Biochemistry* **37**, 777–85.

Hibbet, D. S. (1992). Ribosomal RNA and fungal systematics. *Transactions of the Mycological Society of Japan* **33**, 533–56.

Hirsch, P., Eckhardt, F. E. W. & Palmer, R. J. Jr. (1995). Fungi active in weathering rock and stone monuments. *Canadian Journal of Botany* **73**, 1384–90.

Hochella, M. F. (2002). Sustaining Earth: thoughts on the present and future roles in mineralogy in environmental science. *Mineralogical Magazine* **66**, 627–52.

Hodge, A. (2000). Microbial ecology of the arbuscular mycorrhiza. *FEMS Microbiology Ecology* **32**, 91–6.

Hoffland, E., Giesler, R., Jongmans, T. & van Breemen, N. (2002). Increasing feldspar tunneling by fungi across a north Sweden podzol chronosequence. *Ecosystems* **5**, 11–22.

Horton, T. R. & Bruns, T. D. (2001). The molecular revolution in ectomycorrhizal ecology: peeking into the black-box. *Molecular Ecology* **10**, 1855–71.

Howlett, N. G. & Avery, S. V. (1997). Relationship between cadmium sensitivity and degree of plasma membrane fatty acid unsaturation in *Saccharomyces cerevisiae*. *Applied Microbiology and Biotechnology* **48**, 539–45.

Hugenholtz, P. & Pace, N. R. (1996). Identifying microbial diversity in the natural environment: a molecular phylogenetic approach. *Trends in Biotechnology* **14**, 190–7.

Hughes, K. A. & Lawley, B. (2003). A novel Antarctic microbial endolithic community within gypsum crusts. *Environmental Microbiology* **5**, 555–65.

Hunt, J., Boddy, L., Randerson, P. F. & Rogers, H. L. (2004). An evaluation of 18S rDNA approaches for the study of fungal diversity in grassland soils. *Microbial Ecology* **47**, 385–95.

Jentschke, G. & Godbold, D. L. (2000). Metal toxicity and ectomycorrhizas. *Physiologia Plantarum* **109**, 107–16.

Kahle, C. F. (1977). Origin of subaerial Holocene calcareous crusts: role of algae, fungi and sparmicristisation. *Sedimentology* **24**, 413–35.

Kennedy, N., Conolly, J. & Clipson, N. (2005). Impact of lime, nitrogen and plant species on fungal community structure in grassland microcosms. *Environmental Microbiology* **7**, 780–8.

Kerr, R. A. (1997). Life goes to extremes in the deep Earth – and elsewhere? *Science* **276**, 703–4.

Kikuchi, Y. & Sreekumari, K. R. (2002). Microbially influenced corrosion and biodeterioration of structural metals. *Journal of the Iron and Steel Institute of Japan* **88**, 620–8.

Knorre, Hv. & Krumbein, W. E. (2000). Bacterial calcification. In *Microbial Sediments*, ed. R. E. Riding & S. M. Awramik, pp. 25–39. Berlin: Springer-Verlag.

Krumbein, W. E., Urzi, C. & Gehrmann, C. (1991). On the biocorrosion and biodeterioration of antique and medieval glass. *Geomicrobiology Journal* **9**, 139–60.

Krupa, P. & Kozdroj, J. (2004). Accumulation of heavy metals by ectomycorrhizal fungi colonizing birch trees growing in an industrial desert soil. *World Journal of Microbiology and Biotechnology* **20**, 427–30.

Kumar, R. & Kumar, A. V. (1999). *Biodeterioration of Stone in Tropical Environments: an Overview*. Madison, WI: The J. Paul Getty Trust.

Lapeyrie, F., Picatto, C., Gerard, J. & Dexheimer, J. (1990). TEM study of intracellular and extracellular calcium oxalate accumulation by ectomycorrhizal fungi in pure culture or in association with *Eucalyptus* seedlings. *Symbiosis* **9**, 163–6.

Lapeyrie, F., Ranger, J. & Vairelles, D. (1991). Phosphate-solubilizing activity of ectomycorrhizal fungi *in vitro*. *Canadian Journal of Botany* **69**, 342–6.

Van Leerdam, D. M., Williams, P. A. & Cairney, J. W. G. (2001). Phosphate-solubilizing abilities of ericoid mycorrhizal endophytes of *Woollsia pungens* (Epacridaceae). *Australian Journal of Botany* **49**, 75–80.

Leyval, C. & Joner, E. J. (2001). Bioavailability of heavy metals in the mycorrhizosphere. In *Trace Elements in the Rhizosphere*, ed. G. R. Gobran, W. W. Wenzel & E. Lombi, pp. 165–85. Boca Raton, FL: CRC Press.

Lindahl, B. D. & Olsson, S. (2004). Fungal translocation – creating and responding to environmental heterogeneity. *Mycologist* **18**, 79–88.

Liu, W. T., Marsh, T. L., Cheng, H. & Forney, L. J. (1997). Characterization of microbial diversity by determining terminal restriction fragment length polymorphisms of genes encoding 16SyrRNA. *Applied and Environmental Microbiology* **63**, 4516–22.

Lundstrom, U. S., Van Breemen, N. & Bain, D. (2000). The podzolization process. A review. *Geoderma* **94**, 91–107.

Magyarosy, A., Laidlaw, R. D., Kilaas, R., Echer, C., Clark, D. S. & Keasling, J. D. (2002). Nickel accumulation and nickel oxalate precipitation by *Aspergillus niger*. *Applied Microbiology and Biotechnology* **59**, 382–8.

Manley, E. & Evans, L. (1986). Dissolution of feldspars by low-molecular-weight aliphatic and aromatic acids. *Soil Science* **141**, 106–12.

Manoli, F., Koutsopoulos, E. & Dalas, E. (1997). Crystallization of calcite on chitin. *Journal of Crystal Growth* **182**, 116–24.

Martino, E., Perotto, S., Parsons, R. & Gadd, G. M. (2003). Solubilization of insoluble inorganic zinc compounds by ericoid mycorrhizal fungi derived from heavy metal polluted sites. *Soil Biology and Biochemistry* **35**, 133–41.

Maurice, P. A., McKnight, D. M., Leff, L., Fulghum, J. E. & Gooseff, M. (2002). Direct observations of aluminosilicate weathering in the hyporheic zone of an Antarctic Dry Valley stream. *Geochemica et Cosmochimica Acta* **66**, 1335–47.

May, E. (2003). Microbes on building stone for good or bad? *Culture* **24**, 4–8.

Meharg, A. A. (2003). The mechanistic basis of interactions between mycorrhizal associations and toxic metal cations. *Mycological Research* **107**, 1253–65.

Meharg, A. A. & Cairney, J. W. G. (2000). Co-evolution of mycorrhizal symbionts and their hosts to metal-contaminated environments. *Advances in Ecological Research* **30**, 69–112.

Mehta, A. P., Torma, A. E. & Murr, L. E. (1979). Effect of environmental parameters on the efficiency of biodegradation of basalt rock by fungi. *Biotechnology and Bioengineering* **21**, 875–85.

Mironenko, N. V., Alekhina, I. A., Zhdanova, N. N. & Bulat, S. A. (2000). Intraspecific variation in gamma-radiation resistance and genomic structure in the filamentous fungus *Alternaria alternata*: a case study of strains inhabiting Chernobyl Reactor No. 4. *Ecotoxicology and Environmental Safety* **45**, 177–87.

Money, N. P. (2004). The fungal dining habit – a biomechanical perspective. *Mycologist* **18**, 71–6.

Monger, C. H. & Adams, H. P. (1996). Micromorphology of calcite-silica deposits, Yucca Mountain, Nevada. *Soil Science Society of America Journal* **60**, 519–30.

Morris, C. E., Bardin, M., Berge, O., Frey-Klett, P., Fromin, N., Girardin, H., Guinebretière, Leboron, P., Thièry, J. M. & Troussellier, M. (2002). Microbial diversity: approaches to experimental design and hypothesis testing in primary scientific literature from 1975 to 1999. *Microbiology and Molecular Biology Reviews* **66**, 592–616.

Muller, B., Burgstaller, W., Strasser, H., Zanella, A. & Schinner F. (1995). Leaching of zinc from an industrial filter dust with *Penicillium, Pseudomonas* and *Corynebacterium*: citric acid is the leaching agent rather than amino acids. *Journal of Industrial Microbiology* **14**, 208–12.

Muyzer, G., de Waal, E. C. & Uitterlinden, A. G. (1993). Profiling of complex microbial populations by denaturing gradient gel electrophoresis analysis of polymerase chain reaction amplified genes coding for 16S rRNA. *Applied and Environmental Microbiology* **59**, 695–700.

Nannipieri, P., Ascher, M. T., Ceccherini, M. T., Landi, L., Pietramellara, G. & Renella, G. (2003). Microbial diversity and soil functions. *European Journal of Soil Science* **54**, 655–70.

Nara, K., Nakaya, H., Wu, B., Zhou, Z. & Hogetsu, T. (2003). Underground primary succession of ectomycorrhizal fungi in a volcanic desert on Mount Fuji. *New Phytologist* **159**, 743–56.

Nica, D., Davis, J. L., Kirby, L., Zuo, G. & Roberts, D. J. (2000). Isolation and characterization of microorganisms involved in the biodeterioration of concrete in sewers. *International Biodeterioration and Biodegradation* **46**, 61–8.

Nikolcheva, L. G., Cockshutt, A. M. & Bärlocher, F. (2003). Determining diversity and freshwater fungi in decaying leaves: comparison of traditional and molecular approaches. *Applied and Environmental Microbiology* **69**, 2548–54.

Nikolcheva, L. G., Bourque, T. & Bärlocher, F. (2005). Fungal diversity during initial stages of leaf decomposition in a stream. *Mycological Research* **109**, 246–53.

Olsson, P. A. & Wallander, H. (1998). Interactions between ectomycorrhizal fungi and the bacterial community in soils amended with various primary minerals. *FEMS Microbiology Ecology* **27**, 195–205.

Osborn, A. M., Moore, E. R. B. & Timmis, K. N. (2000). An evaluation of terminal-restriction fragment length polymorphism (T-RFLP) analysis for the study of microbial community structure and dynamics. *Environmental Microbiology* **2**, 9–50.

Perfettini, J. V., Revertegat, E. & Langomazino, N. (1991). Evaluation of cement degradation by the metabolic activities of two fungal strains. *Experientia* **47**, 527–33.

Prosser, J. I. (2002). Molecular and functional diversity in soil-microorganisms. *Plant and Soil* **244**, 9–17.

Purvis, O. W. (1996). Interactions of lichens with metals. *Science Progress* **79**, 283–309.

Purvis, O. W. & Halls, C. (1996). A review of lichens in metal-enriched environments. *Lichenologist* **28**, 571–601.

Putnis, A. (2002). Mineral replacement reactions: from macroscopic observations to microscopic mechanisms. *Mineralogical Magazine* **66**, 689–708.

Ranjard, L., Poly, F., Lata, J.-C., Mougel, C., Thioulouse, J. & Nazaret, S. (2001). Characterization of bacterial and fungal soil communities by automated ribosomal intergenic spacer analysis fingerprints: biological and methodological variability. *Applied and Environmental Microbiology* **67**, 4479–87.

Riding, R. (2000). Microbial carbonates: the geological record of calcified bacterial-algal mats and biofilms. *Sedimentology* **47**, 179–214.

Rivadeneyra, M. A., Delgado, R., Delgado, G., Del Moral, A., Ferrer, M. R. & Ramos-Cormenzana, A. (1993). Precipitation of carbonates by *Bacillus* sp. isolated from saline soils. *Geomicrobiology Journal* **11**, 175–84.

Roberts, D. J., Nica, D., Zuo, G. & Davis, J. L. (2002). Quantifying microbially induced deterioration of concrete: initial studies. *International Biodeterioration and Biodegradation* **49**, 227–34.

Rosling, A., Lindahl, B. D., Taylor, A. F. S. & Finlay, R. D. (2004a). Mycelial growth and substrate acidification of ectomycorrhizal fungi in response to different minerals. *FEMS Microbiology Ecology* **47**, 31–7.

Rosling, A., Lindahl, B. D. & Finlay, R. D. (2004b). Carbon allocation to ectomycorrhizal roots and mycelium colonising different mineral substrates. *New Phytologist* **162**, 795–802.

Rossi, G. (1979). Potassium recovery through leucite bioleaching: possibilities and limitations. In *Metallurgical Applications of Bacterial Leaching and Related Phenomena*, ed. L. E. Murr, A. E. Torma & J. E. Brierley, pp. 279–319. New York: Academic Press.
Sand, W. & Bock, E. (1991). Biodeterioration of mineral materials by microorganisms – Biogenic sulphuric and nitric-acid corrosion of concrete and natural stone. *Geomicrobiology Journal* **9**, 129–38.
Sanderson, M. J. & Shaffer, H. B. (2002). Troubleshooting molecular phylogenetic analysis. *Annual Reviews of Ecology and Systematics* **33**, 49–72.
Sayer, J. A., Raggett, S. L. & Gadd, G. M. (1995). Solubilization of insoluble compounds by soil fungi: development of a screening method for solubilizing ability and metal tolerance. *Mycological Research* **99**, 987–93.
Sayer, J. A., Cotter-Howells, J. D., Watson, C., Hillier, S. & Gadd, G. M. (1999). Lead mineral transformation by fungi. *Current Biology* **9**, 691–4.
Scheublin, T. R., Ridgway, K. P., Young, J. P. W. & van der Heijden, M. G. A. (2004). Nonlegumes, legumes, and root nodules harbor different arbuscular mycorrhizal fungal communities. *Applied and Environmental Microbiology* **70**, 6240–6.
Setala, H., Rissanen, J. & Markkola, A. M. (1997). Conditional outcomes in the relationship between pine and ectomycorrhizal fungi in relation to biotic and abiotic environment. *Oikos* **80**, 112–22.
Sigler, W. V., Bachofen, R. & Zeyer, J. (2003). Molecular characterization of endolithic cyanobacteria inhabiting exposed dolomite in central Switzerland. *Environmental Microbiology* **5**, 618–27.
Smit, E., Leeflang, P., Glandorf, B., van Elsas, J. D. & Wernars, K. (1999). Analysis of fungal diversity in the wheat rhizosphere by sequencing of cloned PCR-amplified genes encoding 18S rRNA and temperature gradient gel electrophoresis. *Applied and Environmental Microbiology* **65**, 2614–21.
Smith, S. E. & Read, D. J. (1997). *Mycorrhizal Symbiosis.* London: Academic Press.
Smits, M. M., Hoffland, E., Jongmans, A. G. & van Breemen, N. (2005). Contribution of mineral tunneling to total feldspar weathering. *Geoderma* **125**, 59–69.
Staley, J. T., Palmer, F. & Adams, J. B. (1982). Microcolonial fungi: common inhabitants on desert rocks. *Science* **215**, 1093–5.
Sterflinger, K. (2000). Fungi as geologic agents. *Geomicrobiology Journal* **17**, 97–124.
Sterflinger, K., Krumbein, W. E. & Schwiertz, A. (1998). PCR in-situ hybridisation of hyphomycetes. *International Microbiology* **1**, 217–20.
Thorn, R. G., Reddy, C. A., Harris, D. & Paul, E. A. (1996). Isolation of saprophytic Basidiomycetes from soil. *Applied and Environmental Microbiology* **62**, 4288–92.
Van Tichelen, K. K., Colpaert, J. V. & Vangronsveld, J. (2001). Ectomycorrhizal protection of *Pinus sylvestris* against copper toxicity. *New Phytologist* **150**, 203–13.
de la Torre, J. R., Goebel, B. M., Friedmann, E. I. & Pace, N. R. (2003). Microbial diversity of cryptoendolithic communities from the McMurdo Dry Valleys, Antarctica. *Applied and Environmental Microbiology* **69**, 3858–67.
Torsvik, V., Øverås, L. & Thingstad, T. F. (2002). Prokaryotic diversity–magnitude, dynamics and controlling factors. *Science* **296**, 1064–6.
Urzi, C., Garcia-Valles, M. T., Vendrell, M. & Pernice, A. (1999). Biomineralization processes of the rock surfaces observed in field and in laboratory. *Geomicrobiology Journal* **16**, 39–54.
Verrecchia, E. P. (2000). Fungi and sediments. In *Microbial Sediments*, ed. R. E. Riding & S. M. Awramik, pp. 69–75. Berlin: Springer-Verlag.
Verrecchia, E. P., Dumont, J. L. & Rolko, K. E. (1990). Do fungi building limestones exist in semi-arid regions? *Naturwissenschaften* **77**, 584–6.

Volkmann, M., Whitehead, K., Rutters, H., Rullkotter, J. & Gorbushina, A. A. (2003). Mycosporine-glutamicol-glucoside: a natural UV-absorbing secondary metabolite of rock-inhabiting microcolonial fungi. *Rapid Communications in Mass Spectrometry* **17**, 897–902.

Wainwright, M., Tasnee, A. A. & Barakah, F. (1993). A review of the role of oligotrophic microorganisms in biodeterioration. *International Biodeterioration and Biodegradation* **31**, 1–13.

Wallander, H., Wickman, T. & Jacks, G. (1997). Apatite as a source in mycorrhizal and non-mycorrhizal *Pinus sylvestris* seedlings. *Plant and Soil* **196**, 123–31.

Wallander, H., Mahmood, S., Hagerberg, D., Johansson, L. & Pallon, J. (2003). Elemental composition of ectomycorrhizal mycelia identified by PCR-RFLP analysis and grown in contact with apatite or wood ash in forest soil. *FEMS Microbiology Ecology* **44**, 57–65.

Warren, L. A., Maurice, P. A., Parmer, N. & Ferris, F. G. (2001). Microbially mediated calcium carbonate precipitation: implications for interpreting calcite precipitation and for solid-phase capture of inorganic contaminants. *Geomicrobiology Journal* **18**, 93–115.

Warscheid, T. & Krumbein, W. E. (1994). Biodeterioration processes on inorganic materials and means of countermeasures. *Materials and Corrosion* **45**, 105–13.

Watts, H. J., Very, A. A., Perera, T. H. S., Davies, J. M. & Gow, N. A. R. (1998). Thigmotropism and stretch-activated channels in the pathogenic fungus *Candida albicans*. *Microbiology* **144**, 689–95.

White, I. D., Mottershead, D. N. & Harrison, S. J. (1992). *Environmental Systems: An Introductory Text*. London: Chapman and Hall.

Whitelaw, M. A. (2000). Growth promotion of plants inoculated with phosphate- solubilizing fungi. *Advances in Agronomy* **69**, 99–151.

Wierzchos, J., De Los Ríos, A., Sancho, L. G. & Ascaso, C. (2004). Viability of endolithic micro-organisms in rocks from the McMurdo Dry Valleys of Antarctica established by confocal and fluorescence microscopy. *Journal of Microscopy* **216**, 57–61.

Wierzchos, J., Sancho, L. G. & Ascaso, C. (2005). Biomineralization of endolithic microbes in rocks from the McMurdo Dry Valleys of Antarctica: implications for microbial fossil formation and their detection. *Environmental Microbiology* **7**, 566–75.

von Wintzingerode, F., Gobel, U. B. & Stackebrandt, E. (1997). Determination of microbial diversity in environmental samples: pitfalls of PCR-based rRNA analysis. *FEMS Microbiology Reviews* **21**, 213–29.

Zhdanova, N. N., Zakharchenko, V. A., Vember, V. V. & Nakonechnaya, L. T. (2000). Fungi from Chernobyl: mycobiota of the inner regions of the containment structures of the damaged nuclear reactor. *Mycological Research* **104**, 1421–6.

6

Mycelial responses in heterogeneous environments: parallels with macroorganisms

LYNNE BODDY AND T. HEFIN JONES

Cardiff School of Biosciences, Cardiff University

Introduction

Terrestrial fungi are commonly studied in the laboratory, growing on artificial media in which nutrients are typically homogeneously distributed and supplied in superabundance, the environment is sterile and microclimate (temperature, moisture, gaseous regime) usually relatively constant. This contrasts with the natural environment, in which: nutrients are often patchily and sparsely distributed or not readily available, because they are locked in recalcitrant material (e.g. lignin); many other organisms are encountered, including other fungi, bacteria and invertebrates; and microclimate is constantly changing, both temporally and spatially. This chapter explores the ways in which fungi cope with environmental heterogeneity. Similar situations are faced by macroorganisms and analogies are drawn. Emphasis is placed on basidiomycetes, not only because they have been studied in most detail, but because of their dominant role as decomposers and mutualistic symbionts (Boddy & Watkinson, 1995; Smith & Read, 1997) and because they are better adapted to respond to environmental heterogeneity over scales ranging from micrometres to many metres than are other fungi.

Both saprotrophic and ectomycorrhizal Basidiomycota form extensive mycelial systems in woodland soil and litter, but it is the former that are the focus of this review. Saprotrophic, cord-forming Basidiomycota that ramify at the soil–litter interface, interconnecting disparate litter components, provide most examples. The key feature of these fungi that fits them for growth in environments where resources are heterogeneously distributed is that they are non-resource-unit restricted, i.e. they can grow out of one resource in search of others. This contrasts with resource-unit restricted fungi that can only be disseminated as spores to new resources,

Fungi in the Environment, ed. G. M. Gadd, S. C. Watkinson & P. S. Dyer. Published by Cambridge University Press.

heterogeneously dispersed in space and time. Both resource-unit restricted and non-resource-unit restricted fungi risk loss of a large amount of biomass (as a result of invertebrate grazing, killing by other microorganisms and death in a harsh environment) in the quest for new resources, but the latter have minimized this by adopting a variety of different strategies for obtaining new resources, and different responses when resources have been encountered. These include operating a 'sit and wait' strategy, whereby a large mycelial network waits for resources to land on it and then actively colonizes those resources, often with responses occurring elsewhere in the system. Another approach, which is not mutually exclusive, is to grow and search actively for new resources, with an array of responses to finding them. With both approaches, mycelia are at risk or have nutritional opportunities from encounter with other organisms, both micro- and macro-, and they have evolved dramatic responses to such encounters.

The mycelial networks that develop are continuously re-modelled in response to local nutritional or environmental cues, interaction with other fungi, grazing by invertebrates or other destructive disturbance events. This occurs through a complex combination of growth, branching, hyphal fusion and regression of different regions of the mycelial system. Not only does morphology change but so also does a complex set of physiological processes associated with uptake, storage and re-distribution of nutrients throughout the network. Morphological and physiological changes are highly coordinated, ensuring that locally perceived environmental changes are responded to locally and over a larger scale. Different fungi respond differently in space and time, leading to different long-term behaviour and foraging strategies. This is reviewed here.

Search strategies

Mycelial pattern formation in different species

The pattern of mycelial outgrowth from a resource into soil and litter in search of new resources varies between species and is modified by the quantity and quality of the initial resource, soil nutrient status and other soil characteristics, and the microclimate (see below). These patterns have been described quantitatively in terms of radial extension rate, hyphal coverage and fractal dimension (D) (Boddy *et al.*, 1999; Boddy, 1999; Boddy & Donnelly, 2007) (Box 6.1). The last effectively describes space-filling and can highlight developmental shifts between diffuse, assimilative and corded, distributive growth, as well as revealing subtle differences that may be missed by simple observation.

Box 6.1

Fractal geometry can be used to describe structures (e.g. many natural objects) that cannot be described in Euclidean whole number dimensions, but rather fractional dimensions, with values ranging between 1 (linear) and 2 (completely plane-filled) in two dimensions, and up to 3 in three dimensions (volumes). Fractal structures are 'self-similar': that is, when they are examined at increasingly closer range more and more, smaller and smaller, identical copies are revealed. This is true over an infinite range of scales for 'ideal' fractals, but natural structures are only self-similar over a finite range of length scales. Several different fractal dimensions can be determined, and two of these – mass fractal dimension D_M and surface/border fractal dimension D_S – are useful for description of mycelia. These two fractal dimensions allow distinction between systems that have gaps in their interiors (mass fractal) and those that are completely plane-filled except at their boundaries (border or surface fractal). With mass fractal systems $D_S = D_M$, whereas with surface fractal systems $D_S < D_M$. There are several ways of determining D, but that most commonly used with mycelia is the box-count method (Boddy *et al.*, 1999; Boddy, 1999; Boddy & Donnelly, 2007), where mass and border fractal dimensions are designated D_{BM} and D_{BS}, respectively.

While growing through soil in search of new resources, some fungi (e.g. *Agrocybe gibberosa*, *Coprinus picaceus*, *Phallus impudicus*, *Phanerochaete velutina* and *Resinicium bicolor*) produce mass fractal mycelia, whereas others (e.g. *Hypholoma fasciculare* and *Stropharia caerulea*) are surface fractal (Fig. 6.1) (Boddy, 1999; Boddy & Donnelly, 2007). The former have open systems characterized by well-defined, rapidly extending cords (high-conductivity channels, formed by aggregation and limited differentiation of hyphae) throughout the system. They might be thought of as long-range foragers that would be unsuccessful at capitalizing on homogeneously supplied nutrients but would successfully discover large, sparsely distributed resources. Although *A. gibberosa* tends to utilize graminaceous litter (Robinson *et al.*, 1993), the others interconnect fallen branches, trunks and stumps with relatively little colonization of leaf litter (Dowson *et al.*, 1988a; Kirby *et al.*, 1990; Owen, 1997). By contrast, surface fractal mycelia are characterized by diffuse, slowly extending search fronts, and might be thought of as short-range foragers. They search areas intensively and would be likely to be successful in discovering abundant, relatively homogeneously distributed resources. Indeed, *Stropharia* species are commonly found on nutrient-rich sites (e.g. compost heaps) and colonize *Urtica dioica* rhizomes and other small woody and non-woody litter (Donnelly,

1995). *Hypholoma fasciculare*, although it does colonize stumps and other large woody resources, frequently colonizes leaves, small twigs, cupules, etc. (Dowson *et al.*, 1988a).

Temporal changes

Mycelial pattern characteristics, not least fractal dimension, change with time. Systems tend to become more open (lower D), and in the long term even mycelia that were initially (for several months) surface fractal become increasingly mass fractal (Donnelly *et al.*, 1995; Boddy *et al.*, 1999). Systems become more open as a result of regression of individual hyphae and minor cords, leaving a persistent network of thick cords behind the foraging margin. Regressing mycelium allows re-distribution of biomass and nutrients for further exploration by foraging fronts. Radiotracer studies have shown that nutrients are also scavenged at foraging fronts and are translocated from the resource base to support growth fronts (Wells & Boddy, 1990; Wells *et al.*, 1990; Boddy, 1999). Modern

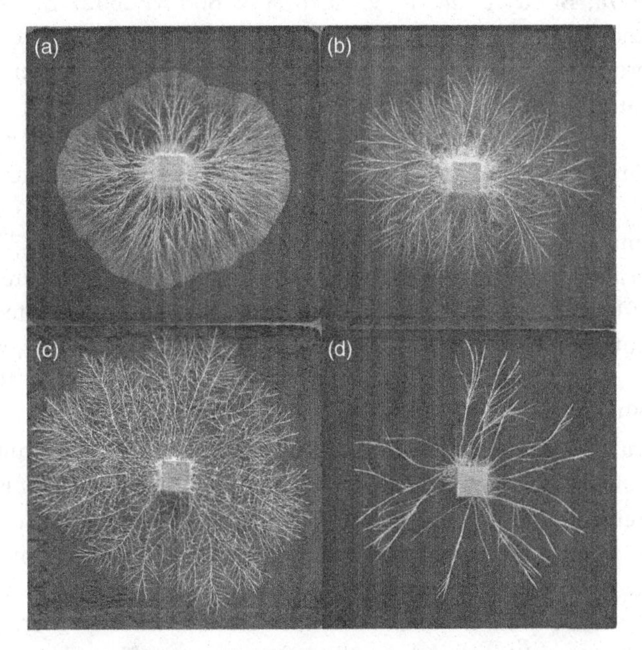

Fig. 6.1. Interspecific differences in foraging mycelial morphology. Mycelia are growing from 4 cm^3 beechwood inocula across trays (24 cm × 24 cm) of compressed non-sterile soil. (a) *Hypholoma fasciculare* after 30 d; (b) *Phanerochaete velutina* after 13 d; (c) *Phallus impudicus* after 46 d; and (d) *Resinicium bicolor* after 31 d. (Digital images courtesy of G. M. Tordoff.)

techniques for imaging nutrient dynamics at smaller spatial scales are now being employed to unravel mechanisms and routes of translocation (Tlalka *et al.*, 2002, 2003; Chapter 1, this volume).

Changes elicited by abiotic factors

Mycelial growth (extension, biomass, *D*) is affected by many environmental variables, including the nutrient status (size, state of decay and other aspects of quality) of the inoculum, nutrient status of newly encountered resources, nutrient status and composition of the soil through which the mycelial system is extending, and non-nutrient abiotic environment (temperature, gaseous regime, water potential and pH). These vary both in time (seasonally, daily) and in space (Boddy 1984, 1986), and hence mycelial pattern and behaviour also vary temporally and spatially.

Most species examined, although not *S. caerulea*, produce mycelia with greater space-filling from larger inocula (Bolton & Boddy, 1993; Donnelly & Boddy, 1997a; Boddy *et al.*, 1999; Zakaria & Boddy, 2002). *Stropharia caerulea* and *R. bicolor* mycelium, though not necessarily that of *Phan. velutina*, produces greater space-filling from resources with a higher nutrient concentration (Boddy *et al.*, 1999; Zakaria & Boddy, 2002); higher *D* values were often correlated with higher extension rates and hyphal coverage, indicating that space-filling is not at the expense of extension. Presumably, denser mycelium reflects the greater availability of carbon and nutrients for generating extra-resource mycelial biomass.

H. fasciculare and *S. caerulea*, with their surface fractal structures, are responsive to changes in the environment and respond to elevated soil nutrient (nitrogen or phosphorus) status by increasing branching, which allows increased uptake of nutrients (Donnelly & Boddy, 1998; Boddy *et al.*, 1999). In contrast, *Phal. impudicus*, with its mass fractal structure and long-range foraging strategy, is less responsive, elevated soil nutrient status having little apparent effect on morphology. *Phan. velutina* is also mass fractal with a long-range foraging strategy, and usually less responsive to environmental change than the short-rangers, but sometimes fills space more when there is a plethora of soil nutrients.

The non-nutrient environment has variable effects; moreover, parameters interact with each other. Temperature can affect morphology (extension, branching, cord-production and space-filling) dramatically (Donnelly & Boddy, 1997b), but what is mediating these effects, and how they relate to foraging, is unclear. Soil pH dramatically affects morphology of some species: *Phan. velutina* had a much higher *D* at pH 7 than at pH 4.4 (Owen, 1997), whereas *C. picaceus* often failed to grow out of wood resources at pH 4.4, but when it did growth was much slower. Nonetheless, with *C. picaceus*, similar amounts of extra-resource biomass were produced at all pH values, owing to much greater space-filling at lower pH. Soil water potential clearly affected morphology of *S. caerulea*, D_{BS} decreasing (*c.* 1.7 to 1.35) with increasing water potential (from −0.02 MPa to −0.002 MPa, i.e. getting wetter), and with greater aggregation into cords at and below −0.2 MPa (Donnelly & Boddy, 1997b). In contrast, D_{BS} of *Phan. velutina*, was not affected, although D_{BM} was sometimes higher on soils drier than −0.006 MPa. These experiments were, however, performed under constant conditions. In the natural environment soil water potential and temperature constantly varies, and such fluctuations have been shown to affect mycelial development, nutrient uptake and movement in systems to which new resources were added behind growth fronts (Wells *et al.*, 2001).

Parallels with macroorganisms

There are distinct parallels between fungal and plant foraging (Hutchings & De Kroon, 1994; Hutchings & Wijesinghe, 1997). In plant ecology two extreme patterns of foraging clonal structures have been described: 'phalanx', in which plants have tight aggregated rosettes, and 'guerrilla' in which, within clones, parts are loosely aggregated with fast-growing branches (Schmid & Harper, 1985). Fungal short-range foraging is equivalent to 'phalangeal' foraging of plants, and fungal long-range foraging equivalent to the 'guerrilla' strategy. In the context of foraging strategies of root systems, three traits can be considered: (1) developing extensive root systems (considered high-scale foraging); (2) concentrating roots in high resource patches (high-precision foraging); (3) rapidly reaching or proliferating in patches (high-rate foraging) (Campbell *et al.*, 1991; Rajaniemi & Reynolds, 2004). These could equally well be applied to mycelia of species described above.

Parallels with animals are also evident, for example, behaviour shown by foraging swarms of *Dorylus* driver ants (Wilson, 1990) and army ants (Rayner & Franks, 1987). Different species of army ant (genus *Eciton*) feed on different types of prey: *E. hamatum* preys upon social insect colonies that are widely dispersed; *E. burchelli* predates social insects and large arthropods that are evenly distributed; *E. repax* has an intermediate diet, feeding on social insects but making rapid raids on their nests, and also predating some large arthropods (Burton & Franks, 1985; Franks *et al.*, 1991). The trail patterns formed by the raid fronts (Fig. 6.2) are remarkably similar to mycelial systems, that of *E. hamatum* looking superficially similar to that of long-range mycelial foraging, *E. burchelli* to short-range foraging, and *E. repax* somewhere in between.

Patterns of trails formed by motile organisms tend to form readily without invoking pre-programming, i.e. they are self-organizing (see, for example, Franks *et al.*, 1991; Rayner, 1997; Couzin & Franks, 2002): trails made by raiding ants are structured by the interaction between inbound and outbound foragers mediated by pheromones. Pioneers make slow progress as they find suitable routes through rough terrain, making easier routes for followers, who reinforce the trails created (Franks *et al.*, 1991; Rayner, 1997; Couzin & Franks, 2002). Followers, which move rapidly along the well-established trails, tend to catch up with the pioneers and then branch off to form trails of their own, creating a structure visually resembling a river delta. This self-organizing pattern generation is enhanced by deposition of trail pheromones that attract other ants, which deposit more pheromones, resulting in positive feedback. When foraging is successful and ants return along the same trails this further amplifies trail formation. Where there is less use, less pheromones are deposited and there is negative feedback. Within trails there is further self-organization (generated by movement rules of individual ants), resulting in the formation of lanes and optimized traffic flow. Differences in the trail patterns produced by the different *Echiton* species can be generated by small differences in the relative attractiveness of the trail pheromones

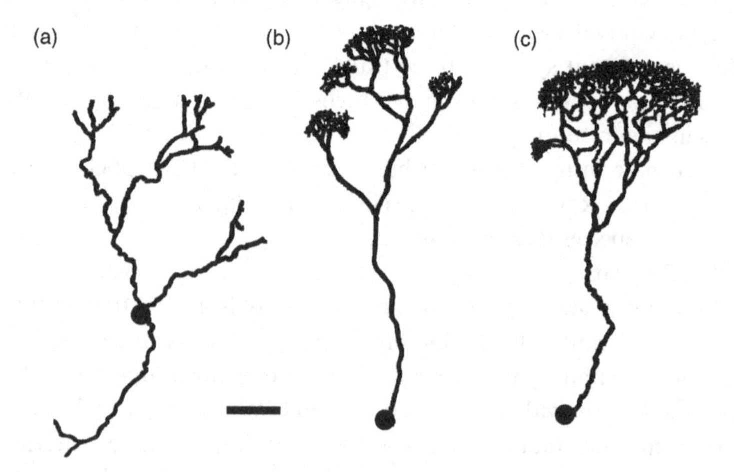

Fig. 6.2. Different trail patterns produced by foraging army ants: (a) *Eciton hamatum*, (b) *E. repax* and (c) *E. burchelli*. The circles indicate the locations of the nests. Bar represents 5 m. (After Burton & Franks, 1985.)

(Burton & Franks, 1985). Moreover, by manipulating the prey distributions for swarms of *E. burchelli* it has been shown that sites of prey capture alter the geometry of the trail patterns, and patterns could be made more typical of other army ant species (Franks *et al.*, 1991). This implies that trail patterns do not result from different search strategies but from different patterns of 'finds', and is therefore not analogous to long- and short-range patterns exhibited by cord-forming fungi. Of course, as with ants, mycelial patterns also change dramatically depending on distribution and abundance of resources.

Mycelial response to discovery of new resources

When foraging mycelia encounter new resources they often exhibit remarkable patterns of re-allocation of biomass and changes in partitioning of mineral nutrients, and further extension through soil may be delayed. Responses differ depending on size and quality of the inoculum, the new resource encountered, the relative sizes (and presumably quality) of the inoculum and the new resource, spatial distribution of the new resource, soil nutrient status and fungal species.

Re-allocation of biomass

When mycelia encounter new resources substantially larger than the resource from which they are growing, changes typically include thickening of resource-connected cords, regression of non-resource-connected mycelium, and cessation or slowing of radial extension, but subsequently with outgrowth from the newly colonized resource (Fig. 6.3) (Dowson

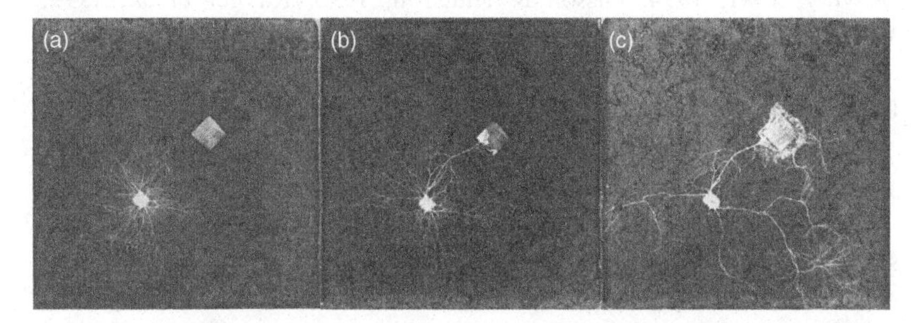

Fig. 6.3. Remodelling of foraging mycelial systems following discovery of a new resource. *Phanerochaete velutina* mycelium extending from beech wood (0.2 cm^3) across trays (24 cm × 24 cm) of sterile soil (a) 9 d, (b) 18 d and (c) 39 d after inoculation. By 18 d a new beechwood block (2 cm^3) had been contacted, and by 39 d all but major cords had regressed, while the latter had continued to extend, and there was outgrowth from the new resource. (Digital images from photographs courtesy of Rory Bolton.)

et al., 1986, 1988b; Bolton *et al.*, 1991; Bolton, 1993; Boddy, 1993, 1999; Donnelly & Boddy, 1997a). With short-range foragers (e.g. *H. fasciculare*) similar though more dramatic visible changes also occur even when newly encountered resources are similar in size to the original resource.

In the field, following outgrowth from new resources, further new resources will be encountered with further mycelial responses. Again, size of resources already within the mycelial system relative to newly encountered resources affects mycelial responses, with large resources having an overriding influence (Fig. 6.4) (Hughes & Boddy, 1996).

As well as encountering resources sequentially, mycelia may simultaneously encounter resources in different locations around the periphery. When *Phan. velutina*, extending from a small (0.5 cm^3) wood resource, encountered four closely spaced (about 3 cm apart) new wood resources or two resources (0.5 cm^3) on either side of the mycelium simultaneously, the mycelium responded by producing fans from points all around the margin (Bolton, 1993; Boddy, 1993, 1999). Such changes did not occur when a single new resource of the same total volume was encountered. Clearly, distribution of resources has a controlling influence on mycelial response.

Colonization of new resources can clearly trigger global, coordinated re-organization of the whole mycelial system, including responses at specific localized sites. Exactly how local sensing occurs or what is the signal that elicits site-specific developmental responses are not known, but several possibilities have been suggested: changes in intracellular hydrostatic pressure, intracellular amino acid concentration and electrical potentials (Rayner, 1991, 1994; Olsson & Hansson, 1995; Rayner *et al.*, 1995; Watkinson, 1999). It may be useful to measure intracellular proteinase expression as a marker indicating reception of a propagated stimulus cued by remote contact of a resource (Watkinson *et al.*, 2006).

Delays to continued foraging

When new resources are encountered mycelial extension often slows or stops completely, at least for a while. The time before mycelial egress from a newly encountered resource is also variable. The size/nutrient status of the newly encountered resource, and that of the original inoculum, dramatically affects the time between contact and egress from a newly encountered resource. With *Phan. velutina*, with a constant inoculum size, the bigger the new resource, the longer the time before egress from the new resource: with a 0.2 cm^3 wood inoculum mycelium grew out of a 0.2 cm^3 new resource after 7 d but not until 63 d with a 16 cm^3 new

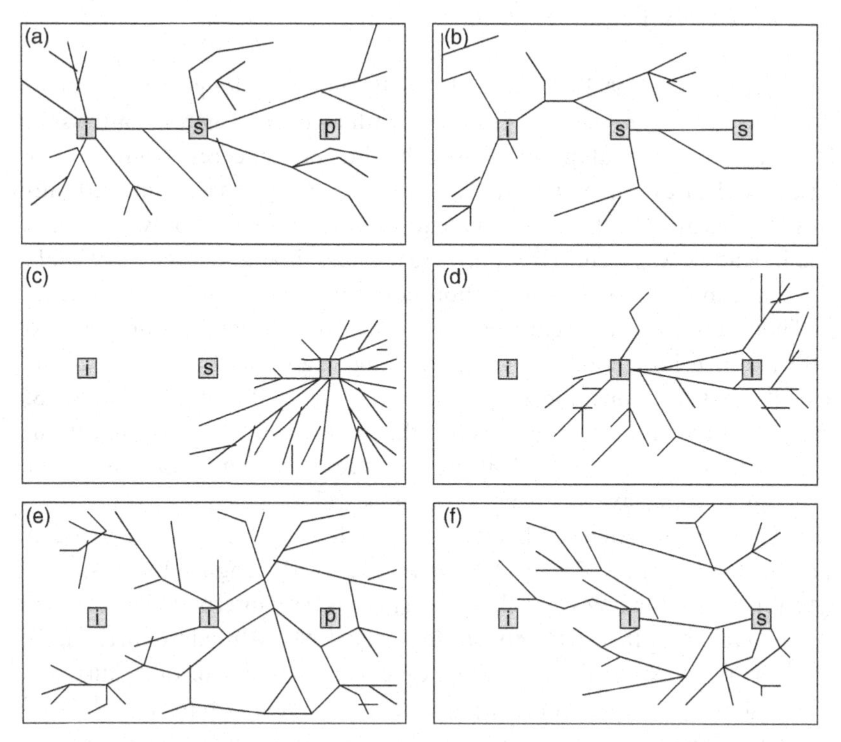

Fig. 6.4. Diagrammatic representation of re-modelling of *Phanerochaete velutina* foraging mycelial systems following sequential discovery of two new resources (based on data in Hughes & Boddy, 1996). Lines represent major cords, much of the finer mycelium being omitted from the diagram. In all cases *Phan. velutina* extended from a small (1 cm^3) beechwood resource (i), and encountered either small (2 cm^3; s) or large (8 cm^3; l) beechwood resource or inert plastic (p). Mycelial systems developed similarly in systems in which the first resource was small wood and the second plastic (a) or small wood (b), although there was less mycelial biomass in the latter systems, i.e. when two small new resources were encountered in succession. Although there was regression of fine mycelium this was not dramatic and large non-resource-connected cords remained. However, when the second new resource was large (c) all mycelium emanating from the original resource, and from the first newly encountered resource, completely regressed, leaving no connection between the large new resource and the smaller ones. When both new resources were large (d) these two remained connected with no major regression of mycelium emanating from them, although there was again complete regression from, and loss of connection with, the small original resource. When the first newly encountered resource was large and the second plastic (e), there was complete regression and loss of connection with the original resource, and outgrowth from the large resource both towards and beyond the plastic, and back towards the original inoculum. When the first newly encountered resource was large and the second small (f), there was (as in (e)) complete regression and loss of connection with the original resource, but following contact with the second new resource the polarity of the system changed such that subsequent growth from the large resource was in the direction of the original resource.

resource (Bolton, 1993). This may be partly due to the longer time taken to colonize the larger resource. However, with a constant new resource size, the smaller the inoculum, the longer the time before egress from the new resource. For example, when a 4 cm^3 new resource was encountered outgrowth was after 53 d when the inoculum was 0.2 cm^3, but only 10 d when the inoculum was 8 cm^3 (Bolton, 1993). Thus, time to egress is probably governed in large part by the carbon/nutrient status of the whole system.

Time between initial contact and egress from a new resource is also affected by prior colonization by other fungi. Egress of *Phan. velutina* was usually first from uncolonized rather than colonized resources (Boddy & Abdalla, 1998), presumably reflecting the ease or difficulty of colonization.

A similar situation is evident among predators feeding on patches of prey: predators leave smaller patches in search of new prey sooner than they leave larger patches. This results in predators tending to aggregate in patches of high prey density: this is known as the aggregative response (Arthur, 1966; Jones & Hassell, 1988). This has been interpreted in terms of optimal foraging theory (Krebs & McCleery, 1984). A predator arriving at a patch first acquires the resource quickly, but the rate of acquisition diminishes with time. The marginal value theorem (Charnov, 1976) of foraging indicates at what time on the curve of diminishing return for this patch the predator should leave the patch and travel to the next one (Fig. 6.5). If travel time between patches increases (patches are far apart)

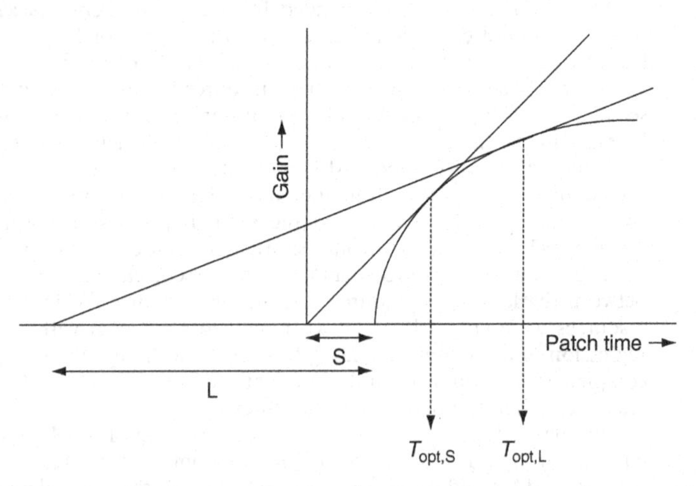

Fig. 6.5. The marginal value theorem. An example of a gain curve arising from resource depression. As travel time between patches increases, so does T_{opt}. For a short time (S), $T_{opt,S}$ is less than $T_{opt,L}$ for a long travel time (L). (Adapted from Krebs & McCleery, 1984.)

the optimal time to stay in a patch also increases: the longer it takes to travel, the lower the value of moving, so the further it is worth going along the curve of diminishing returns in the current patch. If patches within the environment vary in quality, each patch should be exploited until the gain rate within the patch drops to the average for the environment. Many of these arguments are also applicable to the delays before egress from newly encountered resources by foraging mycelia.

'Choice' of resources for colonization

When mycelia more or less simultaneously encounter resources colonized by different fungi those colonized by certain species are, at least initially, favoured over others. For example, when there were three wood blocks, one colonized by *Stereum hirstum*, another by *S. rugosum* and the third by *Kretzschmaria* (= *Ustulina*) *deusta*, *Phan. velutina* preferentially colonized the resource occupied by *K. deusta*, the other two often remaining uncolonized (Boddy & Abdalla, 1998). This is presumably not attraction, but reflects ease of colonization, *K. deusta* being much less combative than the other two species.

The obvious macroorganism parallel is with predators selecting the weakest or slowest prey, for example, lions preying on gazelle. With vertebrates, distribution within a habitat is often affected by the previous feeding activities of conspecifics as well as of other species, and also by the individual's own history of exploitation of food plants. For example, in pastures lightly stocked with sheep, selective grazing is initially in the better parts of the pasture. The vegetation in poorer areas thus becomes taller and older, contains other grass species and becomes less attractive to the sheep, resulting in feeding being more and more aggregated in the better areas as time progresses (Crawley, 1983). Similar factors may drive the pattern of distribution of foraging mycelia, although this will be continually modified by fresh plant litter inputs.

Physiological integration, division of labour and movement of nutrients

Nutrients (e.g. nitrogen and phosphorus) are scavenged as mycelia grow through soil and can be translocated away from the growing front (Cairney, 1992; Boddy, 1993) commonly accumulating in wood resources connected within the mycelial system (Wells & Boddy, 1990; Wells *et al.*, 1990, 1997, 1998, 1999). Likewise nutrients can be transported from wood resources to growing fronts. Rates of translocation can be rapid (sometimes more than 25 cm h^{-1}), the largest fluxes being through cords interconnecting one resource with another (see, for example, Wells & Boddy, 1990). Not surprisingly, temperature and water potential affect nutrient uptake and movement, as with many physiological processes (see, for example, Wells & Boddy, 1995a; Boddy, 1999).

Resources needed for growth are distributed patchily; as already discussed, mycelia respond to resources by re-allocating biomass. Superimposed upon these morphological responses are an equally, if not more, complex set of physiological responses that contribute to uptake, storage and re-distribution of nutrients throughout the network in a highly coordinated way. Many factors will affect the balance between, and the main sites of, uptake, storage and demand for nutrients. Size of new resources is one. For example, when two new resources were encountered simultaneously there was allocation of ^{32}P to these, from the inoculum resource, proportional to size and type (e.g. wood or leaf litter) of litter component (Wells *et al.*, 1990). Likewise, when new resources were encountered in sequence most ^{32}P was allocated to the larger new resource, irrespective of their position in the sequence of encounter (Hughes & Boddy, 1994, 1996). There also tends to be a higher demand for nutrients at relatively early stages of colonization than later on (see, for example, Hughes & Boddy, 1994; Wells *et al.*, 1998).

There is division of labour within the mycelial system in terms of providing carbon and/or energy. In small soil microcosms, resources from which the mycelium originally extended tend to decay more rapidly than newly encountered resources (see, for example, Abdalla & Boddy, 1996; Hughes & Boddy, 1996). This is probably because the former are well colonized, and there is often an increase in decay rate of the original resource when a new resource is met (Wells & Boddy, 1995a; Boddy & Abdalla, 1998). The size of the initial resource relative to the new resource, however, does influence the rate of decay of inoculum, new resource and total resource (Bolton, 1993). When inoculum resources are well decayed, newly encountered resources tend to decay more rapidly (Wells & Boddy, 1990); thus the demand of the system as a whole is being satisfied from elsewhere. The presence of antagonistic decay fungi also affects partitioning of decay (see, for example, Boddy & Abdalla, 1998).

This physiological division of labour and movement of nutrients is analogous to that seen in clonal plants, where there is internal re-distribution of resources from sites of acquisition to parts of clones located where that particular resource is scarce (Hutchings & Wijesinghe, 1997). Ramets of clonal plants also operate local specialization, which enhances acquisition of different resources from sites of greatest abundance. This specialization and resource sharing can result in considerably more plant growth when resources are heterogeneously supplied compared with when an equivalent amount of resources is homogeneously distributed (Hutchings & Wijesinghe, 1997). This physiological division of labour in feeding specialization is also seen in the polyps of some coelenterates, although to a lesser extent (Hyman, 1940).

Mycelial response to encountering other organisms

Antagonistic fungi and bacteria

During mycelial growth through soil, other mycelia are inevitably encountered resulting in direct competition for nutrients in soil solution and/or antagonistic interactions (Boddy, 2000). Little is known about direct competition between mycelia for nutrients in solution in soil and litter. Both the ectomycorrhizal *Paxillus involutus* and the saprotrophic *Phan. velutina* were able to sequester ^{33}P-labelled orthophosphate added to a litter patch in which they were competing in soil, and translocate it elsewhere in their respective mycelia (D. P. Donnelly, L. Boddy & J. R. Leake, unpublished). Phosphorus was transported back to the plant, though interestingly not via the shortest mycelial connection but along what appeared to be a major arterial route.

With solid organic resources, having control of space or territory effectively means having control of the carbon and mineral resources therein. In these situations mycelia are effectively competing for space and it is appropriate to use the term combat (Rayner & Webber, 1984; Boddy, 2000). The outcomes of combative interactions can be: deadlock, where neither fungus gains any territory; replacement, where one fungus replaces the other and gains its territory; partial replacement, where some but not all of the opponent's territory is gained; and mutual replacement, where one fungus gains some of the territory held by the other fungus and vice versa. In a study on interactions between *Stropharia caerulea* mycelium and that of other non-resource-unit restricted wood-decay Basidiomycota in soil trays, changes in mycelial morphology were quantified (D_{BM} and D_{BS}, hyphal coverage and number of cords) (Donnelly & Boddy, 2001). Fractal dimension and rate of biomass increase varied depending on interaction combination (Table 6.1). Importantly, there were often localized changes in mycelial morphology and local fractal dimension, reflecting defensive responses. These changes sometimes resulted in ingress into the opponent's territory.

Outcomes of competition between two plant or two animal species are very similar to those reported for fungi. Theoretically, four outcomes are possible (Lotka, 1932): competitive exclusion of species j by species i,

Table 6.1. *Changes in morphology (expressed by mass fractal dimension (D_{BM}) and biomass) of mycelial systems during interactions between* Stropharia caerulea *and other cord-forming, saprotrophic basidiomycetes in trays of compressed soil*

Other species	Outcome of interaction	Change in D_{BM} and D_{BS} of *S. caerulea*	Change[a] in biomass production rate ($mg\,d^{-1}$) of *S. caerulea*	Change in D_{BM} and D_{BS} of other species	Change in biomass of other species compared with controls with wood or no resource (in parentheses)
Agrocybe gibberosa	Initially deadlock, but *A. gibberosa* completely encircled *S. caerulea*. Subsequently, fans of *S. caerulea* mycelium extended over *A. gibberosa*	Production of dense localized fans (1.89 ± 0.04)	Decrease to 0.07 ± 0.05	No change in D_{BM} or D_{BS}	Significant ($p \leq 0.005$) decrease to 0.48 ± 0.09 (from 0.95 ± 0.14 or 0.71 ± 0.34)
Hypholoma fasciculare	Deadlock, but mutual replacement in one system	Differences not usually significant	Significant ($p \leq 0.05$) decrease to 0.10 ± 0.04	High D_{BM} (1.83 ± 0.03) non-invasive, lateral fans were produced elsewhere	Decrease to 1.35 ± 0.9 (from 1.74 ± 0.32 or 1.7 ± 0.34)
Phanerochaete velutina	*S. caerulea* was rapidly replaced	D_{BM} and D_{BS} were significantly reduced	Significant ($p \leq 0.005$) decrease to 0.09 ± 0.06	D_{BM} was significantly reduced	Significant ($p \leq 0.001$) decrease to 0.73 ± 0.1 (from 1.53 ± 0.12 or 1.03 ± 0.07)
Phallus impudicus	Temporary defensive ridges produced by *S. caerulea* were breached in a few places and the fungus was replaced when *P. impudicus* reached the wood blocks	Differences not usually significant	Significant ($p \leq 0.001$) decrease to 0.04 ± 0.01	No change in D_{BM} or D_{BS}	Decrease to 1.45 ± 0.3 (from 3.25 ± 0.52 or 2.47 ± 0.3)

Source: Data from Donnelly & Boddy (2001)

Notes:

[a] compared with a control with wood resource ($0.19 \pm 0.04\,mg\,d^{-1}$) and a control with no additional resource ($0.34 \pm 0.05\,mg\,d^{-1}$).

[b] 35% less ($p \leq 0.01$) major radial cords produced in interactive system.

competitive exclusion of species *i* by species *j*, an unstable equilibrium between the two species, and a stable equilibrium between the two species. Coexistence will only occur if there is some kind of niche separation, for example by differences in resource use between the species. Gause (1934) in his classic experiment on two laboratory species of *Paramecium* reported how the numbers of both species were lower when they were grown together than when grown individually, and how they eventually reach an approximate equilibrium. Berendse (1981, 1982) found that competition between two grassland species that commonly grow together, the grass *Anthoxanthum odoratum* and *Plantago lanceolata*, was ameliorated by *P. lanceolata* having deeper roots than the grass. When separation of the rooting zones was experimentally prevented in pot or field experiments, the grass had a significantly greater negative effect upon its competitor than in controls where deep rooting by *P. lanceolata* was possible. There was some evidence too that the presence of *A. odoratum* caused *P. lanceolata* to develop deeper roots than it did when growing by itself.

Soil invertebrates

Mycelia are highly nutritious and many invertebrates, including Collembola, termites, nematodes and Diptera, use fungi as a food source, either grazing directly on mycelia or indirectly by ingesting mycelium within decomposing litter (Maraun *et al.*, 2003). High-intensity invertebrate grazing can completely destroy mycelia in the laboratory, but less intense grazing can result in dramatic changes in fungal growth and activity (see, for example, van der Drift & Jansen, 1977; Hedlund *et al.*, 1991; Bengtsson *et al.*, 1993; Hedlund & Augustson, 1995; Harold *et al.*, 2005) (Fig. 6.6). Despite the plethora of studies on fungal–invertebrate interactions, non-resource-unit restricted basidiomycetes have not been studied until recently. Likewise the spatial implications of these interactions that may result in growth responses and movement of biomass, energy and nutrients within fungal mycelia have attracted little consideration. Changes in mycelial morphology and foraging patterns vary in small (0–24 cm diameter) cord-forming mycelial systems growing out of single organic resources depending on resource status (Harold *et al.*, 2005), grazing intensity (density) and invertebrate (Collembola) species (Kampichler *et al.*, 2004). Different cord-forming fungal species are also affected differently by grazing (Tordoff *et al.*, 2006) (Fig. 6.6).

For example, when *Hypholoma fasciculare* growing in trays of non-sterile soil was subject to varying grazing intensities by the Collembola *Folsomia candida*, *Hypogastrura* cf. *tullbergi* and *Proisotoma minuta*, extra-resource mycelium changed from exhibiting slow, dense exploitative growth, to less dense explorative growth during grazing. Effects, however, varied depending on grazing intensity and collembolan species, *F. candida* having the largest effects (Kampichler *et al.*, 2004). Grazing by *F. candida* affected mycelial foraging patterns of *H. fasciculare* differently depending on size, state of decay (time for which wood has been colonized) and nutrient status (inocula colonized on malt agar or nutrient agar) of wood-block inocula from which it was growing (Harold *et al.*, 2005).

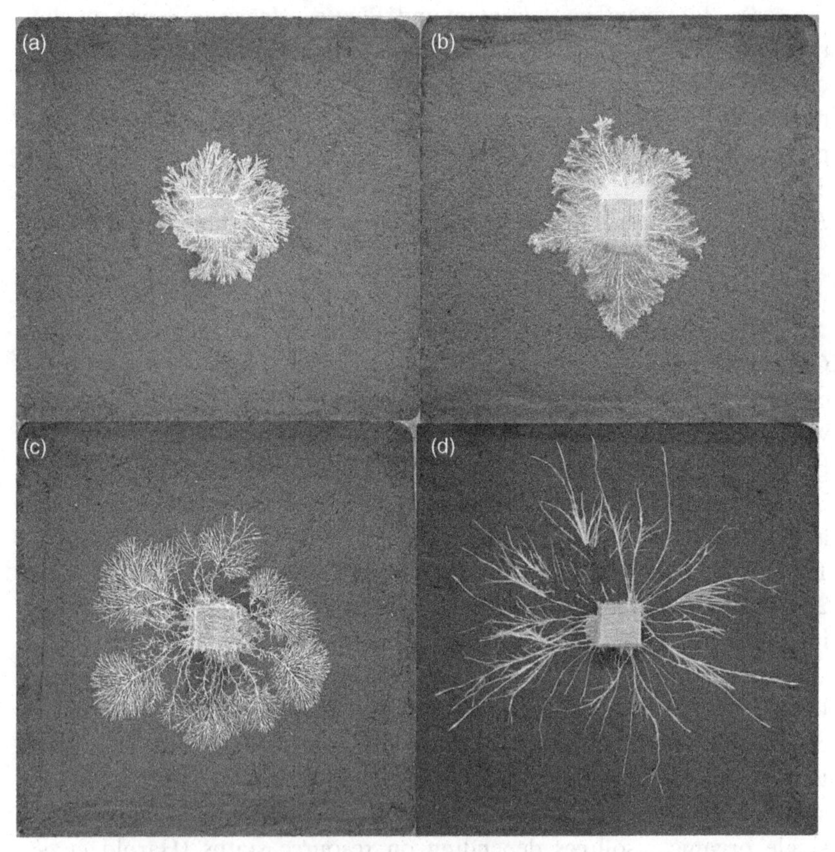

Fig. 6.6. Grazing effects on mycelia. Mycelia are growing from 4 cm^3 beechwood inocula across trays (24 cm × 24 cm) of compressed non-sterile soil. Collembola (2 *Folsomia candida* per square centimetre of mycelium) were added when mycelia had radii of 8 cm. (a) *Hypholoma fasciculare* after 8 d grazing; (b) *Phanerochaete velutina* after 5 d grazing; (c) *Phallus impudicus* after 20 d grazing; (d) *Resinicium bicolor* after 12 d. Ungrazed systems of equivalent age are shown in Fig. 6.1. (Digital images courtesy of G. M. Tordoff.)

Mycelia were most luxuriant, had greater hyphal coverage and extended more rapidly from 2-year-old than from younger inocula, from 4 cm^3 than from 1 cm^3 inocula, and from inocula colonized on malt extract agar rather than on distilled water agar. The less luxuriant systems, characterized by many fine hyphae and fewer mycelial cords, had dramatically reduced coverage and extension. Grazing by Collembola often resulted in points of more rapid outgrowth as cords with a fanned margin (Fig. 6.6).

Different fungi are preferentially grazed in different regions; hyphal tips at growing margins of *H. fasciculare*; indiscriminate grazing on cords and fine mycelium, but generally close to wood inoculum of *R. bicolor*, and fine mycelium within the colony and hyphal tips at growing margin of *Phan. velutina* (Tordoff *et al.*, 2006). Intense (80 Collembola per 24 cm × 24 cm tray) grazing by *F. candida* resulted in complete removal of hyphal tips around large parts of the circumference of *Phan. velutina* mycelia, yielding much smaller systems than in ungrazed controls.

There are many parallels between invertebrate grazing of fungal mycelia and grazing of plants. The 'browse line' around almost the entire margin of *P. velutina* mycelia is analogous to that seen on trees in wood pasture grazed upon by cattle. Insect herbivores cause distortions of tree shape: for example, sawfly, *Perga affinis affinis*, feeding on young *Eucalyptus* results in multiple leading shoots and bushy form (Carne, 1969), whereas forest tent caterpillars *Malacosoma disstria* on aspen (*Populus tremuloides*) result in shorter internodes and smaller leaves which gives rise to tight clusters of foliage (Duncan & Hodson, 1958). Plant shape, like that of mycelia, is determined by the nature and extent of branching. The terminal bud on a branch produces hormones that suppress branching at the lower nodes, in a manner analogous to fungal hyphal apical dominance (Carlile *et al.*, 2001). When herbivores remove or kill the terminal bud, the flow of hormones stops and one or more of the lower buds develops into an elongating shoot. The mechanism behind changes in mycelial morphology following invertebrate grazing is presumably similar.

Although the emphasis of work performed so far has been on invertebrate effects on mycelial morphology, effects on physiology are crucially important. For example, a major question that arises is: in what ways and to what extent do soil mesofauna affect re-allocation of nutrients? Mineral nutrients certainly seem to be released during grazing of litter (Anderson *et al.*, 1983). Effects on translocation within mycelia need urgent investigation. Enzymes associated with ligninolysis were switched on by *Phan. velutina* in the presence of nematodes (Dyer *et al.*, 1992); in the soil-inhabiting zygomycete *Mortierella isabellina* there were differences in expression of proteases during grazing (Hedlund *et al.*, 1991). Moreover, these changes in enzyme expression occurred distant to the site of grazing, and a research priority is examination of effects of grazing in localized areas on morphology, translocation and enzyme production elsewhere in large basidiomycete mycelial systems.

Invertebrates can exert direct nutritional effects on basidiomycete mycelia, through nitrogen inputs in the faeces. In addition, some basidiomycete species can kill invertebrates by immobilizing them with toxins, hyphae subsequently penetrating and feeding on the cadavers: for example, wood-rotting *Pleurotus* species feed on nematodes (Barron & Thorn, 1987) and the ectomycorrhizal *Laccaria bicolor* on Collembola (*Folsomia candida*) (Klironomos & Hart, 2001). Others produce structures that trap nematodes (Tzeam & Liou, 1993; Luo *et al.*, 2004). Obvious analogies with macroorganisms are insectivorous plants (e.g. sundew and Venus flytrap) and insect parasitoids that oviposit on, in, or near to the usually arthropod host and whose larvae then gradually consume the host.

Sit and wait strategy

The mycelial networks described above, growing in laboratory microcosms, have covered areas of at most 0.25 m^2. In forests, however, genets of saprotrophs and ectomycorrhizal Basidiomycota can cover areas from several square metres to many hectares (Thompson & Rayner, 1982; Thompson & Boddy, 1988; Smith *et al.*, 1992; Dahlberg & Stenlid, 1995; Ferguson *et al.*, 2003; Cairney, 2005). Indeed, the largest organisms on Earth may well be *Armillaria* species: a genet of *Armillaria ostoyae* spans 965 hectares, has a maximum separation of 3810 m and is estimated as 1900–8650 years old (Ferguson *et al.*, 2003). However, since parts of mycelia can be separated from each other during development, and of course also rejoin if parts of the same genet meet again, the true extent and degree of interconnectivity is not known. Nonetheless, phosphorus moves to different litter resources over 1 m apart, via saprotrophic mycelial cord systems (Wells & Boddy, 1995b). Likewise, over a smaller scale, carbon transfer certainly occurs from a host tree to neighbouring seedlings via ectomycorrhizal connections, even if the trees are different species (Simard *et al.*, 1997; Leake *et al.*, 2004).

These large saprotrophic mycelial networks are ideally placed to capitalize on new dead organic resources when they arrive on or in the system via litter fall or root death. The same is true for ectomycorrhizal mycelium when new seedlings germinate in their vicinity, and also when appropriate litter resources become available, since they also have saprotrophic ability

(Leake & Read, 1997). When, for example, a woody resource arrives on a saprotrophic mycelial system the network responds by establishing mycelial growth to colonize the new resource, sometimes by growth elsewhere to supplement nutrient supplies, and by movement of carbon and mineral nutrients. Not only do networks form on the forest floor, but also in the canopy of tropical forests rhizomorphic fungi form suspended networks that trap and then colonize falling litter; this strategy has been likened to that of animal 'filter feeders' (Hedger, 1990).

In laboratory microcosms, new resources were added to *Phan. velutina* mycelia of up to 50 cm diameter in pairs, diametrically opposite one another, at 10 d intervals for 60 d (Wells *et al.*, 1998). Ten days after addition of the final pair, [^{32}P]orthophosphate was added to the central inoculum and its translocation to new resources monitored over time. The total ^{32}P allocated to new wood resources increased linearly with time; however, the proportion of the total allocated to each new resource varied depending on the length of time that the resources had been in contact with the mycelial system. It was not the most recently supplied resource that was the main sink for ^{32}P but the better colonized new resources (Fig. 6.7). The rate of ^{32}P acquisition by the most recently added resources increased linearly with time, supported by efflux from the other wood resources, that had earlier been the main sinks for translocated phosphorus. This experiment demonstrates well the spatial and temporal coordination of nutrient partitioning within these complex mycelial networks. The degree of differential partitioning could be explained in terms of net removal of nutrients from a uniform, system-wide translocation stream (which may be analogous to a water ring main) at locations where the nutrient is being actively used or stored, as opposed to demand-driven translocation (Wells *et al.*, 1998). However, mechanisms that allow both a circulatory translocation system (Olsson & Gray, 1998; Wells *et al.*, 1998; Lindahl *et al.*, 2001) and mass flow (Jennings, 1987; Tlalka *et al.*, 2002) are not clear (Watkinson *et al.*, 2006). There is also some evidence, from a similar experiment with two sizes of inocula, that carbon availability affects partitioning of mineral nutrients between newly added resources (Wells *et al.*, 1999). In field mycelial systems centred on large (hundreds of square centimetres) wood resources this may not, however, occur.

When mycelial systems do not have sufficient resources to draw upon to produce mycelial biomass and the necessary enzymes for colonization of newly arriving resources, hyphae develop from cords elsewhere in the system to produce a large surface area for nutrient absorption. Thus, in a microcosm experiment in which a new wood resource was added behind the margin of a *Phan. velutina* system growing on a nutrient-poor soil (made by diluting soil 1:1 with sand), the D_{BM} increased significantly and patches of much-branched fine hyphae developed (Wells *et al.*, 1997). The latter also formed in controls to which inert Perspex was added but were up to 10 times greater in hyphal coverage when an organic resource was added. These patches were ephemeral, regressing more rapidly with age than did mycelial cords. They also ceased to develop when nutrients were added to bring the nutrient status of the sand–soil mix back to that of the 'undiluted' soil. The patches were not sinks but sites of uptake of mineral nutrients ([^{32}P]orthophosphate) from soil.

Again there are parallels with macroorganisms. Clonal species are important components of many natural plant communities, and many

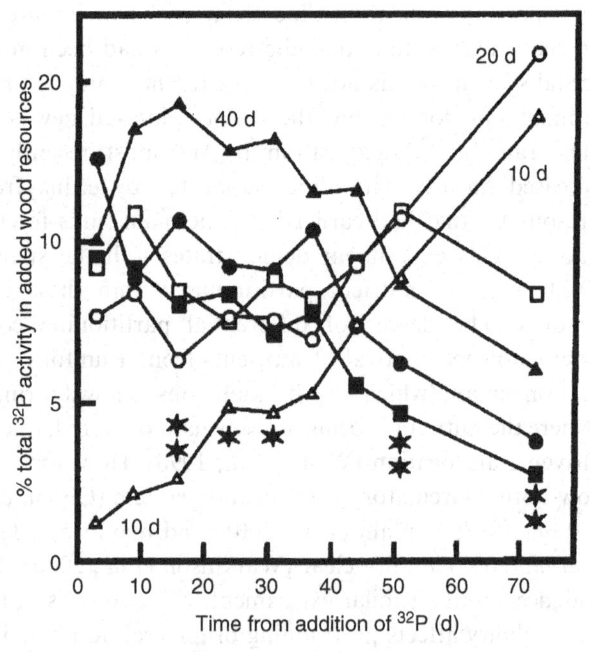

Fig. 6.7. Partitioning of phosphorus amongst wood resources supplied to established *Phanerochaete velutina* mycelial systems at different times, in trays of compressed non-sterile soil. The resources were added to established systems, behind the mycelial margin around the edges of a circle of radius 17 cm centred on the inoculum wood block. Change in the proportion of total ^{32}P activity recorded in new resources with time are shown for resources added after 10, 20, 30, 40, 50 and 60 ± 1 d before addition of the radioisotope to the central inoculum. (After Wells *et al.*, 1998.)

can establish as persistent monocultures. Genets of many clonal species can be extremely long-lived and occupy very large areas (Cook, 1983). Many clones spread laterally by generating an indeterminate number of simple reiterated structures (ramets) at intervals on branched stems. Ramets may bear leaves and roots. The resource-acquiring organs of a clone may therefore be widely distributed in space and time. In most clonal species, resource movement is mainly towards growing apices, but resources can move in either direction, depending on whether young or older ramets are in need of support (Marshall, 1990).

Determinate organisms (*sensu* Andrews, 1991) also operate 'sit and wait' strategies and these also have a dynamic component. Classic examples include *Mantis*, whose bodies remain stationary while their very mobile heads follow prey movements, then strike dramatically when the prey is in range of the forelegs. Ant lions (larval Myrmeleontidae, Neuroptera) dig pits (2–5 cm diameter) with sloping sides in dry sand, and bury themselves at the bottom with only the head exposed. If an ant walks over the edge of the pit it usually falls to the bottom, where it is captured, because of the instability of the sides and the ant lion flicking sand at the ant by sharp movements of the head. Spiders and their webs provide another example.

Importance of network connectivity

Mycelia form complex networks that are constantly being remodelled in response to nutrient discovery and demand, changes in microclimate and destructive disturbance, e.g. by invertebrate grazers. There is considerable scope for communication within the network, because hyphae both maintain continuity with their immediate ancestors and become connected with neighbouring regions via *de novo* formation of cross-links (anastomoses). Thus there is both radial and tangential connectivity, resulting in systems with many connected loops (Fig. 6.8). Tangential connections make communication pathways between peripheral regions shorter. Routes between different regions, shortest path lengths, etc. can be investigated mathematically by graph or network theory (Albert & Barabási, 2001); this has just begun for large mycelial systems (Fricker *et al.*, 2006; Chapter 1, this volume). Different network architectures will impart different degrees of resilience (i.e. the extent to which network properties are changed as links are removed from the network) to disturbance. For example, highly connected networks will be more resilient to invertebrate grazing, because even though some connections will be severed others will remain. Resilience of mycelial networks does not, however, solely depend on the network architecture at the time of disturbance, but also on the ease and efficiency with which the network can re-connect, by re-growth following damage.

Conclusions

The ability to develop and operate complex, extensive and persistent translocation networks that are continually being remodelled in response to the abiotic and biotic environment underlies the unique position of non-resource-unit restricted basidiomycetes in woodland ecosystems. A range of different foraging strategies has evolved, reflecting the broad spectrum of different types and distributions of plant litter resources. The ability of mycelia to respond to new resources is an efficient way of capturing and re-distributing energy and mineral nutrient resources. The production of persistent networks provides a means of 'foraging over time' in areas that have already been 'searched'. The ability to anastomose and form tangential connections gives rise to mycelia containing many loops, which confer resilience on systems, allowing them to be long-lived.

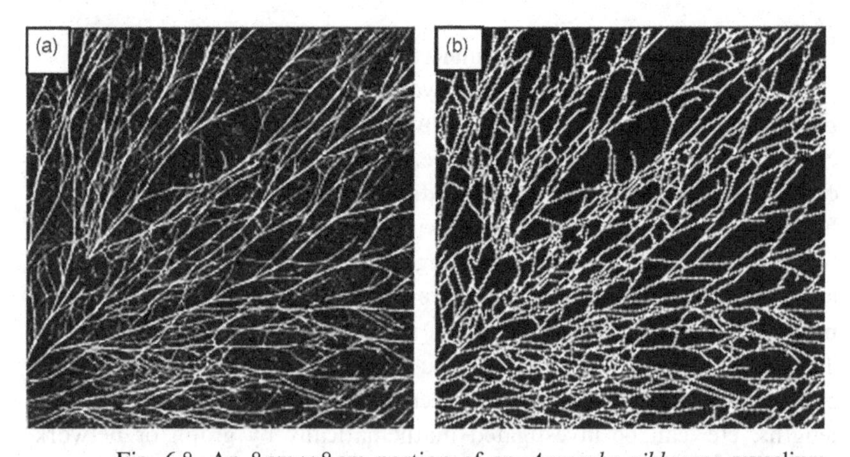

Fig. 6.8. An 8 cm × 8 cm portion of an *Agrocybe gibberosa* mycelium extending for 28 d from a 4 cm^3 beechwood inoculum across a tray (24 cm × 24 cm) of compressed non-sterile woodland soil. (a) Digital image of the system. (b) The same system but with cords represented by lines of constant thickness irrespective of actual diameter of cords, allowing the eye to see the connectivity of the network rather than being focused on the thick radial cords. (Digital image (a) courtesy of Damian P. Donnelly; network diagram courtesy of Juliet Hynes, extracted by using software written by Mark D. Fricker.)

Acknowledgements

We thank N. E. R. C. for funding much of the research reported here. The research on fungal foraging patterns has been conducted over many years and with many collaborators, to whom we extend our thanks: Saleh Abdalla, Dan Bebber, Rory Bolton, Sam Bretherton, Clare Culshaw, Damian Donnelly, Chris Dowson, Mark Fricker, Simon Harold, Melanie Harris, Carolyn Hughes, Juliet Hynes, Christian Kampichler, Jonathan Leake, Louise Owen, Alan Rayner, Johann Rolschewski, Timothy Rotheray, Phillip Springham, Joanna Thomas, Wendy Thompson, Monica Tlalka, George Tordoff, Sarah Watkinson, John Wells, Jonathan Wood and Jamil Abd Zakaria.

References

Abdalla, S. H. M. & Boddy, L. (1996). Effect of soil and litter type on outgrowth patterns of mycelial systems of *Phanerochaete velutina*. *FEMS Microbiology Ecology* **20**, 195–204.

Albert, R. & Barabási, A.-L. (2001). Statistical mechanics of complex networks. *Reviews of Modern Physics* **74**, 47–97.

Anderson, J. M., Ineson, P. & Huish, S. A. (1983). Nitrogen and cation mobilization by soil fauna feeding on leaf litter and soil organic matter from deciduous woodlands. *Soil Biology and Biochemistry* **15**, 463–7.

Andrews, J. H. (1991). *Comparative Ecology of Microorganisms and Macroorganisms*. New York: Springer-Verlag.

Arthur, A. P. (1966). Associative learning in *Itoplectis conquisitor* (Hymenoptera: Ichneumonidae). *Canadian Entomologist* **98**, 213–23.

Barron, G. L. & Thorn, R. G. (1987). Destruction of nematodes by species of *Pleurotus*. *Canadian Journal of Botany* **65**, 774–8.

Bengtsson, G., Hedlund, K. & Rundgren, S. (1993). Patchiness and compensatory growth in a fungus-Collembola system. *Oecologia* **93**, 296–302.

Berendse, F. (1981). Competition between plant populations with different rooting depths. II. Pot experiments. *Oecologia* **48**, 334–41.

Berendse, F. (1982). Competition between plant populations with different rooting depths. III. Field experiments. *Oecologia* **53**, 50–5.

Boddy, L. (1984). The microenvironment of basidiomycete mycelia in temperate deciduous woodlands. In *The Ecology and Physiology of the Fungal Mycelium*, ed. D. H. Jennings & A. D. M. Rayner, pp. 261–89. Cambridge: Cambridge University Press.

Boddy, L. (1986). Water and decomposition processes. In *Water, Fungi and Plants*. ed. P. G. Ayres & L. Boddy, pp. 375–98. Cambridge: Cambridge University Press.

Boddy, L. (1993). Saprotrophic cord-forming fungi: warfare strategies and other ecological aspects. *Mycological Research* **97**, 641–55.

Boddy, L. (1999). Saprotrophic cord-forming fungi: meeting the challenge of heterogeneous environments. *Mycologia* **91**, 13–32.

Boddy, L. (2000). Interspecific combative interactions between wood-decaying basidiomycetes – a review. *FEMS Microbiology Ecology* **31**, 185–94.

Boddy, L. & Abdalla, S. H. M. (1998). Development of *Phanerochaete velutina* mycelial cord systems: effect of encounter of multiple colonised wood resources. *FEMS Microbiology Ecology* **25**, 257–69.

Boddy, L. & Donnelly, D. P. (2007). Fractal geometry and microorganisms in the environment. In *Biophysical Chemistry of Fractal Structures and Process in Environmental Systems*, ed. N. Senesi & K. Wilkinson. Chichester: John Wiley (in press).

Boddy, L. & Watkinson, S. C. (1995). Wood decomposition, higher fungi, and their role in nutrient redistribution. *Canadian Journal of Botany* **73** (Suppl.1), S1377–83.

Boddy, L., Wells, J. M., Culshaw, C. & Donnelly, D. P. (1999). Fractal analysis in studies of mycelium in soil. *Geoderma* **88**, 301–28.

Bolton, R. G. (1993). Resource acquisition by migratory mycelial cord systems of *Phanerochaete velutina* and *Hypholoma fasciculare*. Ph.D. thesis, University of Wales, Cardiff.

Bolton, R. G. & Boddy, L. (1993). Characterisation of the spatial aspects of foraging mycelial cord systems using fractal geometry. *Mycological Research* **97**, 762–8.

Bolton, R. G., Morris, C. W. & Boddy, L. (1991). Non-destructive quantification of growth and regression of mycelial cords using image analysis. *Binary* **3**, 127–32.

Burton, J. L. & Franks, N. R. (1985). The foraging ecology of the army ant *Eciton rapax*: an ergonomic enigma? *Ecological Entomology* **10**, 131–41.

Cairney, J. W. G. (1992). Translocation of solutes in ectomycorrhizal and saprotrophic rhizomorphs. *Mycological Research* **96**, 135–41.

Cairney, J. W. G. (2005). Basidiomycete mycelia in forest soils: dimensions, dynamics and roles in nutrient distribution. *Mycological Research* **109**, 7–20.

Campbell, B. D., Grime, J. P. & Mackey, J. M. L. (1991). A trade-off between scale and precision in resource foraging. *Oecologia* **87**, 532–8.

Carlile, M. J., Watkinson, S. C. & Gooday, G. W. (2001). *The Fungi*. London: Academic Press.

Carne, P. B. (1969). On the population dynamics of the eucalypt-defoliating sawfly *Perga affinis affinis* Kirby (Hymenoptera). *Australian Journal of Zoology* **17**, 113–41.

Charnov, E. L. (1976). Optimal foraging: the marginal value theorem. *Theoretical Population Biology* **9**, 129–36.

Cook, R. E. (1983). Clonal plant populations. *American Scientist* **71**, 244–53.

Couzin, I. D. & Franks, N. R. (2002). Self-organized lane formation and optimized traffic flow in army ants. *Proceedings of the Royal Society of London* B**270**, 139–46.

Crawley, M. J. (1983). *Herbivory. The Dynamics of Animal-Plant Interactions. Studies in Ecology*, vol. 10. Oxford: Blackwell Scientific Publications.

Dahlberg, A. & Stenlid, J. (1995). Spatiotemporal patterns in ectomycorrhizal populations. *Canadian Journal of Botany* **73**, S1222–30.

Donnelly, D. P. (1995). Comparative physiology and ecology of mycelial cord growth of *Stropharia caerulea* and *Phanerochaete velutina*. Ph.D. thesis, University of Wales, Cardiff.

Donnelly, D. P. & Boddy, L. (1997a). Resource acquisition by the mycelial-cord-former *Stropharia caerulea*: effect of resource quantity and quality. *FEMS Microbiology Ecology* **23**, 195–205.

Donnelly, D. P. & Boddy, L. (1997b). Development of mycelial systems of *Stropharia caerulea* and *Phanerochaete velutina* on soil: effect of temperature and water potential. *Mycological Research* **101**, 705–13.

Donnelly, D. P. & Boddy, L. (1998). Developmental and morphological responses of mycelial systems of *Stropharia caerulea* and *Phanerochaete velutina* to soil nutrient enrichment. *New Phytologist* **138**, 519–31.

Donnelly, D. P. & Boddy, L. (2001). Mycelial dynamics during interactions between *Stropharia caerulea* and other cord-forming, saprotrophic basidiomycetes. *New Phytologist* **151**, 691–704.

Donnelly, D. P., Wilkins, M. F. & Boddy, L. (1995). An integrated image analysis approach for determining biomass, radial extent and box-count fractal dimension of macroscopic mycelial systems. *Binary* **7**, 19–28.

Dowson, C. G., Rayner, A. D. M. & Boddy, L. (1986). Outgrowth patterns of mycelial cord-forming basidiomycetes from and between woody resource units in soil. *Journal of General Microbiology* **132**, 203–11.

Dowson, C. G., Rayner, A. D. M. & Boddy, L. (1988a). Inoculation of mycelial cord-forming basidiomycetes into woodland soil and litter. II. Resource capture and persistence. *New Phytologist* **109**, 343–9.

Dowson, C. G., Rayner, A. D. M. & Boddy, L. (1988b). Foraging patterns of *Phallus impudicus*, *Phanerochaete laevis* and *Steccherinum fimbriatum* between discontinuous resource units in soil. *FEMS Microbiology Ecology* **53**, 291–8.

Duncan, D. P. & Hodson, A. C. (1958). Influence of forest tent caterpillar upon the Aspen forests of Minnesota. *Forest Science* **4**, 71–93.

Dyer, H. C., Boddy, L. & Preston-Meek, C. M. (1992). Effect of the nematode *Panagrellus redivivus* on growth and enzyme production by *Phanerochaete velutina* and *Stereum hirsutum*. *Mycological Research* **96**, 1019–28.

Ferguson, B. A., Dreisbach, T. A., Parks, C. G., Filipo, G. M. & Schmitt, C. L. (2003). Coarse-scale population structure of pathogenic *Armillaria* species in a mixed-conifer forest in the Blue Mountains of northeast Oregon. *Canadian Journal of Forest Research* **33**, 612–23.

Franks, N. R., Gomez, N. Goss, S. & Deneubourg, J. L. (1991). The blind leading the blind in army ant raid patterns: testing a model of self-organisation (Hymenoptera: Formicidae). *Journal of Insect Behaviour* **4**, 583–606.

Fricker, M. D., Bebber, D., Darrah, P. R., Tlalka, M., Watkinson, S. C., Boddy, L., Yiasoumis, L., Cartwright, H. M., Meškauskas, A., Moore, D., Smith, M. D., Nakagaki, T., Lee, C. F. & Johnson, N. (2006). Inspiration from microbes: from patterns to networks. In *Complex Systems and Inter-Disciplinary Science*, ed. B. W. Arthur, R. Axtell, S. Bornholdt *et al.* London: World Scientific Publishing Co (in press).

Gause, G. F. (1934). *The Struggle for Existence.* Baltimore: Williams & Wilkins.

Harold S., Tordoff, G. M., Jones, T. H. & Boddy, L. (2005). Mycelial responses of *Hypholoma fasciculare* to collembola grazing: effect of inoculum age, nutrient status and resource quality. *Mycological Research* **109**, 927–35.

Hedger, J. (1990). Fungi in the tropical forest canopy. *The Mycologist* **4**, 200–2.

Hedlund, K. & Augustson, A. (1995). Effects of enchytraeid grazing on fungal growth and respiration. *Soil Biology and Biochemistry* **27**, 905–9.

Hedlund, K., Boddy, L. & Preston, C. M. (1991). Mycelial responses of the soil fungus, *Mortierella isabellina*, to grazing by *Onychiurus armatus* (Collembola). *Soil Biology and Biochemistry* **23**, 361–6.

Hughes, C. L. & Boddy, L. (1994). Translocation of ^{32}P between wood resources recently colonised by mycelial cord systems of *Phanerochaete velutina*. *FEMS Microbiology Ecology* **14**, 201–12.

Hughes, C. L. & Boddy, L. (1996). Sequential encounter of wood resources by mycelial cords of *Phanerochaete velutina*: effect on growth patterns and phosphorus allocation. *New Phytologist* **133**, 713–26.

Hutchings, M. J. & De Kroon, H. (1994). Foraging in plants: the role of morphological plasticity in resource acquisition. *Advances in Ecological Research* **25**, 159–238.

Hutchings, M. J. & Wijesinghe, D. K. (1997). Patchy habitats, division of labour and growth dividends in clonal plants. *Trends in Ecology and Evolution* **12**, 390–4.

Hyman, L. H. (1940). *The Invertebrates*, vol. 1, *Protozoa through Ctenophora.* New York: McGraw-Hill.

Jennings, D. H. (1987). The translocation of solutes in fungi. *Biological Reviews* **62**, 215–43.

Jones, T. H. & Hassell, M. P. (1988). Patterns of parasitism by *Trybliographa rapae*, a cynipid parasitoid of the cabbage root fly, under laboratory and field conditions. *Ecological Entomology* **13**, 309–17.

Kampichler, C., Rolschewski, J., Donnelly, D. P. & Boddy, L. (2004). Collembolan grazing affects the growth strategy of the cord-forming fungus *Hypholoma fasciculare*. *Soil Biology and Biochemistry* **36**, 591–9.

Kirby, J. J. H., Stenlid, J. & Holdenrieder, O. (1990). Population structure and responses to disturbance of the basidiomycete *Resinicium bicolor*. *Oecologia* **85**, 178–84.

Klironomos, J. N. & Hart, H. H. (2001). Animal nitrogen swap for plant carbon. *Nature* **410**, 651–2.

Krebs, J. R. & McCleery, R. H. (1984). Optimization in behavioural ecology. In *Behavioural Ecology: An Evolutionary Approach*, ed. J. R. Krebs & N. B. Davies, 2nd edn, pp. 91–121. Oxford: Blackwell.

Leake, J., Johnson, D., Donnelly, D. P. & Boddy, L. (2004). Networks of power and influence: the role of mycorrhizal mycelium in controlling plant communities and agro-ecosystem functioning. *Canadian Journal of Botany* **82**, 1016–45.

Leake, J. R. & Read D. J. (1997). Mycorrhizal fungi in terrestrial habitats. In: *The Mycota*, vol. 4, *Environmental and Microbial Relationships*, ed. D. T. Wicklow & B. Söderström, pp. 281–301. Heidelberg: Springer-Verlag.

Lindahl, B. O., Finlay, R. D. & Olsson, S. (2001). Simultaneous, bidirectional translocation of ^{32}P and ^{33}P between wood blocks connected by mycelial cords of *Hypholoma fasciculare*. *New Phytologist* **150**, 189–94.

Lotka, A. J. (1932). The growth of mixed populations: two species competing for a common food supply. *Journal of the Washington Academy of Science* **22**, 461–9.

Luo, H., Mo, M., Huang, X., Li, X. & Zhang, K. (2004). *Coprinus comatus*: a basidiomycete fungus forms novel spiny structures and infects nematodes. *Mycologia* **96**, 1218–25.

Maraun, M., Martens, H., Migge, S., Theenhaus, A. & Scheu, S. (2003). Adding to 'the enigma of soil animal diversity': fungal feeders and saprophagous soil invertebrates prefer similar food substrates. *European Journal of Soil Biology* **39**, 85–95.

Marshall, C. (1990) Source-sink relations of interconnected ramets. In *Clonal Growth in Plants – Regulation and Function*, ed. J van Groenendael & H. de Kroon, pp. 23–41. The Hague: SPB Academic Publishing.

Olsson, S. & Gray, S. N. (1998). Patterns and dynamics of ^{32}P phosphate and ^{14}C labeled AIB translocation in intact basidiomycete mycelia. *FEMS Microbiology Ecology* **26**, 109–20.

Olsson, S. & Hansson, B. S. (1995). The action potential-like activity found in fungal mycelium is sensitive to stimulation. *Naturwissenschaften* **82**, 30–1.

Owen, S. L. (1997). Comparative development of the mycelial cord-forming fungi *Coprinus picaceus* and *Phanerochaete velutina*, with particular emphasis on pH and nutrient reallocation. Ph.D. thesis, University of Wales, Cardiff.

Rajaniemi, T. K. & Reynolds, H. L. (2004). Root foraging for patchy resources in eight herbaceous plant species. *Oecologia* **141**, 519–25.

Rayner, A. D. M. (1991). The challenge of the individualistic mycelium. *Mycologia* **83**, 48–71.

Rayner, A. D. M. (1994). Pattern generating processs and fungal communities. In *Beyond the Biomass: Compositional and Functional Analysis of Microbial Communities*, ed. K. Ritz, J. Dighton & K. E. Giller, pp. 247–58. Chichester: Wiley-Sayce.

Rayner, A. D. M. (1997). *Degrees of Freedom: Living in Dynamic Boundaries*. London: Imperial College Press.

Rayner, A. D. M. & Franks, N. R. (1987). Evolutionary and ecological parallels between ants and fungi. *Trends in Ecology and Evolution* **2**, 127–32.

Rayner, A. D. M. & Webber, J. F. (1984). Interspecific mycelial interactions – an overview. In *The Ecology and Physiology of the Fungal Mycelium*, ed. D. H. Jennings & A. D. M. Rayner, pp. 383–417. Cambridge: Cambridge University Press.

Rayner, A. D. M., Griffith, G. S. & Ainsworth, A. M. (1995). Mycelia interconnectedness. In *The Growing Fungus*, ed. N. A. R. Gow & G. M. Gadd, pp. 21–40. London: Chapman & Hall.

Robinson, C. H., Dighton, J. & Frankland, J. C. (1993). Resource capture by interacting fungal colonizers of straw. *Mycological Research*, **97**, 547–58.

Schmid, B. & Harper, J. L. (1985). Clonal growth in grassland perennials. I. Density and pattern-dependent competition between plants with different growth forms. *Journal of Ecology* **73**, 793–808.

Simard, S. W., Perry, D. A., Jones, M. D., Myrold, D. D., Durall, D. M. & Molina, R. (1997). Net transfer of carbon between ectomycorrhizal tree species in the field. *Nature* **388**, 579–82.

Smith, M. L., Bruhn, J. N. & Anderson, J. B. (1992). The fungus *Armillaria bulbosa* is among the largest and oldest living organisms. *Nature* **356**, 428–31.

Smith, S. E. & Read, D. J. (1997). *Mycorrhizal Symbiosis*. London: Academic Press.

Thompson, W. & Boddy, L. (1988). Decomposition of suppressed oak trees in even-aged plantations. II. Colonisation of tree roots by cord and rhizomorph producing basidiomycetes. *New Phytologist* **93**, 277–91.

Thompson, W. & Rayner, A. D. M. (1982). Spatial structure of a population of *Tricholomopsis platyphylla* in a woodland site. *New Phytologist* **92**, 103–14.

Tlalka, M., Watkinson, S. C., Darrah, P. R. & Fricker, M. D. (2002). Continuous imaging of amino acid translocation in intact mycelia of *Phanerochaete velutina* reveals rapid, pulsatile fluxes. *New Phytologist* **153**, 173–84.

Tlalka, M., Hensman, D., Darrah, P. R., Watkinson, S. C. & Fricker, M. D. (2003). Noncircadian oscillations in amino acid transport have complementary profiles in assimilatory and foraging hyphae of *Phanerochaete velutina*. *New Phytologist* **158**, 325–35.

Tordoff, G. M., Jones, T. H. & Boddy, L. (2006). Grazing by *Folsomia candida* (Collembola) differentially affects the mycelial morphology of the cord-forming basidiomycetes *Hypholoma fasciculare*, *Phanerochaete velutina* and *Resinicium bicolor*. Mycological Research **110**, 325–35.

Tzeam, S. S. & Liou, J. Y. (1993). Nematophagous resupinate basidiomycete fungi. *Phytopathology* **83**, 1015–20.

van der Drift, J. & Jansen, E. (1977). Grazing of springtails on hyphal mats and its influence on fungal growth and respiration. In *Soil Organisms as Components of Ecosystems*, ed. U. Lohm & T. Persson, *Ecological Bulletin* **25**, 203–9.

Watkinson, S. C. (1999). Metabolism and differentiation in basidiomycete mycelium. In *The Fungal Colony*, ed. N. A. R. Gow, G. D. Robson & G. M. Gadd, pp. 126–56. Cambridge: Cambridge University Press.

Watkinson, S. C., Bebber, D., Darrah, P., Fricker, M., Tlalka, M. & Boddy, L. (2006). The role of wood decay fungi in the carbon and nitrogen dynamics of the forest floor. In *Fungi in Biogeochemical Cycles*, ed. G. M. Gadd, pp. 151–81. Cambridge: Cambridge University Press.

Wells, J. M. & Boddy, L. (1990). Wood decay, and phosphorus and fungal biomass allocation, in mycelial cord systems. *New Phytologist* **116**, 285–95.

Wells, J. M. & Boddy, L. (1995a). Effect of temperature on wood decay and translocation of soil-derived phosphorus in mycelial cord systems. *New Phytologist* **129**, 289–97.

Wells, J. M. & Boddy, L. (1995b). Phosphorus translocation by saprotrophic basidiomycete mycelial cord systems on the floor of a mixed deciduous woodland. *Mycological Research* **99**, 977–80.

Wells, J. M., Hughes, C. & Boddy, L. (1990). The fate of soil-derived phosphorus in mycelial cord systems of *Phanerochaete velutina* and *Phallus impudicus*. *New Phytologist* **114**, 595–606.

Wells, J. M., Donnelly, D. P. & Boddy, L. (1997). Patch formation and developmental polarity in mycelial cord systems of *Phanerochaete velutina* on nutrient-depleted soil. *New Phytologist* **136**, 653–65.

Wells, J. M., Harris, M. J. & Boddy, L. (1998). Temporary phosphorus partitioning in mycelial systems of the cord-forming basidiomycete *Phanerochaete velutina*. *New Phytologist* **140**, 283–93.

Wells, J. M., Harris, M. J. & Boddy, L. (1999). Dynamics of mycelial growth and phosphorus partitioning in developing *Phanerochaete velutina*: dependence on carbon availability. *New Phytologist* **142**, 325–34.

Wells, J. M., Thomas, J. & Boddy, L. (2001). Soil water potential shifts: developmental responses and dependence on phosphorus translocation by the cord-forming basidiomycete *Phanerochaete velutina*. *Mycological Research* **105**, 859–67.

Wilson, E. O. (1990). *Success and Dominance in Ecosystems: the Case for the Social Insects*. Oldendorf/Luhe, Germany: Ecology Institute.

Zakaria, A. J. & Boddy, L. (2002). Mycelial foraging by *Resinicium bicolor*: interactive effects of resource quantity, quality and soil composition. *FEMS Microbiology Ecology* **40**, 135–42.

7

Natural abundance of ^{15}N and ^{13}C in saprotrophic fungi: what can they tell us?

ANDY F.S. TAYLOR AND PETRA M.A. FRANSSON

Swedish University of Agricultural Sciences, Uppsala

Introduction

The aim of this chapter is to summarize recent developments in the study of the natural abundance of stable isotopes, primarily ^{15}N and ^{13}C, in fungal sporocarps. The main focus will be on saprotrophic fungi but, owing to a scarcity of studies, considerable use is made of the more abundant literature on ectomycorrhizal fungi. A brief introduction to the terminology used in the determination and use of the stable isotopes ^{15}N and ^{13}C is provided. This is followed by a discussion of the most significant findings from investigations into ectomycorrhizal fungi and how these compare with the available data from saprotrophic fungi. Recent results from a study focusing on saprotrophic fungi are then presented. Finally, some suggestions are given as to where isotope data may be useful in investigating the ecology of saprotrophic fungi.

Microorganisms are integral components of most biogeochemical cycles in terrestrial ecosystems (Dighton, 1995). The role of fungi, in particular, in boreal and temperate forests is pivotal. Most of the trees in these forests form mutualistic associations with a wide range of ectomycorrhizal fungi (Smith & Read, 1997) while the breakdown of woody debris is almost exclusively carried out by saprotrophic fungi (Tanesaka *et al.*, 1993). In addition to these essential ecological roles, fungi make up a large proportion of the total organism diversity in these systems. Despite their evident importance, relatively little is known about fungal activities *in situ*. The opaque character of the main growth substrates (woody material and soil) and the cryptic nature of most fungal mycelia have greatly hampered field investigations. Consequently, our knowledge of the involvement of fungi in ecosystem processes is primarily derived from laboratory investigations using a small number of tractable species.

Fungi in the Environment, ed. G. M. Gadd, S. C. Watkinson & P. S. Dyer. Published by Cambridge University Press.

Stable isotopes: use and terminology

Stable isotopes have been used for many years to investigate plant ecology and nutrient cycling in ecosystems (for general review, see Dawson *et al.*, 2002). In particular, ^{15}N and ^{13}C have been used extensively in ecological studies (^{15}N, Högberg, 1997; Robinson, 2001; ^{13}C, Staddon, 2004). The usefulness of these isotopes depends on small but measurable differences in the isotope ratios of ^{13}C:^{12}C and ^{15}N:^{14}N, which result from fractionations during the chemical, physical and biological processing of compounds containing N and C (Dawson *et al.*, 2002). These fractionations result in different resources (e.g. growth substrates) often having distinct isotope signatures that can be traced when the substrate is utilized. For example, the natural abundances of ^{15}N and ^{13}C have been widely used to elucidate pathways of N and C cycling in ecosystems (see, for example, Farquhar *et al.*, 1989; Fung *et al.*, 1997). In recent years, the natural abundance of ^{15}N and ^{13}C has also been investigated as a potential indirect method for assessing the functional roles of the fungi in ecosystems.

The isotope ratios of ^{13}C:^{12}C and ^{15}N:^{14}N in samples are determined by isotope ratio mass spectroscopy and expressed relative to international reference standards. By definition, standards have a δ value of 0‰. For N, atmospheric N_2 is taken as the standard and natural ^{15}N abundance is expressed in delta (δ) units, indicating parts per thousand (‰) deviations from the ratio ^{15}N:^{14}N in atmospheric N_2 (Eq. 7.1):

$$\delta^{15}N(‰) = ((R_{sample}/R_{standard}) - 1) \times 1000, \quad (7.1)$$

where R denotes the ratio ^{15}N:^{14}N (Högberg, 1997). Positive $\delta^{15}N$ values indicate greater abundance of ^{15}N; negative values indicate lower abundance of ^{15}N relative to air. Variations from the standard ratio in biological material are small, with total ranges within ecosystems usually *c.* −10‰ to +20‰ (Handley & Raven, 1992).

The notation ($\delta^{13}C$‰) and the calculations (Eq. 7.1) for natural abundance of ^{13}C are similar, where the reference standard is the Vienna–PeeDee Belemnite (Staddon, 2004). Owing to the high ^{13}C content of the ^{13}C reference standard, biological materials have negative values of $\delta^{13}C$: the more negative the value, the lower the ^{13}C abundance. The $\delta^{13}C$ of plant material varies considerably, from −40 to −9‰, with much of this variation accounted for by differences between C_3 and C_4 plants (Staddon, 2004).

In addition to tracing pathways of N and C movements through ecosystems, the natural abundance of ^{13}C and ^{15}N has been used in a diverse range of other ecological studies: for investigating interactions and trophic status among soil invertebrates (see, for example, Ponsard & Arditi, 2000), and for comparing feeding preferences of arboreal prosimians in forests of East African and Madagascar (Schoeninger *et al.*, 1998).

Stable isotopes and fungal ecology

In the field of mycorrhizal research, the natural abundance of ^{15}N in host plants has been found to be closely linked to the type of mycorrhizal association formed (Michelsen *et al.*, 1998; Hobbie *et al.*, 2000) and it has also been used to indicate the importance of ectomycorrhizal (ECM) fungi in N cycling and in supplying host plants with N derived from organic sources (Hobbie *et al.*, 1998). Remarkably, the abundance of ^{13}C has been determined in single spores of the arbuscular mycorrhizal fungi *Gigaspora marginata* (Nakano *et al.*, 1999). This study used the widely different $\delta^{13}C$ of C_3 and C_4 plants to show that the $\delta^{13}C$ of spores produced by *G. marginata* decreased linearly from the C_4 plant to the proximity of the C_3 plant when grown in dual cultures. Recent work on mycoheterotrophic orchids (Gebauer & Meyer, 2003) and other non-chlorophyllous plants (Trudell *et al.*, 2003) has used ^{13}C and ^{15}N to examine both the degree of dependence of the plants on mycorrhizal fungal associates and the specificity exhibited by the plants to particular fungal taxa.

The study that first raised the potential of using stable isotopes to investigate fungal ecology in the field was by Gebauer & Dietrich (1993). They investigated ^{15}N abundance in different compartments of mixed stands of spruce, larch and beech forests, including a small number of ECM and saprotrophic fungal species. An unexpected result of this study was that ECM fungi were enriched in ^{15}N compared with their host plants. This pattern of enrichment of ECM fungi over the host plants has subsequently been found in a number of other studies (Högberg *et al.*, 1996; Taylor *et al.*, 1997, 2003; Michelsen *et al.*, 1998; Hobbie *et al.*, 1999) and seems to be a general phenomenon. This is surprising because the level of colonization by ECM fungi (percentage of root tips colonized) in most temperate and boreal forests is usually more than 95% (see, for example, Taylor *et al.*, 2000). This means that most of the N entering the host must pass through the fungi first. The fungi and the host plants should therefore have similar ^{15}N signatures. The reason(s) for this difference is (are) still a matter of debate (see Hobbie & Colpaert, 2003). The most likely explanation is that a small discrimination against the heavier isotope ^{15}N during the transfer of N from ECM fungi to the host plant results in an accumulation of ^{15}N within the ECM mycelium. Discrimination for or against the heavier isotope (^{15}N and ^{13}C) is called isotopic fractionation (Dawson *et al.*, 2002).

^{15}N enrichment in ECM fungi may also reflect inherent differences among species in their ability to utilize organic nutrient sources (Smith & Read, 1997). Lilleskov *et al.* (2002) have shown that species of ECM fungi that had high δ^{15}N values in their sporocarps had a greater ability to degrade and utilize organic sources of N than those ECM species with low δ^{15}N values. As litter is degraded its ^{15}N content increases (see below); fungal species utilizing more decomposed material should, in theory, take up breakdown products that are enriched in ^{15}N. However, if the uptake of enriched organic nutrient sources had a major influence on δ^{15}N values, then it could be predicted that most saprotrophic fungi would have high δ^{15}N values, but they do not (see below).

The majority of studies that have examined natural abundance of ^{13}C and ^{15}N in fungi have used sporocarp material, as direct analysis of soil mycelia is rarely practicable. One exception to this is the study by Högberg *et al.* (1996), who determined the ^{15}N abundance in the fungal mantles of beech (*Fagus sylvatica* L.) mycorrhizas: mantle material was enriched by 2.4‰–6.4‰ compared with the enclosed host root.

Although the sample sizes of the fungi analysed by Gebauer & Dietrich (1993) were small, there was a suggestion that ECM fungi and saprotrophic fungi had different ^{15}N signatures. This has also been found in subsequent studies (Gebauer & Taylor, 1999; Kohzu *et al.*, 1999; Taylor *et al.*, 2003; Trudell *et al.*, 2004). In general, ECM fungi as a group are enriched in ^{15}N compared with saprotrophic fungi. Based on the assumption that the δ^{15}N of an organism is correlated with the signature of the substrate, the difference between ECM and saprotrophic fungi would suggest that the two groups are utilizing sources of N in the soil with different ^{15}N contents. Although this may be the case for lignicolous (wood rotting) saprotrophic fungi and ECM fungi, it seems likely that the mycelia of many saprotrophic terricolous (growing on the forest floor) and ECM fungi compete for the same resources in soil organic layers (see Lindahl *et al.*, 2002). The ^{15}N enrichment of ECM fungi compared with saprotrophic fungi may also be explained by the reason outlined above for the differences in ^{15}N signatures between ECM fungi and their host plant (fractionation during N transfer).

Delta ^{13}C has also been shown to differ between saprotrophic and ECM fungi, the latter group being more depleted in ^{13}C than the former (Hobbie *et al.*, 1999; Högberg *et al.*, 1999; Kohzu *et al.*, 1999; Henn & Chapela, 2001). Owing to these differences between saprotrophic and ECM fungi in

$\delta^{13}C$ and $\delta^{15}N$, a joint plot of both parameters usually results in a clear separation between the two trophic groups (see Taylor *et al.*, 2003; Trudell *et al.*, 2004).

The difference between ECM and saprotrophic fungi in isotope data has been called the saprotrophic–mycorrhizal divide (Henn & Chapela, 2001) and has been used to suggest the trophic (ECM or saprotrophic) status of fungi where this was unknown (see, for example, Hobbie *et al.*, 2001). Bomb-carbon or radiocarbon (^{14}C) has also been used in this context. Careful measurements of the concentrations of ^{14}C present in organic matter allow the age of the organic matter to be accurately determined. Hobbie *et al.* (2002) used this method to show that sporocarps of known saprotrophic species contained carbon that had been photosynthetically fixed by plants more than 6 years previously, whereas the carbon in known ECM species was only 0–2 years old. This method gives very strong evidence for trophic status, but the use of radiocarbon for such studies is, unfortunately, prohibitively expensive.

In addition to being useful for distinguishing ECM and saprotrophic fungi, the natural abundance of ^{13}C and ^{15}N has been examined in the latter group in relation to the presumed substrate utilized by the fungi. It is well established that the $\delta^{15}N$ of organic matter in forest soils characteristically increases from the litter layer downwards into more decomposed material (Högberg, 1997 and references therein). The negative values (from *c.* –6‰ to 3‰) in the fresh litter layer reflect those of the plant species present, but the values quickly become positive in the F and H layers and continue to increase in the mineral soil, where $\delta^{15}N$ may be *c.* +6‰. The main reason proposed to explain this pattern is the input of depleted N in plant litter on the soil surface and fractionation against ^{15}N during decomposition (Nadelhoffer & Fry, 1988).

Delta ^{15}N and $\delta^{13}C$ values measured in plant and fungal material are essentially the average of all the N and C contained within the tissue. However, different chemical components within plants and fungi may differ considerably in both $\delta^{15}N$ and $\delta^{13}C$ (Högberg, 1997; Pate & Arthur, 1998). Taylor *et al.* (1997) showed that proteins and amino acids in ECM fungi were enriched by as much as 9‰ compared with the cell wall chitin. Within plants, cellulose is enriched in ^{13}C compared with lignin (Hobbie & Werner, 2004 and references therein). Structural plant components, foliage, stems and roots, may also differ with respect to both ^{13}C and ^{15}N isotopes and these may also differ in $\delta^{13}C$ from the phloem sap carbon. Pate & Arthur (1998) found that mature foliage was markedly depleted in ^{13}C compared with the phloem sap. This observation may explain the *c*. 2‰ difference between ECM trees and the associated host-specific ECM fungi recorded by Högberg *et al.* (1999). When saprotrophic fungi are specialized on or show a preference for particular substrates that differ in isotope ratios, then this could be reflected in the values in the mycelium.

Gebauer & Taylor (1999) examined $\delta^{15}N$ in a range of saprotrophic species in a mixed forest stand in the Fichtelgebirge region of southern Germany. The fungi were separated into taxa (group 1) considered to grow on litter and woody debris and taxa (group 2) that can utilize organic material at more advanced states of decomposition. Species in the group 1 had lower $\delta^{15}N$ signatures than did those in group 2 and this correlated with the $\delta^{15}N$ of the presumed substrates. The sporocarps of group 1 also had higher N concentrations. Enrichment factors for the fungi were calculated by using the equation $\varepsilon_{f\text{-}s} = \delta^{15}N_{fungus} - \delta^{15}N_{substrate}$ (Emmett *et al.*, 1998). The average enrichment factor for fungi growing on litter or wood (group 1) was significantly lower than that of fungi growing on more decomposed material (group 2). Kohzu *et al.* (1999) also examined enrichment in wood-decomposing and litter fungi but unfortunately only provided data for the wood decomposers, which showed little enrichment ($\varepsilon = 0.4 \pm 1.2$‰) between $\delta^{15}N_{substrate}$ and $\delta^{15}N_{fungus}$.

One explanation for the low level of enrichment in fungi growing on wood could be the very low concentrations of N in the woody substrates. In order to grow on these substrates the fungi must extract all of the available N. This would, therefore, result in the ^{15}N content of the wood and the fungi being similar. Those fungi growing on more N-rich substrates and where the N is contained in a wide range of compounds may be able to be more selective in the use of N-containing substrates. This potential for selective utilization of compounds, which may differ in ^{15}N contents, could create greater differences in fungal and substrate signatures. These two different scenarios are analogous to closed and open reaction systems, where no fractionation is measurable in the closed system because all of the N in the substrate is transformed into the product(s) of the reaction (Högberg, 1997).

Kohzu *et al.* (1999) examined enrichment in ^{13}C between substrate and fungi and found a much greater enrichment of 3.5 ± 0.9‰ between $\delta^{13}C_{substrate}$ and $\delta^{13}C_{fungus}$. Based on the previous argument, this could imply selective utilization of compounds with different ^{13}C contents. However, Kohzu *et al.* (1999) also carried out a laboratory experiment in which they inoculated wood blocks with the mycelium of *Trametes versicolor* (L.:Fr.) Quél., a white rot fungus, and measured $\delta^{13}C$ of the wood, the fungus and the CO_2 respired off. Over a period of 181 days, the value of the wood ($-26.3 \pm 0.3‰$) was remarkably constant, even with a mass loss of 70%! The mycelium was consistently enriched in ^{13}C compared with the wood ($\varepsilon = 3.5 \pm 0.5$), i.e. the same value recorded in the study for the field samples. The $\delta^{13}C$ of the CO_2 was similar to that of the wood, suggesting that respiration had little effect upon the fungal enrichment. It was suggested that the enrichment in the fungal material in ^{13}C compared with the wood might not be due to utilization of different substrates but relate to differences in cellular components within the fungal tissue. It is known that chitin may be enriched in ^{13}C relative to woody substrates (Gleixner *et al.*, 1993). Because much of the N in fungi is contained within chitin, measurements of vacuolated mycelia could result in $\delta^{13}C$ values higher than those of the wood.

Fungal species potentially differ in how they fractionate for or against the heavier isotope within each isotope pair during uptake and subsequent metabolic transformations. This phenomenon has so far received little attention but could have a significant influence on the $\delta^{15}N$ and $\delta^{13}C$ values of fungi. Fractionations occur because heavier isotopes react more slowly and are more strongly bound in chemical reactions than lighter isotopes (Högberg, 1997). Abraham & Hesse (2003) examined fractionation by four 'mould' fungal species (two zygomycetes and two ascomycetes) growing on a range of C and N sources. Differences were found between the fungi both in fractionation on uptake and, more interestingly, in the ^{13}C enrichment of many amino acids. A discriminant analysis of the isotope ratios in four amino acids (Thr, Ile, Phe, Val) divided the fungi into their respective groups. This is an important finding as it could potentially be used to investigate selective feeding on fungal species by soil fungivores. In addition, these findings may complicate the interpretation of $\delta^{15}N$ and $\delta^{13}C$ values. If fungi differ markedly in the degree of fractionation during uptake of nutrients and carbon compounds, then without precise knowledge of the fractionation pattern little may be assumed from $\delta^{15}N$ and $\delta^{13}C$ values concerning the original substrates utilized.

Henn & Chapela (2000) examined fractionation against ^{13}C by three fungi (a wood decomposer, a litter decomposer and an ECM fungal species) growing on sucrose from C_3 and C_4 plants. They found a taxon-specific fractionation pattern that was apparently related to the distribution of the heavier isotope within the sucrose molecule. Based on these results, the authors suggested that the usefulness of stable isotopes for determining resources of fungi might be limited owing to fractionation and the differential use of different C atoms within organic C sources.

However, a recent study by Hobbie *et al.* (2004), which specifically examined fractionation and differential use of C during uptake and incorporation of N and C by a selection of ECM and saprotrophic fungi growing on different artificial media, found little influence of fractionation on δ values. In the closed Petri dish systems utilized in this study, there were no differences among the fungi with respect to $\delta^{15}N$ values. However, saprotrophic fungi were depleted in ^{13}C relative to ECM fungi. This was attributed to the utilization of ^{13}C-depleted agar in the medium by the saprotrophic fungi. The fungi obtained up to 45% of their carbon from the agar. This possibility was not taken into account in the study by Henn & Chapela (2000) and could partly explain their results.

What is the potential for extracting species-level information from natural abundance of ^{13}C and ^{15}N stable isotopes? At present, we know little about species-level patterns in saprotrophic fungi: sample sizes have, in general, been too small. However, studies examining species-related differences in ECM fungi have suggested that species-level patterns do exist and that they potentially reflect the ecophysiology of the species (Taylor *et al.*, 2003; Trudell *et al.*, 2004). Trudell *et al.* (2004) also found some evidence for species-level patterns in the saprotrophic genus *Pholiota*. However, the influence of site cannot be ignored, particularly in analyses of ^{15}N values. On sites subject to elevated levels of nitrogen deposition, the ^{15}N content of the deposition may have a significant impact on the average $\delta^{15}N$ values in different ecosystem compartments, including fungi (Bauer *et al.*, 2000). In addition, when N deposition is high, the range of potential $\delta^{15}N$ values will be narrower than on unaffected sites (Gebauer & Taylor, 1999).

Summary of a recent study in central Sweden

A recent study on stable isotopes and field-collected fungi, with a special focus on saprotrophic fungi, was carried out in forests in and around Uppsala, central Sweden (59°52' N, 17°13' E, 35 m a.s.l.). Most of these forests are similar in structure and typical of the area, with Scots pine (*Pinus sylvestris* L.) and Norway spruce (*Picea abies* [L.] Karst.) as the overstorey and with understorey deciduous tree species that include aspen (*Populus tremula* L.), birch (*Betula pendula* Roth), willow (*Salix caprea* L.), alder (*Alnus incana* [L.] Moench), and oak (*Quercus robur*).

The main forest site is the Stadsskogen, situated within Uppsala, and is ideal for intensive studies as the ECM and saprotrophic fungal community is very diverse (see Taylor *et al.*, 2003) and there is an abundance of woody substrates for the growth of saprotrophic species. A previous study showed that the foliage of the trees in the different canopy layers differed in their $\delta^{13}C$ (Högberg *et al.*, 1999). This phenomenon was used to show that the $\delta^{13}C$ of the host-specific ECM fungi was determined by that of their host plant. This study also demonstrated that saprotrophic fungi and ECM fungi in the forest differed markedly in $\delta^{13}C$ values.

Regular visits were made to the forests during the fruiting season (June–October) over a three-year period. Mature sporocarps of saprotrophic fungi and from fungi of unknown trophic status were collected and identified to species. Nomenclature follows that of Hansen & Knudsen (1992, 1997). Species were classified into terricolous species (growing on the forest floor) and lignicolous species (growing on wood). A small sample of the woody substrate was collected in the proximity of the lignicolous fungi whenever possible. All samples were dried (70 °C, 24 h) and ground to a powder in a ball mill; $\delta^{13}C$ and $\delta^{15}N$ were then determined (see Taylor *et al.*, 2003 for details). Mean values are usually given ± standard error (se). A total of *c.* 200 samples were collected representing 74 species, 45 of which were lignicolous. This latter group could be further divided into taxa causing white or brown rot of woody substrates, by using relevant literature (Gilbertson, 1981; Worrall *et al.*, 1997). Because white-rot fungi utilize all cell-wall components and brown-rot fungi utilize carbohydrates but degrade lignin to only a limited extent, differences in ^{15}N and, in particular, ^{13}C could be expected between these groups of fungi.

A comparison was made with a large data set available on ECM fungi from the Stadsskogen (data from Taylor *et al.*, 2003). The total range of δ values measured in the ECM and saprotrophic fungal sporocarps was remarkable: $\delta^{15}N$ from –4.60‰ to +17.52‰; $\delta^{13}C$ from −29.20‰ to −19.79‰. Kohzu *et al.* (1999) recorded an even greater range of $\delta^{15}N$ values (−8‰ – +22‰) in fungal sporocarps, but this included material from a number of different ecosystems. Bearing in mind that the total range of $\delta^{15}N$ values in most ecosystems is usually only from *c.* −10‰ to +20‰ (Handley & Raven, 1992), it is clear that fungi cover much of this range.

It was mentioned previously that $\delta^{15}N$ and $\delta^{13}C$ values have been used in soil faunal ecology to investigate the trophic status of soil invertebrates (see, for example, Ponsard & Arditi, 2000). Fungi form the basis of many soil food webs and the observed variation in $\delta^{15}N$ and $\delta^{13}C$ values in fungi, assuming that the values in sporocarps reflect those of the mycelia, must raise a question over the usefulness of using ^{15}N and ^{13}C natural abundance as indicators of the trophic status of soil animals. If fungivores demonstrate any feeding specificity, then it is difficult to see how differences in $\delta^{15}N$ can be used to infer trophic status when the values may be more a reflection of feeding preferences for different fungal species or groups.

Table 7.1. *Mean values for percentage nitrogen (%N), $\delta^{15}N$, percentage carbon (%C) and $\delta^{13}C$ in sporocarps of saprotrophic and ectomycorrhizal (ECM) fungi collected in forests around Uppsala in central Sweden*

	Saprotrophic ($n = 74$)	ECM fungi ($n = 109$)	p^a
$\delta^{15}N$ (‰)	1.86 ± 0.4	5.83 ± 0.3	<0.001
$\delta^{13}C$ (‰)	-23.2 ± 0.1	-25.6 ± 0.1	<0.001
%N	4.35 ± 0.2	3.96 ± 0.1	ns
%C	41.5 ± 0.3	43.4 ± 0.1	<0.001

[a] Significance of differences between fungal life strategies, analysed by using Student's *t*-test; ns, not significant.

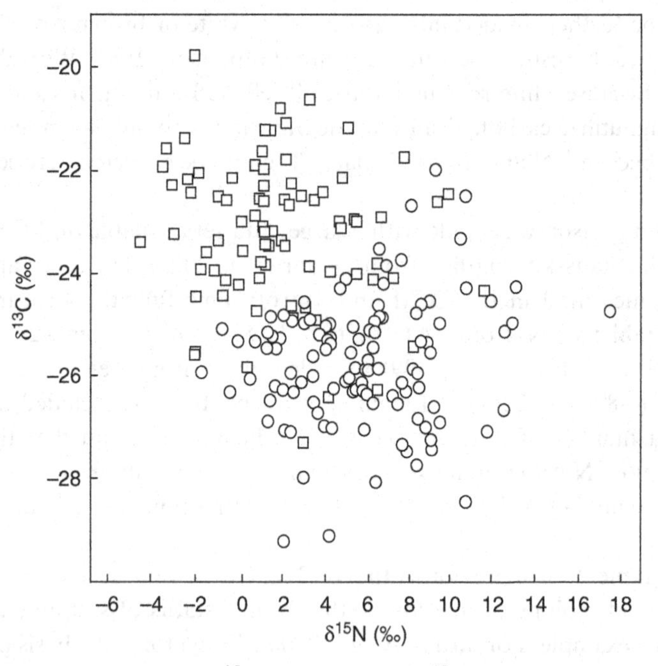

Fig. 7.1. Nitrogen (^{15}N) and carbon (^{13}C) stable isotope values for sporocarps of saprotrophic and ectomycorrhizal fungi collected in forests in and around Uppsala, central Sweden. Data points represent species means. Open squares, saprotrophic fungi; open circles, ectomycorrhizal fungi.

The ECM–saprotroph divide outlined above was also clear in this study (Table 7.1; Fig. 7.1). However, several of the terricolous saprotrophic fungi included in this study had high $\delta^{15}N$ values that are more indicative of ECM fungi. This group of fungi included, among others, four *Agaricus* species (*A. aestivalis* (Møll.) Pilát, *A. arvensis* Schaeff., *A. bitorquis* (Quél.) Sacc., *A. silvaticus* Schaeff.), *Entoloma clypeatum* (L.:Fr.) Kumm. and *Clavariadelphus pistillaris* (L.:Fr.) Donk with $\delta^{15}N$ values of 5.56‰–9.92‰, 8.17‰ and 11.52‰, respectively. It is unlikely that the *Agaricus* species are mutualistic with plant species and this is supported by the high $\delta^{13}C$ values (−23.04‰ – −21.74‰) that were similar to or even higher than those from the other saprotrophic fungi. However, *E. clypeatum*, like a small number of other *Entoloma* species, has been noted as consistently growing under plants of the family Rosaceae (Hansen & Knudsen, 1992) and a number of ectomycorrhizal structures formed by *Entoloma* species have been described (Agerer, 1987–2002). The low $\delta^{13}C$ value (−25.43‰) would support the idea that this species is a mutualist. The very high $\delta^{15}N$ value of the remaining species, *C. pistillaris*, is highly indicative of an ECM species and the taxonomic placing of the genus near *Gomphus* and *Ramaria*, known ECM formers, would support this idea. The $\delta^{13}C$ value (−24.34‰) is, however, ambiguous and further studies are needed to clarify the trophic status of this fungus.

The high $\delta^{15}N$ values of the *Agaricus* species either suggest taxon-specific fractionations that enrich the mycelium of this group or, more likely, they reflect the utilization of ^{15}N-enriched organic substrates, possibly in the mineral soil layers. As mentioned above, as decomposition progresses the remaining material becomes progressively enriched in ^{15}N. The soil horizon in which the major proportion of a mycelium exists is likely to have significant influence upon the overall δ values of the mycelium. It is now possible to localize the mycelium of individual fungi in soil (see Dickie *et al.*, 2002; Landeweert *et al.*, 2003) and it would be very instructive to use these methodologies and relate mycelial position to isotope signatures of the organic material in that layer and in the sporocarps.

The wood decay fungus *Fistulina hepatica* (Schaeff.:Fr.) Fr. had the lowest recorded $\delta^{13}C$ value (−26.424‰) of the saprotrophic fungi included in the study. This species is considered to be a weak parasite of oak trees (*Quercus* spp.) (Hansen & Knudsen, 1997) and the low $\delta^{13}C$ value may indicate that the fungus, at least for sporocarp formation, can use sugars from the phloem tissue of the host plant. The sporocarps of this species are large, fleshy and produced annually, and the mycelium has the unusual ability of very rapidly re-growing new sporocarps should the existing ones be removed (A. Taylor, personal observation). This ability could be linked to the fungus having access to readily available carbon sources in the phloem.

In the study, terricolous and lignicolous fungi differed with respect to N% and $\delta^{15}N$ (Fig. 7.2), with terricolous fungi having, on average, higher N contents (5.87 ± 0.3 compared with $3.25 \pm 0.2\%$, $p < 0.001$, Student's *t*-test) and, on average, higher $\delta^{15}N$ ($2.58 \pm 0.6‰$ compared with $0.74 \pm 0.4‰$, $p = 0.018$, Student's *t*-test). There was a suggestion that carbon contents (C%) were higher in lignicolous fungi ($41.9 \pm 0.3\%$ compared with $40.9 \pm 0.5\%$, $p = 0.077$, Student's *t*-test), but there were no differences between the two groups with respect to $\delta^{13}C$ (lignicolous, $-23.06 \pm 0.18‰$; terricolous, $-23.23 \pm 0.29‰$). Kohzu *et al.* (1999) and Trudell *et al.* (2004) found similar patterns in N% and $\delta^{15}N$ values between terricolous (called litter fungi in these studies) and lignicolous fungi.

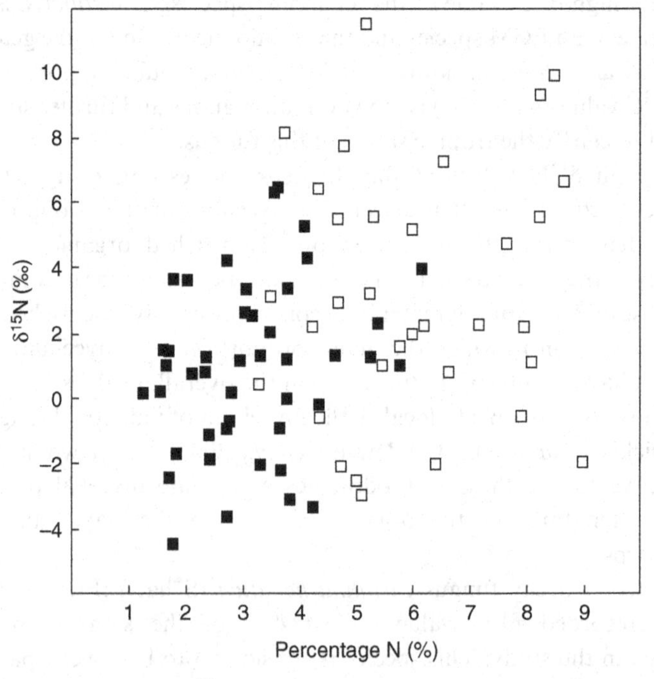

Fig. 7.2. Nitrogen (^{15}N) stable isotope and nitrogen concentration (%) values in sporocarps of terricolous and lignicolous saprotrophic fungi collected in forests in and around Uppsala, central Sweden. Data points represent species means. Open squares, terricolous saprotrophic fungi; filled squares, lignicolous saprotrophic fungi.

However, these studies recorded significant differences between the two groups with respect to $\delta^{13}C$. It seems likely that, just as with ECM fungi, the taxonomic composition of the saprotrophic fungi analysed will have a significant impact on the average isotope values recorded at a site.

Seven species of brown-rot fungus were included in the study. These could, in theory, differ from white-rot fungi with respect to both ^{13}C and ^{15}N because the latter group has access to a wider range of substrates within wood. However, with the exception of C% (brown rot, $43.5 \pm 0.9\%$; white rot, $41.6 \pm 0.3\%$, $p = 0.009$, Student's *t*-test), no differences were found between these two groups of fungi. Increasing the sample size may uncover differences, or it may be that differences are masked by differences in the ^{13}C and ^{15}N contents of woody materials from different plant species. Growing isolates of brown- and white-rot fungi on the same woody substrate under laboratory conditions would enable this question to be addressed systematically.

Conclusions

(1) In general, the mean $\delta^{13}C$ and $\delta^{15}N$ values from saprotrophic and ECM fungal sporocarps differ significantly and, at a given site, can be used to indicate the likely trophic status of unknown taxa.
(2) Terricolous and lignicolous saprotrophic fungi seem to differ with respect to N contents and $\delta^{15}N$, but differences in $\delta^{13}C$ may be site-specific.
(3) At any given site, the mean $\delta^{15}N$ and $\delta^{13}C$ values and N and C contents for saprotrophic fungi are, just as with ECM fungi, dependent upon the taxonomic composition of the fungal community sampled.
(4) Extreme values of $\delta^{15}N$ and $\delta^{13}C$ may be useful for indicating taxon- or group-specific aspects of nutrient acquisition (e.g. *Fistulina hepatica* or *Agaricus* spp.).
(5) We still know too little concerning the determinants of $\delta^{15}N$ and $\delta^{13}C$ in fungi, in particular fractionation during uptake and metabolic processing, to fully utilize the potential of using stable isotopes to investigate fungal ecology.

Future potential applications of stable isotope signatures to fungal ecology

One aspect of fungal ecology that has received considerable attention is that of species or community succession in saprotrophic fungi

(Frankland, 1998). The traditional hypothesis to explain succession has been based on the principle that each successive colonizer utilizes substrates that the preceding species either could not use or could do so only partly (Frankland, 1998). However, it seems likely that at least some of the later colonizers are specialized in using the mycelium of earlier colonizers (Holmer *et al.*, 1997). It should be possible to determine the extent to which these late colonizers are actually utilizing woody components by using $\delta^{15}N$ and $\delta^{13}C$, as the mycelium of the earlier colonizer and the wood should differ significantly.

It is well established that many insects, particularly some groups of wood-boring beetles, inoculate wood with spores of wood-decay fungi during the process of egg laying (Paine *et al.*, 1997). The fungal spores germinate and the mycelium rapidly degrades the wood. The insects may benefit from this association with fungi by feeding on the more palatable, decayed woody tissue and/or by the consumption of fungal mycelium. The relative importance of each of these potential benefits is unclear. A detailed analysis of $\delta^{15}N$ and $\delta^{13}C$ in the larvae, the wood and the mycelium, could help provide an answer to this question.

References

Abraham, W-.R. & Hesse, C. (2003). Isotope fractionations in the biosynthesis of cell components by different fungi: a basis for environmental carbon flux studies. *FEMS Microbiology Ecology* **46**, 121–8.

Agerer, R. (1987–2002). *Colour Atlas of Ectomycorrhizae*. 1st–12th delivery. Schwäbisch Gmünd: Einhorn Verlag.

Bauer, G. A., Andersen, B., Gebauer, G., Harrison, T., Högberg, P., Högbom, L., Taylor, A. F. S., Novak, M., Harkness, D., Schulze, E.-D. & Persson, T. (2000). Biotic and abiotic controls over ecosystem cycling of stable natural isotopes for nitrogen, carbon and sulfur. In *Carbon and Nitrogen Cycling in European Forest Ecosystems*, ed. E.-D. Schulze, *Ecological Studies*, vol. 142. Heidelberg: Springer-Verlag.

Dawson, T., Mambelli, S., Plamboeck, A. H., Templer, P. H. & Tu, K. P. (2002). Stable isotopes in plant ecology. *Annual Review of Ecology and Systematics* **33**, 507–59.

Dickie, I. A., Xu, B. & Koide, R. T. (2002). Vertical niche differentiation of ectomycorrhizal hyphae in soil as shown by T-RFLP analysis. *New Phytologist* **156**, 527–35.

Dighton, J. (1995). Nutrient cycling in different terrestrial ecosystems in relation to fungi. *Canadian Journal of Botany* **73**, 1349–60.

Emmett, B. A., Kjonaas, O. J., Gundersen, P., Koopmans, C., Tietema, A. & Sleep, D. (1998). Natural abundance of ^{15}N in forests across a nitrogen deposition gradient. *Forest Ecology and Management* **101**, 9–18.

Farquhar, G. D., Ehleringer, J. R. & Hubick, K. T. (1989). Carbon isotope discrimination and photosynthesis. *Annual Review of Physiology Plant and Molecular Biology* **40**, 503–37.

Frankland, J. C. (1998) Fungal succession – unravelling the unpredictable. *Mycological Research* **102**, 1–15.

Fung, I., Field, C. B., Berry, J. A., Thompson, M. V., Randerson, J. T., Malmström, C. M., Vitousek, P. M., Collatz, G. J., Sellers, P. J., Randall, D. A., Denning, A. S., Badeck, F. & John, J. (1997). Carbon 13 exchanges between the atmosphere and biosphere. *Global Biogeochemical Cycles* **11**, 507–33.

Gebauer, G. & Dietrich, P. (1993). Nitrogen isotope ratios in different compartments of a mixed stand of spruce, larch and beech trees and of understorey vegetation including fungi. *Isotopenpraxis Environmental Health Studies* **29**, 35–44.

Gebauer, G. & Meyer, M. (2003). ^{15}N and ^{13}C natural abundance of autotrophic and mycoheterotrophic orchids provides insight into nitrogen and carbon gain from fungal association. *New Phytologist* **160**, 209–23.

Gebauer, G. & Taylor, A. F. S. (1999). ^{15}N natural abundance in fruit bodies of different functional groups of fungi in relation to substrate utilization. *New Phytologist* **142**, 93–101.

Gilbertson, R. L. (1981). North American wood-rotting fungi that cause brown rots. *Mycotaxon* **12**, 372–416.

Gleixner, G., Danier, H.-J., Werner, R. A. & Schmidt, H.-L. (1993). Correlations between the ^{13}C of primary and secondary plant products in different cell compartments and that in decomposing Basidiomycetes. *Plant Physiology* **102**, 1287–90.

Handley, L. L. & Raven, J. A. (1992). The use of natural abundance of nitrogen isotopes in plant physiology and ecology. *Plant, Cell and Environment* **15**, 965–85.

Hansen, L. & Knudsen, H. (1992). *Nordic Macromycetes*, vol. 2, *Polyporales, Boletales, Agaricales, Russulales.* Copenhagen, Denmark: Nordsvamp.

Hansen, L. & Knudsen, H. (1997). *Nordic Macromycetes*, vol. 3, *Heterobasidioid, Aphyllophoroid and Gasteromycetoid Basidiomycetes.* Copenhagen, Denmark: Nordsvamp.

Henn, M. R. & Chapela, I. H. (2000). Differential C isotope discrimination by fungi during decomposition of C-3- and C-4-derived sucrose. *Applied and Environmental Microbiology* **66**, 4180–6.

Henn, M. R. & Chapela, I. H. (2001). Ecophysiology of ^{13}C and ^{15}N isotopic fractionation in forest fungi and the roots of the saprotrophic-mycorrhizal divide. *Oecologia* **128**, 480–7.

Hobbie, E. A. & Colpaert, J. V. (2003). Nitrogen availability and colonization by mycorrhizal fungi correlate with nitrogen isotope patterns in plants. *New Phytologist* **157**, 115–26.

Hobbie, E. A., Macko, S. A. & Shugart, H. H. (1998). Patterns in N dynamics and N isotopes during primary succession in Glacier Bay, Alaska. *Chemical Geology* **152**, 3–11.

Hobbie, E. A. & Werner, R. A. (2004). Intramolecular, compound-specific, and bulk carbon isotope patterns in C3 and C4 plants: a review and synthesis. *New Phytologist* **161**, 371–85.

Hobbie, E. A., Macko, S. A. & Shugart, H. H. (1999). Insights into nitrogen and carbon dynamics of ectomycorrhizal and saprotrophic fungi from isotopic evidence. *Oecologia* **118**, 353–60.

Hobbie, E. A., Macko, S. A. & Williams, M. (2000). Correlations between foliar $\delta^{15}N$ and nitrogen concentrations may indicate plant-mycorrhizal interactions. *Oecologia* **122**, 273–83.

Hobbie, E. A., Weber, N. S. & Trappe, J. M. (2001). Mycorrhizal vs saprotrophic status of fungi: the isotopic evidence. *New Phytologist* **150**, 601–10.

Hobbie, E. A., Weber, N. S., Trappe, J. M. & van Klinken G. J. (2002). Using radiocarbon to determine the mycorrhizal status of fungi. *New Phytologist* **156**, 129–36.

Hobbie, E. A., Sánchez, F. S. & Rygiewicz, P. T. (2004). Carbon use, nitrogen use, and isotope fractionation of ectomycorrhizal and saprotrophic fungi in natural abundance and ^{13}C-labelled cultures. *Mycological Research* **108**, 725–36.

Högberg, P. (1997). Tansley Review No 95 – N-15 natural abundance in soil-plant systems. *New Phytologist* **137**, 179–203.

Högberg, P., Högbom, L., Schinkel, H. *et al.* (1996). ^{15}N abundance of surface soils, roots and mycorrhizas in profiles of European forest soils. *Oecologia* **108**, 207–14.

Högberg, P., Plamboeck, A. H., Taylor, A. F. S & Fransson, P. M. A. (1999). Natural C-13 abundance reveals trophic status of fungi and host-origin of carbon in mycorrhizal fungi in mixed forests. *Proceedings of the National Academy of Sciences of the USA* **96**, 8534–9.

Holmer, L., Renvall, P. & Stenlid, J. (1997). Selective replacement between species of wood-rotting basidiomycetes, a laboratory study. *Mycological Research* **101**, 714–20.

Kohzu, A., Yoshioka, T., Ando, T., Takahashi, M., Koba, K. & Wada, E. (1999). Natural ^{13}C and ^{15}N abundance of field-collected fungi and their ecological implications. *New Phytologist* **144**, 323–30.

Landeweert, R., Leeflang, P., Kuyper, T. W., Hoffman, E., Rosling, A., Werners, K. & Smit, E. (2003). Molecular identification of ectomycorrhizal mycelium in soil horizons. *Applied Environmental Microbiology* **69**, 327–33.

Lilleskov, E. A., Hobbie, E. A. & Fahey, T. J. (2002). Ectomycorrhizal fungal taxa differing in response to nitrogen deposition also differ in pure culture organic nitrogen use and natural abundance of nitrogen isotopes. *New Phytologist* **154**, 219–31.

Lindahl, B. O., Taylor, A. F. S. & Finlay, R. D. (2002). Defining nutritional constraints on carbon cycling in boreal forests – towards a less 'phytocentric' perspective. *Plant and Soil* **242**, 123–35.

Michelsen, A., Quarmby, C., Sleep, D. & Jonasson, S. (1998). Vascular plant N-15 natural abundance in heath and forest tundra ecosystems is closely correlated with presence and type of mycorrhizal fungi in roots. *Oecologia* **115**, 406–18.

Nadelhoffer, K. J. & Fry, B. (1988). Controls on natural nitrogen-15 and carbon-13 abundances in forest soil organic matter. *Soil Science Society of America Journal* **53**, 1633–40.

Nakano, A., Takahashi, K. & Kimura, M. (1999). The carbon origin of arbuscular mycorrhizal fungi estimated from δ^{13}C values of individual spores. *Mycorrhiza* **9**, 41–7.

Paine, T. D., Raffa, K. F. & Harrington T. C. (1997). Interactions among scolytid bark beetles, their associated fungi, and live host conifers. *Annual Review of Entomology* **42**, 179–206.

Pate, J. & Arthur, D. (1998). δ^{13}C analysis of phloem sap carbon: novel means of evaluating seasonal water stress and interpreting carbon isotope signatures of foliage and trunk wood of *Eucalyptus globulus*. *Oecologia* **117**, 301–11.

Ponsard, S. & Arditi, R. (2000). What can stable isotopes (δ^{15}N and δ^{13}C) tell about the food web of soil macro-invertebrates? *Ecology* **81**, 852–64.

Robinson, D. (2001). δ^{15}N as an integrator of the nitrogen cycle. *Trends in Ecology and Evolution* **16**, 153–62.

Schoeninger, M. J., Iwaniec, U. T. & Nash, L. T. (1998). Ecological attributes recorded in stable isotope ratios of arboreal prosimian hair. *Oecologia* **113**, 222–30.

Smith, S. E. & Read, D. J. (1997). *Mycorrhizal Symbiosis*. London: Academic Press.

Staddon, P. L. (2004). Carbon isotopes in functional soil ecology. *Trends in Ecology and Evolution* **19**, 148–54.

Tanesaka, E., Masuda, H. & Kinugawa, K. (1993). Wood degrading ability of basidiomycetes that are wood decomposers, litter decomposers, or mycorrhizal symbionts. *Mycologia* **85**, 347–54.

Taylor, A. F. S., Högbom, L., Högberg, M. *et al.* (1997). Natural ^{15}N abundance in fruit bodies of ectomycorrhizal fungi from boreal forests. *New Phytologist* **136**, 713–20.

Taylor, A. F. S., Martin, F. & Read, D. J. (2000). Fungal diversity in ectomycorrhizal communities of Norway spruce (*Picea abies* [L.] Karst.) and beech (*Fagus sylvatica* L.) along north-south transects in Europe. In *Carbon and Nitrogen Cycling in European Forest Ecosystems*, ed. E.-D. Schulze, pp. 343–65. Berlin: Springer-Verlag.

Taylor, A. F. S., Fransson, P. M., Högberg, P., Högberg, M. & Plamboeck, A. H. (2003). Species level patterns in ^{13}C and ^{15}N abundance of ectomycorrhizal and saprotrophic fungal sporocarps. *New Phytologist* **159**, 757–74.

Trudell, S. A., Rygiewicz, P. T. & Edmonds, R. L. (2003). Nitrogen and carbon stable isotope abundances support the myco-heterotrophic nature and host-specificity of certain achlorophyllous plants. *New Phytologist* **160**, 391–401.

Trudell, S. A., Rygiewicz, P. T. & Edmonds, R. L. (2004). Patterns of nitrogen and carbon stable isotope ratios in macrofungi, plants and soils in two old-growth conifer forests. *New Phytologist* **164**, 317–35.

Worrall, J. J., Anagnost, S. E. & Zabel, R. A. (1997). Comparison of wood decay among diverse lignicolous fungi. *Mycologia* **89**, 199–219.

III

Mutualistic interactions in the environment

8

Mycorrhizas and the terrestrial carbon cycle: roles in global carbon sequestration and plant community composition

JONATHAN R. LEAKE
Department of Animal & Plant Sciences, University of Sheffield

Introduction

The mycorrhizal symbiosis is characterized by a reciprocal exchange of photosynthetically-fixed plant carbon in return for the main plant-growth-limiting nutrients, nitrogen or phosphorus. Studies of mycorrhizal functioning have focused on their roles in providing nutrients to plants, but their importance as a significant component of the terrestrial carbon (C) cycle has generally been overlooked. However, over 80% of plant species invest substantial amounts of their below-ground C flow into these fungal symbionts (Smith & Read, 1997; Leake *et al.*, 2004). At the global scale, the annual C flux through soil respiration is ten times greater than fossil fuel combustion and recycles *c.* 10% of atmospheric CO_2 (Raich *et al.*, 2002). Roots and associated mycorrhizas are the single most important component of this flux. Knowledge of the mycorrhizal contribution to the C cycle is of increasingly paramount importance, as this component is likely to be among the most sensitive to ongoing anthropogenic disturbance of both C (Staddon *et al.*, 2002) and nitrogen (N) biogeochemical cycles (Nilsson & Wallander, 2003).

The major biomes are dominated by plants with one of three kinds of mycorrhiza (Fig. 8.1), each of which is adapted to the particular vegetation and soil characteristics of the bioclimatic regions in which it is most important (Read *et al.*, 2004). Central to understanding the contributions of mycorrhizas to the plant–soil–atmosphere continuum of the C cycle is appreciation of the properties and functions of these three major types of mycorrhiza and the ecosystems in which they are of greatest importance. In the heathlands and northern tundra, ericaceous plants (*Cassiope, Vaccinium, Erica, Calluna, Rhododendron*) with ericoid mycorrhizas (ERM) are frequently dominant. To the south, the tundra grades into the

Fungi in the Environment, ed. G. M. Gadd, S. C. Watkinson & P. S. Dyer. Published by Cambridge University Press.

taiga or boreal coniferous forest, where ericaceous plants remain major but sub-dominant components, mainly under the cover of ectomycorrhizal (EM) coniferous trees (*Pinus, Picea*) and some very hardy ectomycorrhizal species in the deciduous genera *Larix, Betula* and *Salix*. The boreal forest forms the world's largest vegetation system, stretching as a continuous circumpolar belt 1000–2000 km wide around the Northern Hemisphere. The southern margins of this vast region grade into temperate deciduous forest, which includes both ectomycorrhizal (e.g. *Fagus, Quercus, Betula*) and arbuscular (AM) mycorrhizal (e.g. *Fraxinus, Acer*) trees, with an understorey increasingly dominated to the south by herbs and grasses with AM. The latter kind of mycorrhiza is of greatest importance throughout temperate grasslands and sub-tropical and tropical regions in savannas and rainforests, although some of the locally dominant tree species in these regions have EM.

The extent of soil carbon sequestration varies widely and systematically across these major biomes (Fig. 8.2), being closely correlated with a number of variables including vegetation, mycorrhiza, soil and microbial characteristics as well as climate. The most important biomes for soil C sequestration are the heathland and boreal forest regions that together cover approximately 70% of the vegetated land mass of the Northern Hemisphere and contain the majority of the world soil C stocks, mainly

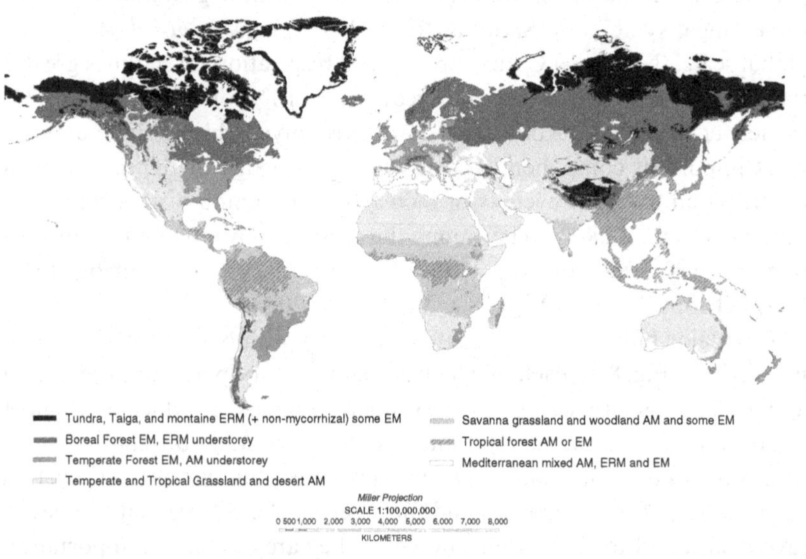

Fig. 8.1. The distribution and extent of the three dominant types of mycorrhiza in the major biomes. Adapted from Read *et al.* (2004) and USDA Natural Resources and Conservation Service maps (http://soils.usda.gov/use/worldsoils/mapindex/soc.html).

in peat and raw humus. There is now increasing evidence that variations in the dominant types of mycorrhizal association in the major biomes are important drivers of ecosystem processes, and of C cycling and storage in particular (Read *et al.*, 2004).

This chapter reviews the evidence that mycorrhizal fungi make significant direct and indirect contributions to soil C stores. It affirms their importance in C cycling by consuming 10–30% of net photosynthesis in some grassland and forest ecosystems, and the importance of mycorrhizal mycelia as a major, and distinct, component of below-ground carbon fluxes. Mycorrhizas are shown to make large direct contributions to soil carbon pools and are implicated in the processes of carbon sequestration in boreal forest and peatland ecosystems that are major terrestrial carbon sinks. It highlights the positive feedback mechanisms whereby mycorrhizas affect both biogeochemical cycles and plant community composition. It also shows that the magnitude of the below-ground C energy flow from recent photosynthate through mycorrhizal mycelial networks in soil has been so large that it has driven the evolution of many plants that exploit this source of C to support their establishment and growth. More than 10% of plant species depend upon 'cheating' mycorrhizal fungal partners out of carbon in order to establish from seed. The importance of mycorrhizas in carbon cycling demands radical re-appraisal of ecosystem C flux models and concepts of plant competition.

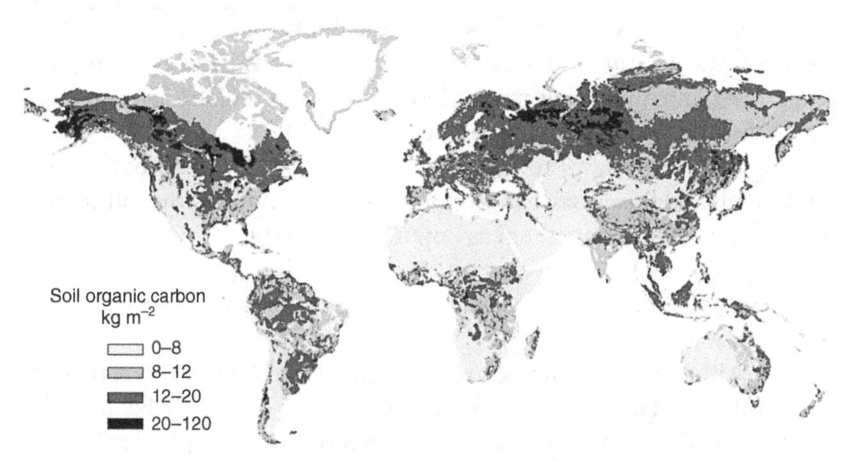

Fig. 8.2. The distribution and amounts of soil organic carbon stocks. Adapted and simplified from the United States Department of Agriculture Natural Resources Conservation Service, Soil Survey Division World Soil Organic Carbon Map (http://soils.usda.gov/use/worldsoils/mapindex/soc.html).

Direct effects of mycorrhizas on soil C inputs and fluxes

Several factors have constrained progress in distinguishing between mycorrhizal and saprotrophic microbial biomass in soil and the contribution of mycorrhizal mycelium to soil C fluxes. The standard substrate-induced respiration technique used to quantify microbial biomass (Anderson & Domsch, 1978) is applied to sieved soil samples in which mycorrhizal mycelia are fragmented and detached from their life-supporting plant carbohydrate supply. The latter alone, within a few hours, causes the mycorrhizal contribution to soil respiration to decrease by 60%–95% (Söderström & Read, 1987). External AM mycelium is unable to assimilate exogenous sugar (Pfeffer *et al.*, 1999) so the substrate-induced respiration systematically discriminates against detecting AM contributions to microbial biomass.

Mycorrhizal mycelia are, nonetheless, major components of soil microbial biomass, being supported by receiving a quality and quantity of carbohydrate supply directly from plants that is unparalleled among soil microbial populations (Leake *et al.*, 2004). The flow of recent assimilate into and through these fungi makes them a primary source of C input into soil, but one that is chemically distinct from that of their host plants (e.g. cell walls of chitin rather than cellulose, melanin rather than lignin, etc.). By adding new C to the soil, mycorrhizal mycelia are also functionally distinct from saprotrophs, even if their necromass is chemically similar: saprotrophs only decrease and alter the existing detrital C stocks in the soil rather than adding to them. These functional differences are reflected in distinct $\delta^{13}C$ signature differences between mycorrhizal and saprotrophic fungi, the symbionts being much closer to the $\delta^{13}C$ values of their host plants than the saprotrophs (Högberg *et al.*, 1999).

A key question central to understanding mycorrhizal contributions to soil C pools and fluxes is: what proportion of the C entering the soil comes through mycorrhizal mycelium as opposed to roots?

Contribution of ericoid mycorrhizal fungi to soil C pools and fluxes

Although the combined respiration of roots and mycorrhizas of plants with ERM dominates the respiratory fluxes from soils in heathlands and tundra (Fig. 8.3a), the relative contributions of these physically and functionally linked components remains uncertain. The hair roots of ericaceous plants contain extensive fungal coils of their mycorrhizal symbionts, which typically occupy 10%–80% of the volume of these roots. However, because the uninfected root cortical cells are largely vacuolate, and the fungal structures turn over faster than do the roots (Read, 1996), the relative importance of the fungus as a sink for plant photosynthate is likely to be much larger than suggested by either fractional volume or biomass measures of roots. Estimates of C allocation to fine roots and mycorrhizas in the field by using ^{14}C pulse-labelling of subarctic tundra dominated by ericaceous plants have shown very marked seasonality with peak values in the autumn, the mycorrhizal hair roots being the dominant sink at that time (Olsrud & Christensen, 2004).

The importance of ERM in soil C fluxes remains uncertain and serves to illustrate the difficulties of making estimates of the mycorrhiza component. Measurement of ^{14}C allocation into ERM hair roots is particularly difficult because in cleaning and extracting these delicate roots from soil the thin-walled root cortical cells containing the dense coils of mycorrhizal mycelium are often detached from roots more easily than from the adhering litter. Olsrud *et al.* (2004) found a very strong linear correlation between ecosystem net rates of C fixation and the concentration of ergosterol (a fungal sterol) in hair roots one week later through the growing season. This indicates a strong dependency of the active mycorrhizal fungal biomass upon rates of C fixation by plants, and implies that the C demand by the fungi must be sufficiently large to be not fully met when rates of photosynthesis are sub-optimal. By using field ^{14}C-pulse labelling of the ericaceous plants, only 20%–40% of ^{14}C allocation below ground was found in ericoid mycorrhizal hair roots (Olsrud *et al.*, 2004). At an adjacent mire site, Olsrud & Christensen (2004), using the same approach, had found that approximately 3%–10% of the total ^{14}C assimilated by the plants was found in hair roots located in the top 5 cm of soil.

Based on a number of assumptions, Olsrud (2004) suggested that approximately 50% of the assimilated C reaching hair roots of ericaceous plants passes to the fungi, and that at its seasonal peak the allocation to mycorrhiza accounted for up to 4.8% of net fixation. These estimates are not based on the actual ^{14}C flux through the roots and do not include any allocation to external mycelium, but are based only on the peak amount of ^{14}C present in roots over a series of post-label harvests. The estimates must significantly under-estimate the true C costs of the mycorrhiza. The proportion of C incorporated in mycelial biomass or lost to respiration and exudation is not known for ericoid mycorrhizas under field conditions, but studies of AM have found that the respiratory losses of C can be several times larger than instantaneous amounts of C in mycelium and their exudates (Johnson *et al.*, 2002a,b). The high rate of C turnover seen in

AM (Staddon *et al.*, 2003) and EM (Ek, 1997) fungi are almost certainly a feature too of ERM. Coupled to the dynamic seasonal patterns of mycorrhiza activities that peak at the end of the growing season, and which are time-lagged from rates of photosynthesis in the preceding weeks (Olsrud, 2004), estimation of the annual C flux budgets for mycorrhizas present a formidable challenge.

Contribution of ectomycorrhizal fungi to soil C pools and fluxes

The quantities of C passing into EM typically range from 10% to 30% of net photosynthate in many forest ecosystems (Leake *et al.*, 2004). Given the much greater productivity of boreal forests than of Arctic heathlands, the total quantity of mycorrhizal C inputs into the former is much larger than that into the latter. Because ectomycorrhizas entirely sheath root tips, over 90% of which are normally symbiotic at any time, virtually all the labile C released from living roots in the major forest regions dominated by ectomycorrhizal plants passes through the fungi. A conservative estimate of the contribution of ectomycorrhizas to soil microbial biomass in boreal forest soil is 32%, based on field observations following a large-scale girdling experiment (Högberg & Högberg, 2002). Soil samples from the forest floors of stands *Picea abies* and mixed *Picea–Quercus robur*, incubated for 5 weeks, showed a decrease of between 47% and 84% of the total fungal biomass. This decrease is assumed to be primarily due to the senescence of mycorrhizal mycelium no longer supported by plant photosynthate (Bååth *et al.*, 2004). This result suggests that EM mycelium dominates fungal biomass in these forests. At least 50% of the soil respiration and half the dissolved organic C in soil solution measured by Högberg & Högberg (2002) was attributed to mycorrhizal mycelium plus roots, all of these findings being consistent with the trend for root and mycorrhizal components to dominate soil respiration in the more northern latitudes such as boreal forests (Fig. 8.3a). Using sand-filled hyphal in-growth bags inserted into coniferous and mixed deciduous/coniferous forest plots, Wallander *et al.* (2004) estimated the production of external EM in soil (i.e. not including the sheath on the roots) to be 420–590 kg $ha^{-1}year^{-1}$. In a Swedish *Pinus sylvestris* and *Picea abies* forest, Wallander *et al.* (2001) estimated the combined biomass of EM mycelia and roots to be 700–900 kg ha^{-1}, and on the basis of previous estimates of the biomass of EM mantles (Kårén and Nylund, 1997) it was concluded that approximately 80% of the EM biomass was extra-radical mycelium. EM mycelium was estimated to account for over 70% of the total soil microbial biomass.

It is not only the quantity of C passing into and through EM that is important for the global C cycle; it is the quality and persistence of C inputs that determines the extent of sequestration in soil. Some EM fungi, particularly in boreal forests with a summer dry season, have been found to secrete copious amounts of nutrient-mobilizing low-molecular-mass organic acids such as oxalic acid and citric acid, and calcium oxalate can form extensive crystalline deposits coating mycorrhizal hyphae and rhizomorphs (Wallander *et al.*, 2002). Such exudates contribute to the supply of low-molecular-mass organic compounds in boreal forest soils that are dominated by root- and mycorrhiza-derived inputs (Högberg & Högberg, 2002).

Persistence of mycorrhizal fungal C after death will be facilitated by the very high C:N ratios of EM mycelium, which are similar to, or even greater than, that of Norway spruce forest organic matter and range from 18.5 to 21.9 (Wallander *et al.*, 2003). These values are unusually high for a component of soil microbial biomass. Live and dead mycelium was not distinguished in this study and it is possible that the latter is very highly depleted in N owing to efficient internal recycling of this critical element, which typically limits plant and microbial activity in boreal forests. Many EM fungi produce robust hydrophobic multicellular hyphal cords that are persistent for months, and some produce highly melanized hyphae that decay very slowly. The best example of the latter is *Cenococcum geophillum*, whose heavily melanized hyphae and sclerotia are produced in abundance under dry soil conditions and are implicated directly in soil C sequestration (Meyer, 1964, 1987).

There is also evidence that mycelial systems of EM fungi are better defended against being eaten by fungal-feeding animals than are many of the saprotrophs, and this may relate both to properties such as melanin and polyphenolic compounds found in mycelia of some EM fungi, and also to more specific toxins. Some EM fungi have been shown to kill fungivorous microarthropods, apparently by an immobilizing toxin, and predate them as a source of nitrogen (Klironomus & Hart, 2001). Such activities will serve to reduce the rate of decomposition and cycling of C from EM mycelium in comparison with less well-defended soil fungi. Factors such as these may together directly affect the persistence of mycorrhiza-derived C in soil. There is now evidence from a short-term (2 weeks) litter-bag incubation study with *Pinus edulis* roots that mycorrhizal colonization resulted in a 30% lower rate of decomposition of C than found for non-mycorrhizal roots, thereby increasing short-term soil C storage (Langley & Hungate, 2003).

Contribution of AM fungi to soil C pools and fluxes

Rates of soil organic matter decomposition tend to be higher in ecosystems dominated by AM plants than in those dominated by plants with ERM or EM (Fig. 8.3b). As a consequence, the combined contributions of roots plus mycorrhizas of AM plants to soil respiration are generally less than for the two other kinds of mycorrhiza (Fig. 8.3a).

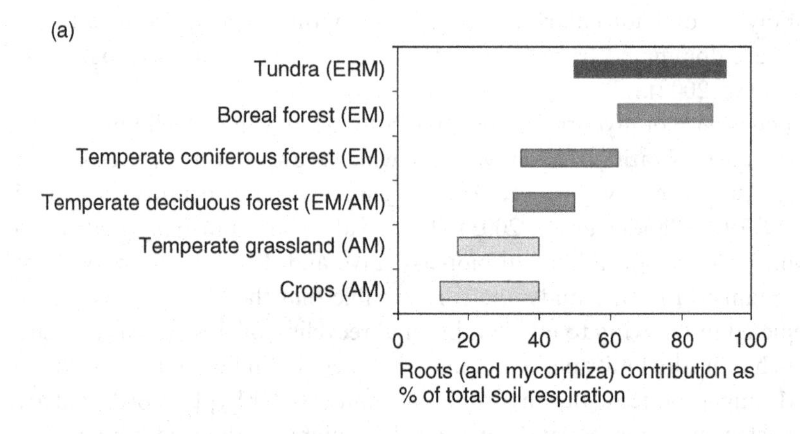

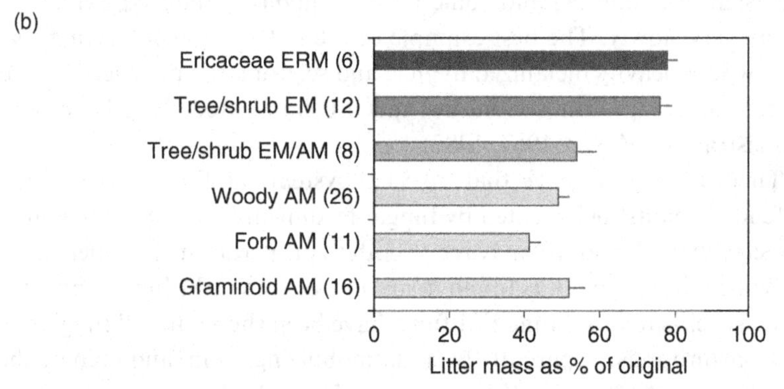

Fig. 8.3. Associations between biome and plant functional groups, their types of mycorrhiza and C cycling processes. (a) Variation in the typical range of the percentage total soil respiration that is due to roots (and associated mycorrhizas) across the main biome types. Data plotted from Raich & Tufekcioglu (2000). (b) The percentage of original mass remaining in litter bags containing litter from plants of different functional types and different kinds of mycorrhiza after a 20 week incubation under standard conditions in the same soil. Numbers in parentheses refer to the number of replicate species in each category. Error bars show 1 standard error of the mean. Data re-calculated from Cornelissen *et al.* (2001).

The pioneering laboratory studies of Jakobsen & Rosendahl (1990) and Pearson & Jakobsen (1993), in which they quantified C allocation from pot-grown cucumber plants into mycorrhiza and its external mycelium by using pulse-labelling, suggested that although the total C cost of mycorrhiza accounted for up to 20% of net fixation less than 1% was found in external mycelium. Such findings could be used to justify the omission of mycorrhizas from current grassland C flux models that assume that root exudation, sloughed cells and dead roots provide the only significant pathways flow of plant-fixed C into soil (see, for example, Toal *et al.*, 2000; Kuzyakov *et al.*, 2001; Jones & Donnelly, 2004). It might also be used to justify farmland food-web diagrams that assume that the only significant fungal components in the soil are the saprotrophs (de Ruiter *et al.*, 2002).

The contribution of AM to soil C fluxes should not be so quickly dismissed. It is now apparent that for various reasons the early measurements of C allocation to AM mycelium in soil will have significantly under-estimated this component, as the rates of C flux through AM hyphae are very high but not easily quantified. Up to 16% of the C content of AM mycelia can be replaced within 24 h by new assimilate from a 5 h labelling (Staddon *et al.*, 2003). The abundance of AM fungal mycelia in soil strongly supports the view that they are functionally important components of the C cycle. Their mycelia have been estimated to make up 50% of the fungal length in soil under herbaceous plants (Rillig *et al.*, 2002). They account for 20%–30% of the total soil microbial biomass in temperate grassland, being particularly important in soils of inherently low organic matter content (Miller & Kling, 2000; Olsson & Wilhelmsson, 2000).

Total C allocation to AM external mycelium in upland temperate grassland has recently been quantified by pulse-labelling of the turf, both in the field with ^{13}C and in turf monoliths transported to a laboratory with ^{14}C (Johnson *et al.*, 2002a, b). These studies suggest that the C allocation through AM mycelia in permanent grassland is an order of magnitude higher than the earlier estimates of Pearson & Jakobsen (1993) based on pot-grown cucumbers. Combining estimates of $^{13}CO_2$ respiration from AM mycelium with ^{14}C allocated to AM biomass indicated that external mycelium received more than 9% of net C fixation by grassland in summer, most of this C being received within the first 3 days after plant photosynthesis (Johnson *et al.*, 2002a, b). This is significantly greater than the typical C inputs to grasslands through herbivore dung (Jones & Donnelly, 2004) and does not include C allocation to fungal mycelium within the plant roots, which will typically add an additional 4%–10% of net fixation (Smith & Read, 1997). Although the respiratory release of labelled C from AM hyphae is now known to be maximal in 12–16 h after pulse-labelling of the plants (Johnson *et al.*, 2002a), the optimal time for sampling mycorrhizal mycelium for *total* labelled C remains uncertain, but appears to be soon after labelling, probably within the first 24 h (see Staddon *et al.*, 2003). It will be governed by the balance between the rates of C transfer, the rates of respiration and the rates of C incorporation into hyphal C storage and growth. Many pulse-labelling studies either have used relatively long fixation periods, which make it difficult to establish the chronology of carbon fluxes, or have taken the first samples many days after labelling (see, for example, Stewart & Metherell, 1999) and so may miss the importance of rapid carbon transfers and fluxes such as those through AM and roots that happen sooner than this. This problem is not unique to AM; a fuller knowledge of the rates, routes and fate of C fluxes through mycelium of the three major types of mycorrhiza is required to avoid these difficulties in future.

As with EM, there are recalcitrant components of AM fungi that contribute significantly to soil C sequestration. The main arterial hyphae in the mycelium of AM fungi in soil have highly thickened cell walls (Nicholson, 1959). They are much less palatable to fungal-feeding collembolans than are saprotrophic fungi (Gange, 2000) and their C inputs to soil appear to be particularly persistent. One apparently unique and quantifiable product of AM hyphae is the hydrophobic protein glomalin, which is secreted into soil and has a major role in soil aggregation. This protein strongly accumulates in soil with a residence time of 6–42 years, thereby contributing as much as 15% of the total stable soil organic carbon pool in grasslands (Miller & Kling, 2000) and 4%–5% of that in tropical rainforest soils (Rillig *et al.*, 2001; Zhu & Miller, 2003).

Indirect effects of mycorrhizas on soil C inputs and fluxes

The indirect effects of mycorrhizas on C cycling through soil are likely to be of even greater importance than their direct effects, but are more difficult to quantify. There are important functional distinctions between ERM, EM and AM and their host plants that have direct bearing upon soil C sequestration (Figs. 8.1–8.4). Plants with ERM produce litter that is among the most extremely recalcitrant, being highly enriched in polyphenolic compounds, low-molecular-mass phenolics and aliphatic acids that are fungistatic and fungitoxic (Jalal *et al.*, 1982; Jalal & Read, 1983). This litter has extremely high C:N ratios and decomposes very slowly (Figs. 8.3b, 8.4) (Cornelissen *et al.*, 2001; Read, 1991). ERM are highly adapted to the litter produced by their host plants, not only in tolerating and being able to metabolize some of the most toxic phenolic compounds (Leake & Read, 1991a) but also in the range of oxidative and hydrolytic enzymes they express to mobilize N and P from litter (Leake & Read, 1997). The enzymatic activities of these mycorrhizas enable them to effectively mobilize N from host plant litter (Kerley & Read, 1998) and to recycle N contained in the structural polymers of their own fungal cell walls (Leake & Read, 1991b; Kerley & Read, 1998). The key features of these processes are that they involve selective and highly effective exploitation of both N and P (Myers & Leake, 1996), with very incomplete decomposition of lignocellulose, resulting in the accumulation of recalcitrant humic compounds. The effectiveness of N exploitation by ERM is often reflected in an absence of detectable mineral N (nitrate or ammonium) in soils under ericaceous plants (see, for example, Adams, 1986).

These same processes also occur under trees in the boreal forest, with EM mycelia providing intensive (Donnelly *et al.*, 2004), spatially precise nutrient foraging (Perez-Moreno & Read, 2000) through decaying litter that is rich in polyphenolics and tannins (Northup *et al.*, 1995). The litter of boreal and temperate forest trees is generally less recalcitrant and nutrient-poor than the litter of ericaceous plants (Figs. 8.3b, 8.4); these differences are reflected in the more limited abilities of EM compared with ERM fungi to use the protein–phenol complexes and other semi-recalcitrant sources of nutrients (Leake & Read, 1997). Nonetheless, through use of organic N such as amino acids and proteins, EM can also short-circuit ammonification and nitrification pathways of the N cycle (Leake *et al.*, 2004) resulting in very low rates of inorganic N release into

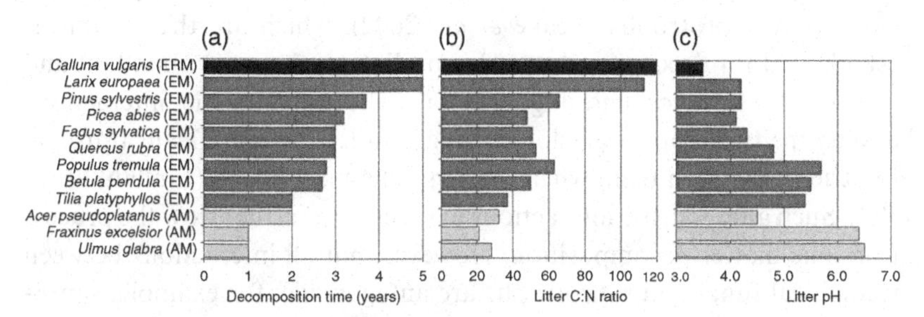

Fig. 8.4. The association between mycorrhiza type (ERM, black; EM, dark grey; and AM, light grey) and plant litter decomposition rates and measures of litter quality. (a) Time taken for litter to decompose for plants with different mycorrhiza types. (b) The association between C:N content of plant litter and mycorrhiza type. (c) The association between litter pH and mycorrhiza type. Data plotted from Read (1991).

the soil. In mycorrhizas competing with saprotrophs for the limiting resource of labile nutrient sources, their large supply of host photosynthate, especially in EM fungi, gives them a major advantage in both biomass (see earlier) and absorptive surface area (Leake *et al.*, 2004). By selective nutrient removal (Nilsson & Wallander, 2003) and their limited abilities to degrade the main plant litter components of lignocellulose or the accumulating soil humus, they increase the already high C:N and C:P ratios of the remaining residues (Bending & Read, 1995) and further their recalcitrance. By these positive feedback mechanisms both ERM and EM restrict the supplies of the mineral nutrients that are most limiting to the activities of the saprotroph decomposer communities, and thereby contribute significantly to the processes of soil C sequestration in heathland and boreal forest ecosystems (Read *et al.*, 2004). In addition to competition for nutrients, the drying of soil under trees by water extraction by EM has recently been implicated in reduced litter decomposition in the presence of EM networks (Koide & Wu, 2003).

AM fungi also affect processes of soil C sequestration, but in a different way. Although there is currently no evidence that AM fungi directly affect decomposer organisms, their effects on stabilization of soil macroaggregates (more than 2 mm in diameter) through both the activities of hyphae and their glomalin secretions (Rillig *et al.*, 2002) is implicated in increased soil C storage (Zhu & Miller, 2003). Organic matter entrapped inside macroaggregates appears to be biologically and chemically stabilized to a greater degree than that in the bulk soil (Jones & Donnelly, 2004).

In addition to the competitive interactions for nutrients between mycorrhizas and saprotrophs, there is evidence of direct antagonism between the two groups of fungi. EM mycelia can sometimes territorially exclude wood-decay saprotrophs (Leake *et al.*, 2002), which are the organisms best adapted for decomposition of lignocellulose, the most abundant and recalcitrant C source entering the soil. The antagonistic interactions between the two trophic groups of fungi can lead to significant transfers of nutrients between them when their mycelia meet in soil (Lindahl *et al.*, 1999). Such antagonistic interactions may serve to further slow the process of organic matter decomposition. However, not all interactions between mycorrhizal fungi and saprotrophs are antagonistic: for example, saprotrophs facilitate N recovery by EM from otherwise recalcitrant protein–phenol complexes (Wu *et al.*, 2003).

Little appears to be known about antagonism between ERM and saprotrophs, although there is much circumstantial evidence that ERM fungi in combination with the 'allelopathic' and nutrient-poor decaying litter of their host plants are directly antagonistic to some EM fungi and may contribute to suppression of EM trees in the boreal regions (Mallik, 2003). It is inconceivable that the exceptionally effective capture of N and P by ERM in nutrient-impoverished heathland and tundra environments does not affect the functioning of the saprotrophic decomposer communities.

The host plants of AM plants typically produce litter of low recalcitrance and comparatively high N content that decomposes relatively quickly (Figs. 8.3b, 8.4). As a consequence, there is generally much less accumulation of C in soils under AM plants (Figs. 8.1, 8.2) and no evidence of strong antagonisms between AM fungi and saprotrophs. The biomass of AM mycelia in soil is typically much lower than that of EM, so their effects on other organisms would be expected to be less; this expectation is supported by studies showing little or no effects of AM mycelia on bacteria and other fungi (see, for example, Olsson & Wilhelmsson, 2000).

Together, the observations on the three main types of mycorrhiza that dominate the major biomes and the properties of their host plants suggest strong functional interdependencies between the properties of plant litter, the adaptive features of the mycorrhizal fungi and the nature of the soils that develop in different bioclimatic regions that define the world biomes. The extent of soil C sequestration (Fig. 8.2) is strongly correlated with these factors, and mycorrhizas are now doubly implicated in soil C storage through both direct inputs of C and indirect effects on decomposition processes (ERM and EM) and soil aggregation (AM).

Effects of mycorrhizal C fluxes on plant community composition

A substantial proportion of the large C-energy flux that passes from plants to mycorrhizas passes out of roots into mycorrhizal mycelial networks (Leake *et al.*, 2004); these networks often interconnect roots of different species (Simard & Durall, 2004). The basis of the mycorrhizal symbiosis is the reciprocal exchange of plant carbohydrate in return for soil nutrients absorbed by the fungi. In linking to communal mycorrhizal mycelial networks, individual plants may be in a position to cheat mycorrhizal fungi and neighbouring plants by gaining more nutrients than they have ‘paid for’ in C, and in some cases may even steal C from the fungi. There is now compelling evidence that such cheating and stealing does occur to varying degrees in natural ecosystems and has important implications for plant community composition and ecosystem functioning.

Direct effects of C flow through mycorrhizal mycelial networks on plant communities

The magnitude of the C flux through mycorrhizal mycelial networks, particularly in deeply shaded forest floor environments, has driven the evolution of holoparasitic plants that exploit mycorrhizal fungi as their sole source of C. In total, over 400 achlorophyllous plants, termed mycoheterotrophs, are known that depend exclusively on fungi for their C by exploiting mycorrhiza-like relationships in their underground roots and tubers (Leake, 1994). These plants have originated from multiple independent lineages of green plants and are found in approximately 90 genera that range from liverworts to dicotyledons and monocotyledons (Leake, 2004). By using DNA-based identification, the fungal partners of some of these plants have been recently identified for the first time; most have proved to be associated, with exceptional specificity, with EM (Taylor *et al.*, 2002) or AM fungi (Bidartondo *et al.*, 2002) that are co-infecting the roots of adjacent green plants. Pulse-labelling studies, and biomass measurements in experimental microcosms with and without mycelial interconnections between autotrophs and mycoheterotrophs, have confirmed that carbon is transported via co-linking mycorrhizal mycelium (McKendrick *et al.*, 2000; Bidartondo *et al.*, 2003).

Plant exploitation of fungal C has been a defining feature of the establishment phases of many plants that produce extremely large numbers of tiny dust-seeds, best exemplified by the hugely successful family Orchidaceae, which contains the largest number of species (*c.* 17 500) and is distributed globally through virtually all terrestrial habitats. However, it is also a feature of the establishment of many ferns and clubmosses that have achlorophyllous subterranean gametophytes that are nourished by mycoheterotrophy (Leake, 1994). In total about 10% of plant species depend on exploitation of fungal C (Leake, 2004), but the proportion of these that exploit the mycorrhizal fungi of autotrophs is uncertain. Nonetheless, for the majority of species whose fungal partners have been determined by DNA-based analysis, or definitive experimentation, we see an emerging picture of associations with mycorrhizal fungi of autotrophic plants as the rule rather than the exception (Table 8.1). Mycoheterotrophs have been found to exploit a very wide phylogenetic breadth of mycorrhizal fungi, ranging from Glomeromycota to Basidiomycota and Ascomycota (Table 8.1).

Although all orchids require fungal C for germination and establishment, and the family Orchidaceae accounts for nearly half the species of mycoheterotroph that never photosynthesize, most orchids produce green

Table 8.1. *Fungal partners of mycoheterotrophs and their sources of carbon*

AM, arbuscular mycorrhiza; EM, ectomycorrhiza. In cases of fungi known to be obligate mycorrhizal symbionts of autotrophs these are assumed (*) to provide the carbon sources for these fungi.

Fungus	Fungus group	Mycoheterotrophic plant (M, Monocotyledonae; D, Dicotyledonae; H, Hepaticae)	Plant family	Source of carbon for the fungus	Reference
Arbuscular mycorrhizal fungi					
Glomus Group A	Glomeromycota	*Voyria tenuiflora* (D) *V. aurantiaca* (D) *V. corymbosa* (D) *V. rosea* (D) *V. caerulea* (D)	Gentianaceae	AM with tropical forest trees*	Bidartondo *et al.* (2002)
Glomus Group A	Glomeromycota	*Voyriella parviflora* (D)	Gentianaceae	AM with tropical forest trees*	Bidartondo *et al.* (2002)
Glomus Group A	Glomeromycota	*Arachnitis uniflora* (M)	Corsiaceace	AM with tropical forest trees*	Bidartondo *et al.* (2002)
Glomus sp.	Glomeromycota	*Sciaphila tosaensis* (M)	Triuridaceae	AM with tropical forest trees*	Yamato (2001)
Glomus Group A	Glomeromycota	*Gymnosiphon minutus* (M)	Burmanniaceae	AM with tropical forest trees*	Dirk Redecker (personal communication)
Heterobasidiomycetes					
Sebacina sp.	Sebacinales	*Neottia nidus-avis* (M)	Orchidaceae	EM with a range of temperate trees including *Fagus sylvatica* and *Corylus*	Selosse *et al.* (2002); McKendrick *et al.* (2002)
Sebacina spp.	Sebacinales	*Hexalectris revoluta* (M)	Orchidaceae	Probably EM on temperate trees	Taylor *et al.* (2003)

Table 8.1. (cont.)

Fungus	Fungus group	Mycoheterotrophic plant (M, Monocotyledonae; D, Dicotyledonae; H, Hepaticae)	Plant family	Source of carbon for the fungus	Reference
Sebacina spp.	Sebacinales	*Hexalectris spicata* (M)	Orchidaceae	Probably EM on temperate trees	Taylor *et al.* (2003)
Thanatephorus ochraceus	Ceratobasidiales	*Hexalectris spicata* (M)	Orchidaceae	Unknown; the fungus may be a casual invader of the orchid	Taylor *et al.* (2003)
Thanatephorus gardneri	Ceratobasidiales	*Rhizanthella gardneri* (M)	Orchidaceae	EM on a myrtaceous shrub	Warcup (1991)
Tulasnella spp.	Tulasnellales	*Cryptothallus mirabilis* (H)	Aneuracaceae	EM with *Betula* and other trees	Bidartondo *et al.* (2003)
Homobasidiomycetes					
Tomentella/ Thelephoraceae: c. 14 spp.	Thelephorales	*Cephalanthera austinae* (M)	Orchidaceae	EM with surrounding trees	Bidartondo *et al.* (2003)
Tomentella spp.	Thelephorales	*Corallorhiza striata* (M)	Orchidaceae	EM	Taylor *et al.* (2002)
Tomentella spp.	Thelephorales	*Corallorhiza trifida* (M)	Orchidaceae	EM with *Salix repens* and *Betula pendula*	McKendrick *et al.* (2000)
Russula: c. 20 spp.	Russulales:	*Corallorhiza maculata* (M)	Orchidaceae	EM with trees	Taylor & Bruns (1997)
Russula: c. 3 spp.	Russulales	*Corallorhiza mertensiana* (M)	Orchidaceae	EM with trees	Taylor & Bruns (1999)
Russulaceae: c. 2–3 spp.	Russulales	*Monotropa uniflora* (D)	Ericacaceae	EM with deciduous and coniferous trees	Young *et al.* (2002)
Gauteria monticola group	Gautieriales	*Pleuricospora fimbriolata* (D)	Ericacaceae	EM with trees	Bidartondo & Bruns (2002)

Rhizopogon salebrosus group	Boletales	*Pterospora andromedea* (D)	Ericacaceae	EM with trees	Bidartondo & Bruns (2002)
Rhizopogon ellenae group	Boletales	*Sarcodes sanguinea* (D)	Ericacaceae	EM with trees	Bidartondo & Bruns (2002)
Tricholoma magnivelare	Agaricales	*Allotropa virgata* (D)	Ericacaceae	EM with trees	Bidartondo & Bruns (2002)
Tricholoma: c. 2 spp.	Agaricales	*Pityopus californicus* (D)	Ericacaceae	EM with trees	Bidartondo & Bruns (2002)
Tricholoma: more than 12 spp.	Agaricales	*Monotropa hypopitys* (D)	Ericacaceae	EM with trees	Bidartondo & Bruns (2002)
Mycena osmundicola	Agaricales	*Gastrodia elata* (M)	Orchidaceae	Litter decay	Lan *et al.* (1994)
Armillaria mellea	Agaricales	*Gastrodia elata* (M)	Orchidaceae	Wood decay and parasitism of trees	Lan *et al.* (1994)
Armillaria jezoensis	Agaricales	*Galeola septentrionalis* (M)	Orchidaceae	Wood decay and parasitism of trees	Rasmussen (2002)
Erythromycetes crocicreas	Hymenochaetales	*Galeola altissima* (M)	Orchidaceae	Wood decay	Rasmussen (2002)
Erythromycetes crocicreas	Hymenochaetales	*Erythrorchis ochobiense* (M)	Orchidaceae	Wood decay	Rasmussen (2002)
Euascomycetes					
Tuber excavatum	Tuberales	*Epipactis microphylla* (M)	Orchidaceae	EM with *Corylus* and *Fagus sylvatica*	Selosse *et al.* (2004)

leaves and may no longer depend on fungal C when established. Green orchids during their initial mycoheterotrophic phase of establishment normally depend upon exploitation of soil fungi that include saprophytes and root-pathogens of autotrophic plants such as fungi in the form genus *Rhizoctonia*, and are not normally associated with mycorrhizal symbionts of other plants. However, evidence has accumulated in the past few years that most fully mycoheterotrophic orchids have switched fungal partners to associate with fungi forming EM on green plants (McKendrick *et al.*, 2002; Taylor *et al.*, 2002) that are therefore able to supply large amounts of C over long periods of time. This is not unique to orchids. The mycoheterotrophic liverwort *Cryptothallus mirabilis* depends upon a *Tulasnella* species, a genus of fungus often found in mycorrhizal partnership with green orchids, but in this tripartite association it forms ectomycorrhizal associations with birch trees (Bidartondo *et al.*, 2003). The achlorophyllous gametophytic stages of many ferns and clubmosses appear to depend upon AM fungal symbionts that must obtain their C from adjacent autotrophic plants (Leake, 1994). Both in these plants, and green orchids, once their long-lived autotrophic adult plants are established, it is possible that the initial C investment by the fungi might be 'paid back' many times over by new photosynthate.

There is increasing evidence that obligate mycoheterotrophy represents one extreme end of a continuum of carbon transfer from fungi to plants, and the multiple evolutionary origins of this form of nutrition are consistent with the view that a large number of plants can gain C, at least transiently, by this pathway. Fully mycoheterotrophic orchids that exploit EM mycelial networks have distinct ^{15}N and ^{13}C enrichment (Trudell *et al.*, 2003). This isotopic 'fingerprint' of mycoheterotrophs associated with EM fungi has exposed examples of green orchids that have also switched fungal partners to EM associates of trees and which gain most of their N and C from exploiting these partners (Gebauer & Meyer, 2003; Bidartondo *et al.*, 2004). Isotopic differences between plants with C4 and C3 photosynthesis has also been used to demonstrate putative C transfer via AM mycorrhiza linking two green plants: a grass and an invasive weed (Carey *et al.*, 2004). The distinction between autotrophy and mycoheterotrophy clearly cannot be made purely on the basis of presence or absence of chlorophylls alone.

In the absence of their critical fungal partners, mycoheterotrophic plants typically fail to germinate and certainly fail to establish (McKendrick *et al.*, 2000, 2002). In the mycoheterotrophic plants that are linked into mycorrhizal mycelial networks it is these that directly and strictly control the distributions of the plants. Mycoheterotrophs thus exemplify the ability of mycorrhizal networks to transport substantial quantities of carbon and nutrients, and the potential importance of these networks for the control of species composition of natural communities. These processes have potentially far-reaching implications for evolutionary theories applied to plant communities (Simard & Durall, 2004) and for our understanding of plant–fungus co-evolution in mycorrhizal symbiosis.

Indirect effects of C flow through mycorrhizal mycelial networks on plant communities

The C costs and functional 'benefits' to plants of linking to mycorrhizal networks are fungus-specific and, owing to variations in physiology and host-specificity, are not shared equally among plants (Hartnett & Wilson, 2002). Grasses, for example, with their extensive fibrous root systems with long root hairs, are much less mycorrhiza-dependent than are coarser-rooted forbs, but may contribute disproportionately to the C cost of supporting a mycorrhizal mycelial network that most benefits the forbs (Grime *et al.*, 1987). For small-seeded plants like *Centaurium erythraea* growing in a highly P-deficient medium, being linked, shortly after germination, into the established mycorrhizal network already paid for by C supplied by other plants is a prerequisite for survival (Francis & Read, 1994).

One further important indirect effect of C flow through mycorrhizas on plant community composition is the effect of the fungal uptake of soil nutrients that is empowered by host photosynthate. In heathland and boreal forest ecosystems, the almost complete monopoly of labile organic nutrients by ERM and EM may play an important role in suppressing the growth of species with AM that are less well adapted to compete for these nutrients (Weber *et al.*, 2005). These factors contribute to the patterns of global distribution of biomes and mycorrhiza types (Fig. 8.1).

Conclusions

Carbon flow through mycorrhizal mycelial networks is substantial, makes important direct and indirect contributions to soil C sequestration, particularly in boreal forest and heathland biomes, and affects plant community composition and functioning. Although the mycorrhizal contribution to the terrestrial C cycle has proved to be one of the most difficult components to quantify, there is increasing evidence that it is both chemically distinct and quantitatively important and can no longer be ignored.

Although little attention has been directed at studying the C flux through mycorrhizal mycelial networks, the quantities of C-energy passing through these fungi is so large that it has driven the evolution of plants that exploit the networks of both AM and EM fungi as a source of both C and nutrients. These observations challenge conventional views of mycorrhizal functioning and emphasize the need to further our understanding of the complex interlinkages between mycorrhizas, plants, soil and C cycling processes.

Acknowledgements

I gratefully acknowledge funding for the research underpinning this review provided by the NERC: grant nos: NER/A/S/2000/00411 and Soil Biodiversity NER/T/2001/00177.

References

Adams, J. A. (1986). Nitrification and ammonification in acid forest litter and humus as affected by peptone and ammonium N amendment. *Soil Biology and Biochemistry* **18**, 45–51.

Anderson, J. P. E. & Domsch, K. H. (1978). A physiological method for the quantitative measurement of microbial biomass in soils. *Soil Biology and Biochemistry* **10**, 215–21.

Bååth, E., Nilsson, L. O., Goransson, H. & Wallander, H. (2004). Can the extent of degradation of soil fungal mycelium during soil incubation be used to estimate ectomycorrhizal biomass in soil? *Soil Biology and Biochemistry* **36**, 2105–9.

Bending, G. D. & Read, D. J. (1995). The structure and function of the vegetative mycelium of ectomycorrhizal plants. V. The foraging behaviour of ectomycorrhizal mycelium and the translocation of nutrients from exploited litter. *New Phytologist* **130**, 401–9.

Bidartondo, M. I. & Bruns, T. D. (2002). Fine-level mycorrhizal specificity in the Monotropoideae (Ericaceae): specificity for fungal species groups. *Molecular Ecology* **11**, 557–69.

Bidartondo, M. I., Redecker, D., Hijri, I., Wiemken, A., Bruns, T. D., Dominguez, L., Sérsic, A., Leake, J. R. & Read, D. J. (2002). Epiparasitic plants specialized on arbuscular mycorrhizal fungi. *Nature* **419**, 389–92.

Bidartondo, M. I., Bruns, T. D., Weiss, M., Sérgio, C. & Read, D. J. (2003). Specialized cheating of the ectomycorrhizal symbiosis by an epiparasitic liverwort. *Proceedings of the Royal Society of London* **B270**, 835–42.

Bidartondo, M. I., Burghardt, B., Gebauer, G., Bruns, T. D. & Read, D. J. (2004). Changing partners in the dark: isotopic and molecular evidence of ectomycorrhizal liaisons between forest orchids and trees. *Proceedings of the Royal Society of London* **B271**, 1799–806.

Carey, E. V., Marler, M. J. & Callaway, R. M. (2004). Mycorrhizae transfer carbon from a native grass to an invasive weed: evidence from stable isotopes and physiology. *Plant Ecology* **172**, 133–41.

Cornelissen, J. H. C., Aerts, R., Cerabolini, B., Werger, M. J. A. & Van der Heijden, M. G. A. (2001). Carbon cycling traits of plant species are linked with mycorrhizal strategy. *Oecologia* **129**, 611–19.

de Ruiter, P. C., Griffiths, B. & Moore, J. C. (2002). Biodiversity and stability in soil ecosystems: patterns, processes and the effects of disturbance. In *Biodiversity and Ecosystem Functioning, Synthesis and Perspectives*, ed. M. Loreau, S. Naeem & P. Inchausti, pp. 102–13. Oxford: Oxford University Press.

Donnelly, D. P., Boddy, L. & Leake, J. R. (2004). Development, persistence and regeneration of ectomycorrhizal mycelial systems in soil microcosms. *Mycorrhiza* **14**, 37–45.

Ek, H. (1997). The influence of nitrogen fertilization on the carbon economy of *Paxillus involutus* in ectomycorrhizal association with *Betula pendula*. *New Phytologist* **135**, 133–42.

Francis, R. & Read, D. J. (1994). The contributions of mycorrhizal fungi to the determination of plant community structure. *Plant and Soil* **159**, 11–25.

Gange, A. (2000). Arbuscular mycorrhizal fungi, Collembola and plant growth. *Trends in Ecology and Evolution* **15**, 369–72.

Gebauer, G. & Meyer, M. (2003). N-15 and C-13 natural abundance of autotrophic and mycoheterotrophic orchids provides insight into nitrogen and carbon gain from fungal association. *New Phytologist* **160**, 209–23.

Grime, J. P., Mackey, J. M. L., Hillier, S. H. & Read, D. J. (1987). Floristic diversity in a model system using experimental microcosms. *Nature* **328**, 420–2.

Hartnett, D. C. & Wilson, G. W. T. (2002). The role of mycorrhizas in plant community structure and dynamics: lessons from grasslands. *Plant and Soil* **244**, 319–31.

Högberg, M. N. & Högberg, P. (2002). Extramatrical ectomycorrhizal mycelium contributes one- third microbial biomass and produces, together with associated roots, half the dissolved organic carbon in a forest soil. *New Phytologist* **154**, 791–5.

Högberg, P., Plamboeck, A. H., Taylor, A. F. S. & Fransson, P. M. A. (1999). Natural ^{13}C abundance reveals trophic status of fungi and host-origin of carbon in mycorrhizal fungi in mixed forests. *Proceedings of the National Academy of Sciences of the USA* **96**, 8534–9.

Jakobsen, I. & Rosendahl, L. (1990). Carbon flow into soil and external hyphae from roots of mycorrhizal cucumber plants. *New Phytologist* **115**, 77–83.

Jalal, M. A. F. & Read, D. J. (1983). The organic acid composition of *Calluna* heathland soil with special reference to phytotoxicity and fungitoxicity. 2. Monthly quantitative determination of the organic acid content of *Calluna* and Spruce dominated soils. *Plant and Soil* **70**, 273–86.

Jalal, M. A. F., Read, D. J. & Haslam, E. (1982). Phenolic composition and its seasonal variation in Calluna vulgaris. *Phytochemistry* **21**, 1397–401.

Johnson, D., Leake, J. R., Ostle, N., Ineson, P. & Read, D. J. (2002a). *In situ* $^{13}CO_2$ pulse-labelling of upland grassland demonstrates a rapid pathway of carbon flux from arbuscular mycorrhizal mycelia to the soil. *New Phytologist* **153**, 327–34.

Johnson, D., Leake, J. R. & Read, D. J. (2002b). Transfer of recent photosynthate into mycorrhizal mycelium of an upland grassland: short-term respiratory losses and accumulation of ^{14}C. *Soil Biology and Biochemistry* **34**, 1521–4.

Jones, M. B., & Donnelly, A. (2004). Carbon sequestration in temperate grassland ecosystems and the influence of management, climate and elevated CO_2. *New Phytologist* **164**, 423–39.

Kårén, O. & Nylund, J. E. (1997). Effects of ammonium sulphate on the community structure and biomass of ectomycorrhizal fungi in a Norway spruce stand in South West Sweden. *Canadian Journal of Botany* **75**, 1628–43.

Kerley, S. J. & Read, D. J. (1998). The biology of mycorrhiza in the Ericaceae. XX. Plant and mycorrhizal necromass as nitrogenous substrates for the ericoid mycorrhizal fungus *Hymenoscyphus ericae* and its host. *New Phytologist* **139**, 353–60.

Klironomus, J. N. & Hart, M. (2001). Animal nitrogen swap for plant carbon. *Nature* **410**, 651–2.

Koide, R. T. & Wu, T. (2003). Ectomycorrhizas and retarded decomposition in a *Pinus resinosa* plantation. *New Phytologist* **158**, 401–7.

Kuzyakov, Y., Ehrensberger, H. & Stahr, K. (2001). Carbon partitioning and below-ground translocation by *Lolium perenne. Soil Biology and Biochemistry* **33**, 61–74.

Lan, J., Xu, J. T. & Li, J. S. (1994). Study on symbiotic relation between *Gastrodia elata* and *Armillariella mellea* by autoradiography. *Acta Mycologica Sinica* **13**, 219–22.

Langley, J. A. & Hungate, B. A. (2003). Mycorrhizal controls on belowground litter quality. *Ecology* **84**, 2302–12.

Leake, J. R. (1994). The biology of myco-heterotrophic ('saprophytic') plants. *New Phytologist* **127**, 171–216.

Leake, J. R. (2004). Myco-heterotroph/epiparasitic plant interactions with ectomycorrhizal and arbuscular mycorrhizal fungi. *Current Opinion in Plant Biology* **7**, 422–8.

Leake, J. R. & Read, D. J. (1991a). Experiments with ericoid mycorrhiza. *Methods in Microbiology* **23**, 435–59.

Leake, J. R. & Read, D. J. (1991b). Chitin as a nitrogen source for mycorrhizal fungi. *Mycological Research* **94**, 993–5.

Leake, J. R. & Read, D. J. (1997). Mycorrhizal fungi in terrestrial habitats. In *The Mycota*, vol. IV, *Environmental and Microbial Relationships*, ed. D. T. Wicklow & B. Söderström, pp. 281–301. Berlin: Springer-Verlag.

Leake, J. R., Donnelly, D. P. & Boddy, L. (2002). Interactions between ectomycorrhizal fungi and saprotrophic fungi. *Ecological Studies* **157**, 346–72.

Leake, J. R., Johnson, D., Donnelly, D., Muckle, G. E., Boddy, L. & Read, D. J. (2004). Networks of power and influence: the role of mycorrhizal mycelium in controlling plant communities and agro-ecosystem functioning. *Canadian Journal of Botany* **82**, 1016–45.

Lindahl, B., Stenlid, J., Olsson, S. & Finlay, R. D. (1999). Translocation of ^{32}P between interacting mycelia of a wood-decomposing fungus and ectomycorrhizal fungi in microcosm systems. *New Phytologist* **144**, 183–93.

Mallik, A. U. (2003). Conifer regeneration problems in boreal and temperate forests with ericaceous understory: role of disturbance, seedbed limitation, and keystone species change. *Critical Reviews in Plant Sciences* **22**, 341–66.

McKendrick, S. L., Leake, J. R. & Read, D. J. (2000). Symbiotic germination and development of myco-heterotrophic plants in nature: transfer of carbon from ectomycorrhizal *Salix repens* and *Betula pendula* to the orchid *Corallorhiza trifida* through shared hyphal connections. *New Phytologist* **145**, 539–48.

McKendrick, S. L., Leake, J. R., Taylor, D. L. & Read, D. J. (2002). Symbiotic germination and development of the myco-heterotrophic orchid *Neottia nidus-avis* in nature and its requirement for locally distributed *Sebacina* spp. *New Phytologist* **154**, 233–47.

Meyer, F. H. (1964). The role of the fungus *Cenococcum graniforme* (Sow.) Ferd. *et* Winge in the formation of mor. In *Soil Micromorphology*, ed. J. Jongerius, pp. 67–87. Amsterdam: Elsevier.

Meyer, F. H. (1987). Extreme site conditions and ectomycorrhizas (especially *Cenococcum geophilum*). *Angewandte Botanik* **61**, 39–46.

Miller, R. M. & Kling, M. (2000). The importance of integration and scale in the arbuscular mycorrhizal symbiosis. *Plant and Soil* **226**, 295–309.

Myers, M. D. & Leake, J. R. (1996). Phosphodiesters as mycorrhizal P sources II. Ericoid mycorrhiza and the utilization of nuclei as a phosphorus source by *Vaccinium macrocarpon*. *New Phytologist* **132**, 445–51.

Nicholson, T. H. (1959). Mycorrhiza in the Gramineae. I. Vesicular-arbuscular endophytes with special reference to the external phase. *Transactions of the British Mycological Society* **42**, 421–38.

Nilsson, L. O. & Wallander, H. (2003). Production of external mycelium by ectomycorrhizal fungi in a Norway spruce forest was reduced in response to nitrogen fertilization. *New Phytologist* **158**, 409–16.

Northup, R., Yu, Z., Dahlgren, R. A. & Vogt, K. (1995). Polyphenol control of nitrogen release from pine litter. *Nature* **377**, 227–9.

Olsrud, M. (2004). Mechanisms of below-ground carbon cycling in subarctic ecosystems. Ph.D thesis, Lund University.

Olsrud, M. & Christensen, T. R. (2004). Carbon cycling in subarctic tundra; seasonal variation in ecosystem partitioning based on *in situ* ^{14}C pulse-labelling. *Soil Biology and Biochemistry* **36**, 245–53.

Olsrud, M., Melillo, J. M., Christensen, T. R., Michelsen, A., Wallender, H. & Olsson, P. A. (2004). Response of ericoid mycorrhizal colonization and functioning to global change factors. *New Phytologist* **162**, 459–69.

Olsson, P. A. & Wilhelmsson, P. (2000). The growth of external AM fungal mycelium in sand dunes and in experimental systems. *Plant and Soil* **226**, 161–9.

Pearson, J. N. & Jakobsen, I. (1993). Symbiotic exchange of carbon and phosphorus between cucumber and three arbuscular mycorrhizal fungi. *New Phytologist* **124**, 481–8.

Perez-Moreno, J. & Read, D. J. (2000). Mobilization and transfer of nutrients from litter to tree seedlings via the vegetative mycelium of ectomycorrhizal plants. *New Phytologist* **145**, 301–9.

Pfeffer, P. E., Douds, D. D. Jr., Bécard, G. & Shachar-Hill, Y. (1999). Carbon uptake and the metabolism and transport of lipids in an arbuscular mycorrhiza. *Plant Physiology* **120**, 587–98.

Raich J. W. & Tufekcioglu, A. (2000). Vegetation and soil respiration: correlations and controls. *Biogeochemistry* **48**, 71–90.

Raich, J. W., Potter, C. S. & Bhagawati, D. (2002). Interannual variability in global soil respiration, 1980–94. *Global Change Biology* **8**, 800–12.

Rasmussen, H. N. (2002). Recent developments in the study of orchid mycorrhizas. *Plant and Soil* **244**, 149–63.

Read, D. J. (1991). Mycorrhizas in ecosystems. *Experientia* **47**, 376–91.

Read, D. J. (1996). The structure and function of the ericoid mycorrhizal root. *Annals of Botany* **77**, 365–91.

Read, D. J., Leake, J. R. & Perez-Moreno, J. (2004). Mycorrhizal fungi as drivers of ecosystem processes in heathland and boreal forest biomes. *Canadian Journal of Botany* **82**, 1243–63.

Rillig, M. C., Wright, S. F., Nichols, K. A., Schmidt, W. F. & Torn, M. S. (2001). Large contribution of arbuscular mycorrhizal fungi to soil carbon pools in tropical forest soils. *Plant and Soil* **233**, 167–77.

Rillig, M. C., Wright, S. F. & Eviner, V. T. (2002). The role of arbuscular mycorrhizal fungi and glomalin in soil aggregation: comparing effects of five plant species. *Plant and Soil* **238**, 325–33.

Selosse, M.-A., Weiss, M., Jany, J. L. & Tillier, A. (2002). Communities and populations of sebacinoid basidiomycetes associated with the achlorophyllous orchid *Neottia nidus-avis* (L.) LCM Rich. and neighbouring tree ectomycorrhizae. *Molecular Ecology* **11**, 1831–44.

Selosse, M.-A., Faccio, A., Scappaticci, G. & Bonfante, P. (2004). Chlorophyllous and achlorophyllous specimens of *Epipactis microphylla* (Neotteae, Orchidaceae) are associated with ectomycorrhizal septomycetes, including truffles. *Microbial Ecology* **47**, 416–26.

Simard, S. W. & Durall, D. M. (2004). Mycorrhizal networks: a review of their extent, function and importance. *Canadian Journal of Botany* **82**, 1140–65.

Smith, S. E. & Read, D. J. (1997). *Mycorrhizal Symbiosis*. London: Academic Press.

Söderström, B. & Read, D. J. (1987). Respiratory activity of intact and excised ectomycorrhizal mycelial systems growing in unsterilised soil. *Soil Biology and Biochemistry* **19**, 231–6.

Staddon, P. L., Heinemeyer, A. & Fitter, A. H. (2002). Mycorrhizas and global environmental change: research at different scales. *Plant and Soil* **244**, 253–61.

Staddon, P. L., Ramsey, C. B., Ostle, N., Ineson, P. & Fitter, A. H. (2003). Rapid turnover of hyphae of mycorrhizal fungi determined by AMS microanalysis of ^{14}C. *Science* **300**, 1138–40.

Stewart, D. P. C. & Metherell, A. K. (1999). Carbon (^{13}C) uptake and allocation in pasture plants following field-pulse-labelling. *Plant and Soil* **210**, 61–73.

Taylor, D. L., & Bruns, T. D. (1997). Independent, specialized invasions of ectomycorrhizal mutualism by two nonphotosynthetic orchids. *Proceedings of the National Academy of Sciences of the USA* **94**, 4510–15.

Taylor, D. L. & Bruns, T. D. (1999). Population, habitat and genetic correlates of mycorrhizal specialization in the cheating orchids *Corallorhiza maculata* and *C. mertensiana. Molecular Ecology* **8**, 1719–32.

Taylor, D. L., Bruns, T. D., Leake, J. R. & Read, D. J. (2002). Mycorrhizal specificity and function in myco-heterotrophic plants. *Ecological Studies* **157**, 375–413.

Taylor, D. L., Bruns, T. D., Szaro, T. M. & Hodges, S. A. (2003). Divergence in mycorrhizal specialization within *Hexalectris spicata* (Orchidaceae), a nonphotosynthetic desert orchid. *American Journal of Botany* **90**, 1168–79.

Toal, M. E., Yeomans, C., Killham, K. & Meharg, A. A. (2000). A review of rhizosphere carbon flow modelling. *Plant and Soil* **222**, 263–81.

Trudell, S. A., Rygiewicz, P. T. & Edmonds, R. L. (2003). Nitrogen and carbon stable isotope abundances support the myco-heterotrophic nature and host specificity of certain myco-heterotrophic plants. *New Phytologist* **160**, 391–401.

Wallander, H., Nilsson L. O., Hagerberg, D. & Bååth, E. (2001). Estimation of the biomass and seasonal growth of external mycelium of ectomycorrhizal fungi in the field. *New Phytologist* **151**, 753–60.

Wallander, H., Johansson, L. & Pallon, J. (2002). PIXE analysis to estimate the elemental composition of ectomycorrhizal rhizomorphs grown in contact with different minerals in forest soil. *FEMS Microbiology Ecology* **39**, 147–56.

Wallander, H., Nilsson L. O., Hagerberg, D. & Rosengren, U. (2003). Direct estimates of C:N ratios of ectomycorrhizal mycelia collected from Norway spruce forest soils *Soil Biology and Biochemistry* **35**, 997–9.

Wallander, H., Göransson, H. & Rosengren, U. (2004). Production, standing biomass and natural abundance of ^{15}N and ^{13}C in ectomycorrhizal mycelia collected at different soil depths in two forest types. *Oecologia* **139**, 89–97.

Warcup, J. H. (1991). The *Rhizoctonia* endophytes of *Rhizanthella* (Orchidaceae). *Mycological Research* **95**, 656–9.

Weber, A., Karst, J., Gilbert, B. & Kimmins, J. P. (2005). *Thuja plicata* exclusion in ectomycorrhiza-dominated forests: testing the role of inoculum potential of arbuscular mycorrhizal fungi. *Oecologia* **143**, 148–56.

Wu, T. H., Sharda, J. N. & Koide, R. T. (2003). Exploring interactions between saprotrophic microbes and ectomycorrhizal fungi using a protein-tannin complex as an N source by red pine (*Pinus resinosa*). *New Phytologist* **159**, 131–9.

Yamato, M. (2001). Identification of a mycorrhizal fungus in the roots of achlorophyllous *Sciaphila tosaensis* Makino (Triuridaceae). *Mycorrhiza* **11**, 83–8.

Young, B. W., Massicotte, H. B., Tackaberry, L. E., Baldwin, Q. F. & Egger, K. N. (2002). *Monotropa uniflora*: morphological and molecular assessment of mycorrhizae retrieved from sites in the Sub-Boreal Spruce biogeoclimatic zone in central British Columbia. *Mycorrhiza* **12**, 75–82.

Zhu, Y.-G. & Miller, R. M. (2003). Carbon cycling by arbuscular mycorrhizal fungi in soil-plant systems. *Trends in Plant Sciences* **8**, 407–9.

9
Water relations in lichens

ROSMARIE HONEGGER
Institute of Plant Biology, University of Zurich

Lichen symbiosis

Lichen-forming fungi are a polyphyletic group of nutritional specialists, which derive fixed carbon from a population of living cyanobacteria and/or green algal cells. Every fifth fungus (approximately 14 000 species), or every second ascomycete, respectively, is a lichen. Species names of lichens refer to the fungal partner, the photoautotrophic symbionts having their own names and phylogenies. Most lichen-forming fungi are physiologically facultatively biotrophic, but occur in nature almost exclusively in the symbiotic state.

The majority of lichen-forming fungi form crustose, often quite inconspicuous thalli on or within the substratum where they meet their photoautotrophic partners, but about 25% of lichen mycobionts differentiate morphologically and anatomically complex 3-D thalli, either shrubby, leaf- or band-shaped, erect or pendulous, which are the result of an amazing hyphal polymorphism. Morphologically and anatomically complex lichen thalli are sophisticated culturing chambers, built up by the fungal partner, for a population of minute photobiont cells. Most lichen-forming fungi grow at or even above the surface of the substratum in order to keep their photoautotrophic partner adequately illuminated. Thus they are exposed to solar radiation, drought and temperature extremes. Lichen-forming ascomycetes produce a wide range of polyphenolic secondary metabolites, which crystallize at hyphal surfaces in the medullary layer and/or within the peripheral cortex, giving the thalli a characteristic coloration (Huneck & Yoshimura, 1996). Most of the cortical secondary compounds absorb ultraviolet (UV) light and transmit longer wavelengths, thus protecting fungal and photobiont cells from radiation damage.

Fungi in the Environment, ed. G. M. Gadd, S. C. Watkinson & P. S. Dyer. Published by Cambridge University Press.

The main building blocks of heteromerous (internally stratified) lichen thalli are tissue-like, conglutinate pseudoparenchyma, which may be found in the peripheral cortical layers and/or as internal conglutinate strands. These border upon loosely interwoven plectenchyma such as are found in the medullary and algal layers. Lichen thalli derive their mechanical stability from conglutinate pseudoparenchyma, which are hydrophilic and passively absorb water and dissolved nutrients. Plectenchyma are gas-filled zones built up by aerial hyphae with hydrophobic surfaces, as recognized by Goebel (1926). Within the lichen thallus the fungal hyphae may change their growth patterns and wall surface properties over distances of a few micrometres (Figs. 9.1b–c, 9.2a, c). The position of photobiont cells is determined by the fungal partner over short distances within the algal layer, where optimal conditions for photosynthesis and gas exchange are found (Honegger, 1991, 1997, 2001). Even in the juvenile state the photobiont wall surfaces begin to be coated with the same mycobiont-derived, water-repellent compounds as are typically found on medullary hyphae (Honegger, 1984, 1991). Thus the apoplastic continuum between both partners of the symbiosis is sealed with a mycobiont-derived hydrophobic coat, which prevents the algal and medullary layers from becoming waterlogged at high levels of thalline hydration (Figs. 9.1a–c, 9.2a–d). During the continuous wetting and drying cycles, passive fluxes of solutes from the surface to the algal layer in the thallus interior and vice versa occur underneath this hydrophobic lining. The photobiont cell population has access to water and dissolved nutrients only via the passive apoplastic fluxes within the quantitatively dominant fungal exhabitant (Honegger, 1991, 1997).

Hydrophobin-derived cell wall surface hydrophobicity

The functionally important wall surface hydrophobicity in lichen thalli has been assumed to be generated by mycobiont-derived secondary metabolites, which crystallize on hyphal surfaces within the medullary layer (Goebel, 1926). Intercellular spaces in medullary and algal layers with heavy incrustations of secondary compounds (Fig. 9.1a) are indeed very difficult to infiltrate with aqueous solutions. However, water-repellency is also a characteristic feature of species with no medullary lichen compounds such as *Xanthoria* and *Peltigera* spp. (lichen-forming ascomycetes) (Fig. 9.1b). In freeze–etch preparations a semicrystalline rodlet layer was resolved on wall surfaces of medullary hyphae of vegetative thalli of *Xanthoria* and *Peltigera* spp. (Honegger, 1982, 1984; Scherrer *et al.*, 2000) and on aerial hyphae in the photobiont stratum of the lichenized basidiocarps of *Dictyonema glabratum* (syn. *Cora pavonia*), a tropical basidiomycete (Trembley *et al.*, 2002a). These rodlet layers are formed by class 1 hydrophobins (Wessels, 1999; Wösten, 2001), as characterized in various *Xanthoria* spp. (haploid mycelia, one hydrophobin per species

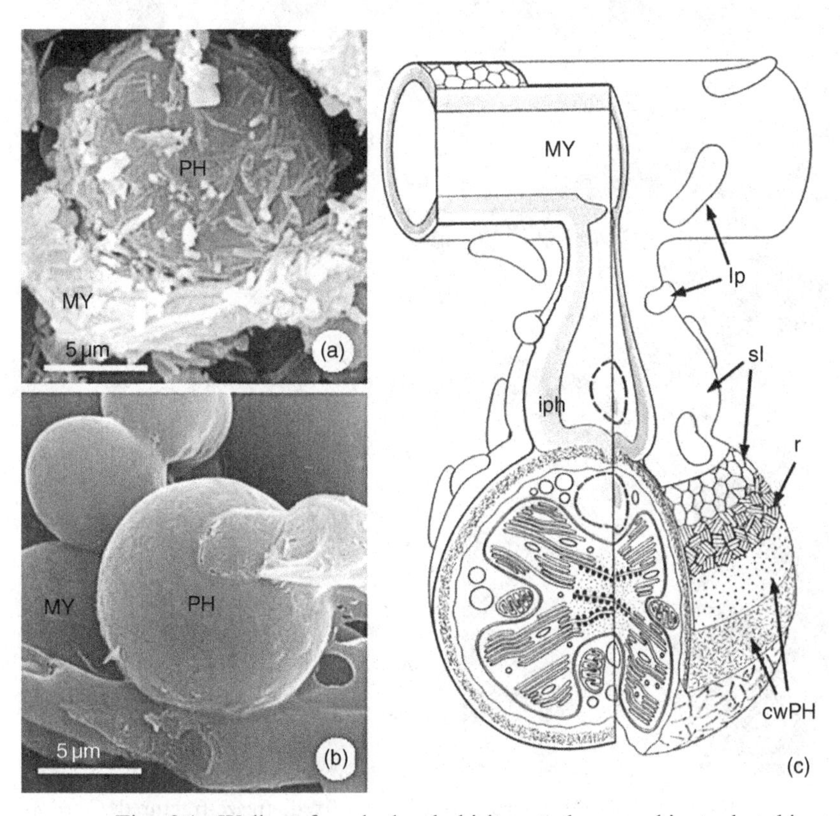

Fig. 9.1. Wall surface hydrophobicity at the mycobiont–photobiont interface in fully hydrated specimens of morphologically advanced lichen-forming ascomycetes with internally stratified thallus and *Trebouxia* sp. as green algal photobiont. (a) LTSEM micrograph of *Hypogymnia physodes* with crystals of mycobiont-derived, phenolic secondary metabolites, which enhance the protein-based wall surface hydrophobicity. (b) LTSEM micrograph of *Xanthoria parietina* with hydrophobin-based wall surface hydrophobicity (XPH1) (Scherrer *et al.*, 2000, 2002). (c) Diagram showing wall layers at the mycobiont–photobiont interface (after Honegger 1991; line drawing by Sybille Erni). MY, mycobiont; PH, photobiont; cwPH, cell wall layers of the photobiont; iph, intraparietal haustorium; lp, mycobiont-derived, crystalline secondary metabolites (polyphenolic); r, mycobiont-derived rodlet layer (hydrophobin or hydrophobin-like); sl, mycobiont-derived compounds overlying the hydrophobic rodlet layer in some, but not all, interfaces of lichen-forming fungi and their photobionts.

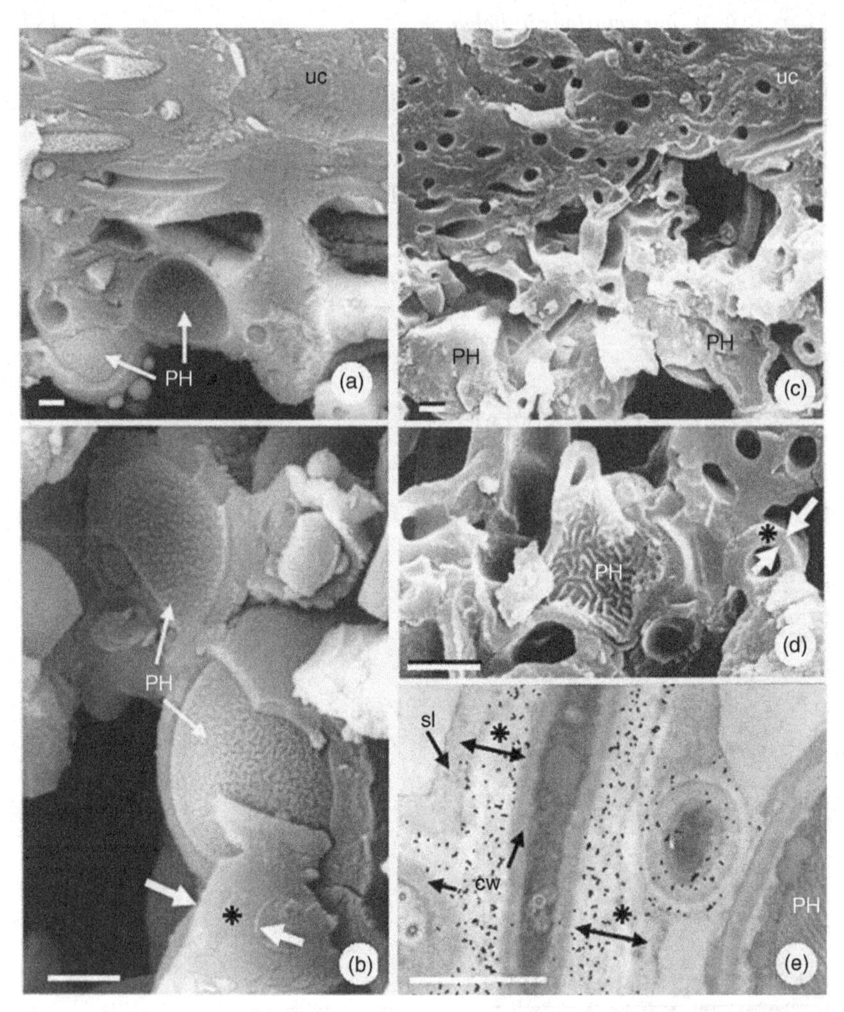

Fig. 9.2. LTSEM micrographs of cryofixed, freeze-fractured *Cetraria islandica* and its photobiont (PH), *Trebouxia* sp., in the fully hydrated (a)–(b) and desiccated state (c)–(d), and immunocytochemical location of lichenin on ultrathin sections (e). Same magnification in (a) and (c), or (b) and (d), respectively. Transition zone of the upper cortex (uc) to the algal layer in (a) and (c), and mycobiont–photobiont interface in (b) and (d). Fully hydrated fungal and algal cells in (a)–(b) are compared with cavitated fungal and shrivelled algal cells in (c)–(d). Differences in fungal wall thickness are marked with thick arrows and asterisks in fully hydrated (b) and desiccated (d) cells. Passive water translocation and consequent swelling and shrinkage occur primarily in the thick, amorphous glucan layer, marked with double arrows and asterisks in (e), overlying the cell wall (cw); electron-dense gold granules visualize the binding sites of a monoclonal antibody against the hydrophilic (1→3), (1→4)-β-glucan lichenin. The hydrophobic surface layer (sl) channels the flow of solutes within the apoplast and prevents free water from accumulating on wall surfaces. Scale bars, 2 μm. From Honegger & Haisch (2001), with kind permission of Blackwell Publishing Ltd.

(Scherrer *et al.*, 2000; Scherrer & Honegger, 2003)) and in *Dictyonema glabratum* (dikaryotic hyphae, three different hydrophobins (Trembley *et al.*, 2002a)). As in many non-lichenized fungi, hydrophobin gene expression in lichenized asco- and basidiomycetes is developmentally regulated (Scherrer *et al.*, 2002; Trembley *et al.*, 2002b). Mycobiont-derived medullary secondary compounds of lichen-forming ascomycetes crystallize on and within the water-repellent, proteinaceous rodlet layer that spreads over the photobiont cells, thus enhancing the hydrophobicity of the medullary and algal layers (Fig. 9.1c). Water-repellent wall surface layers overlie highly hydrophilic, often very thick glucan layers of hyphal walls (Honegger & Haisch, 2001) and channel the flow of solutes from the thallus surface to the interior and vice versa during the wetting and drying cycles (Fig. 9.2a–e).

Ecophysiology

Lichens dominate *c.* 10% of terrestrial ecosystems. These encompass regions in which higher plants, being at their physiological limits, exert little or no competitive pressure: polar, alpine and desert ecosystems. Lichen mats covering thousands of square kilometres of Arctic tundra (built up by reindeer lichens, *Cladonia* spp., 'Icelandic moss', *Cetraria islandica*, etc.), or the extensive lichen fields of the southern Namib Desert, in which dozens of hectares of one the driest deserts worldwide are covered by an intensely orange-coloured carpet of shrubby, finely branched *Teloschistes capensis* (Lange *et al.*, 1991; Jürgens & Niebel-Lohmann, 1995), are indeed very impressive examples of fungal adaptations to extreme environments and also to the symbiotic way of life. Moreover, lichens may locally dominate the often inconspicuous biological soil crust communities in arid, whether cold, temperate or hot, ecosystems (various authors in Belnap & Lange, 2003), which cover very large areas worldwide but are difficult to estimate quantitatively at a global scale. Biological soil crusts are ecologically important stabilizers, which reduce wind erosion in steppe, semi-desert and desert ecosystems and thus lower the severity of large-scale (often intercontinental) dust storm events (Belnap & Lange, 2003).

Generations of ecophysiologists have explored the metabolic activities of lichens under extreme climatic conditions (Kappen, 1988, 1993). In field and laboratory studies on gas exchange the lichen thallus is usually taken as an entity and fungal gas exchange is assumed to be negligible. Even nowadays many investigators consider lichens as plants, although the fungal partner usually amounts to more than 80% of thalline biomass.

The key to understanding the impressive environmental stress tolerance of lichen-forming fungi and their photobionts lies in their poikilohydric water relations and the ability of their desiccated cells to survive temperature extremes unharmed in a state of dormancy. Desiccated thalli of the golden yellow, almost ubiquitous *Xanthoria parietina* grew normally after cryoimmobilization in sub-cooled liquid nitrogen (*c.* −200 °C), freeze–fracturing, sputter coating and examination in the low-temperature scanning electron microscope (LTSEM) under high vacuum conditions at 60 kV, but hydrated thalli did not survive this treatment (Honegger, 1995, 1998). Most lichen-forming fungi and their photobionts are fully adapted to continuous wetting and drying cycles, as experienced during diurnal and seasonal fluctuations in water availability in their ecosystems, but die off under continuously moist conditions or when they are prevented from drying out at elevated temperatures. Consequently, abundant epiphytic lichen growth is typically found in boreal to temperate climates, but bryophytes dominate epiphytic communities in temperate rainforests and tracheophytes (ferns, bromeliads, orchids, etc.) in tropical rainforests (Sillett & Antoine, 2004). Long-living leaves in tropical rainforests carry species-rich, taxonomically diverse communities of foliicolous lichens and their free-living photobionts (*Trentepohlia*, *Phycopeltis* and *Cephaleuros* spp., etc.) (Hawksworth, 1988; Kappen, 1988; Lücking, 1995, 1997; de los Angeles Herrera-Campos *et al.*, 2004; Lücking *et al.*, 2003). Although most species grow on the cuticle and do not penetrate the leaf proper, growers consider foliicolous lichens as a pest because they also colonize the leaves of economic plants such as coffee, tee, cacao, rubber, etc. (Hawksworth, 1988). Foliicolous lichens even grow on plastic (Sipman, 1994; Sanders, 2002).

Approximately 85% of lichen-forming ascomycetes have green algal photobionts (chlorolichens), 10% have cyanobacterial photobionts (cyanolichens) and *c.* 4% have both simultaneously, a green algal partner, which provides fixed carbon, and a diazotrophic cyanobacterial partner delivering fixed nitrogen to the symbiotic system. In such triple symbioses, so-called cephalodiate lichens, the fungal partner creates microaerobic conditions around the cyanobacterial colony, resulting in increased heterocyst frequencies and thus enhanced nitrogen fixation compared with the free-living state (Englund, 1977; Honegger, 1993, 2001; Honegger & Hugelshofer, 2000). Drought-stressed chlorolichens with photobionts of the genus *Trebouxia* (Trebouxiophyceae; symbionts of over 50% of lichen-forming ascomycetes) were shown to rehydrate and to re-establish at least partial metabolic activities with water vapour, but cyanolichens require liquid water for full recovery (Lange *et al.*, 1986, 1988; Schroeter, 1994). Accordingly, chlorolichens with *Trebouxia* spp. as photobionts dominate in arid, hot or cold ecosystems, but cyanolichens with internally stratified thalli (*Peltigera*, *Sticta* and *Nephroma* spp.) are abundant in moist habitats (Lange *et al.*, 1986, 1988; Kappen, 1988; Sillett & Antoine, 2004). In contrast, gelatinous cyanolichens with non-stratified thalli (*Collema* spp.), in which the fungal partner lives within the thick, gelatinous sheaths of *Nostoc* colonies (examples in Honegger, 2001) dominate, together with free-living cyanobacteria (*Nostoc* and *Microcoleus* spp., etc.), in soil crust communities in arid ecosystems with erratic rainfall events (Lange *et al.*, 1998; various authors in Belnap & Lange, 2003). Numerous tropical chlorolichens, foliicolous species included, contain representatives of Trentepohliales (e.g. *Trentepohlia*, *Phycopeltis*, etc.) as photobiont; these are less tolerant of cold stress than are *Trebouxia* species.

With regard to growth rate, polar, alpine and desert lichens all face principally the same problem: photosynthetic activity of their photoautotrophic partner is restricted to a narrow time window when thalli are at least partly hydrated and solar radiation is available at temperatures within the range suitable for photosynthesis. Thalline growth rates depend on the frequency and length of such time windows per day and year. In the Negev and Namib Deserts lichen photobionts photosynthesize during a short period at sunrise, when thalli are partly hydrated from fog and/or nocturnal dewfall before both symbionts dry out (that is, achieve water content of less than 10% of dry mass) and fall into dormancy until water becomes available (Kappen *et al.*, 1979; Kappen, 1988; Lange *et al.*, 1970, 1991). Annual rates of gas exchange were measured in *Ramalina maciformis* (photobiont, *Trebouxia* sp.) in the Negev Desert. A positive balance was recorded on 218 days, a negative balance (due to too high a water content at night or too low a water content in the morning) on 88 days and balance zero (due to drought-induced dormancy) on 49 days (Kappen *et al.*, 1979; Kappen, 1988). A radial size increase around 0.4 mm yr^{-1} was recorded in the crustose *Caloplaca aurantia* in the Negev desert (Lange, 1990). The most conspicuous chlorolichen communities develop in desert areas with very low precipitation, but frequent periods of high atmospheric moisture as found in coastal fog oases (Atacama Desert in South America, Namib Desert in South Africa, etc.). These harbour high proportions of 'fog comb lichens': band-shaped (*Ramalina* or *Roccella* spp.) to finely branched species (*Usnea* or *Teloschistes* spp.) with high surface to volume ratios. Water can temporarily be retained in the thallus by hydrophilic, amorphous fungal wall components such as lichenin, a linear

(1→3), (1→4)-β-glucan, which overlie the cell wall proper (Fig. 9.2e) (Honegger & Haisch, 2001). Lichen-forming fungi do not protect their photobiont from desiccation, as is often assumed. Both symbionts are fully adapted to continuous wetting and drying cycles, which are an integral part of their lifestyle and the driving force behind passive fluxes of solutes between the partners (Honegger, 1991, 1997).

Temperature and water availability are also the limiting factors in polar and alpine lichens. Frost is the main water source available to lichens of high Andean Paramos (Pérez, 1997a,b). *Trebouxia* photobionts of polar lichens were shown to be photosynthetically active even at sub-zero temperatures, at low photon flux densities and in a partly hydrated state (Kappen, 1988, 1993; Kappen *et al.*, 1995). In extensive field measurements, various investigators recorded metabolic activity in lichen thalli at sub-zero temperatures, when snow was the only water source available. Covered by snow, ideally with ice windows due to melting and freezing cycles above darkly pigmented thalli, lichens live as in a greenhouse, well protected from the harsh climate, and achieve a positive balance (Kappen, 1988, 1993). Schroeter & Scheidegger (1995) visualized the level of rehydration from snow in desiccated, deep-frozen thalli of *Umbilicaria aprina* from the continental Antarctic at various sub-zero temperatures with LTSEM techniques and explored their metabolic activity in parallel experiments. Very low growth rates were recorded among lichens of the continental Antarctic with its very cold and dry climate, but surprisingly 'fast' growth was documented in lichen communities in the maritime Antarctic (Smith, 1995; Sancho & Pintado, 2004).

Drought-stress-induced structural changes

Drought-stressed lichen thalli shrink without wilting, but reveal a different consistency and, if thallus coloration derives primarily from the photobiont cell population, a colour different from that of the fully hydrated state. The cortex of fully hydrated thalli is elastic and glassy-translucent, thus allowing light to reach the algal layer and the green (chlorolichens, e.g. lungwort, *Lobaria pulmonaria*) or grey colour (cyanolichens, e.g. dog lichen, *Peltigera canina*) of the photobiont to be seen from outside. The cortex is opaque and brittle, usually greyish in the desiccated state. The severity of drought-stress-induced shrinkage depends on thalline morphology: crustose and squamulose thalli, which are affixed with their whole lower surface to the substratum, shrink mainly in the vertical axis, but thalli with no tight contact to the substratum shrink in all dimensions (Figs. 9.2a–d, 9.3a–b). Ascomal shrinkage during desiccation

(Fig. 9.3a–b) is a functionally important element, which generates the pressure on the flanks of mature asci required for ascospore ejection.

Symplastic cavitation Drought-stress-induced cytoplasmic gas bubbles in fungal cells were first described by de Bary (1866), who realized that wall thickness and stiffness have an impact on spatial changes during desiccation. Cells with thin, hyaline walls undergo cytorrhysis (collapse): they shrivel and shrink during desiccation, globose or ovoid spores often achieving a boat-like shape. This applies also to the green algal and cyanobacterial photobiont cells of lichen thalli (Fig. 9.2c–d) (Brown *et al.*, 1987; Honegger, 1995, 1998; Scheidegger *et al.*, 1995; Schroeter & Scheidegger 1995; Honegger *et al.*, 1996). Cells with thick, rigid walls cannot be sufficiently deformed during water loss and therefore cavitate (implode). Reversible symplastic cavitation (Figs. 9.2c–d, 9.4a–b) is a regularly occurring feature in lichen-forming fungi under drought stress (Honegger, 1995; Scheidegger *et al.*, 1995; Schroeter & Scheidegger, 1995). In material sciences cavitation is feared, because metal parts of pumps and piping systems may become pitted by cavitation-derived shock waves. The warning of an engineer: 'One thing you can count on: wherever the cavitation implosion occurs there will be damage' (DeClerc D., unpublished). In biological systems drought-stress-induced cavitation has been intensely investigated in water-conducting tissues, i.e. in xylem vessels of plants (tracheophytes) under negative pressure due to high transpiratory losses. These are dead cells with lignified walls whose

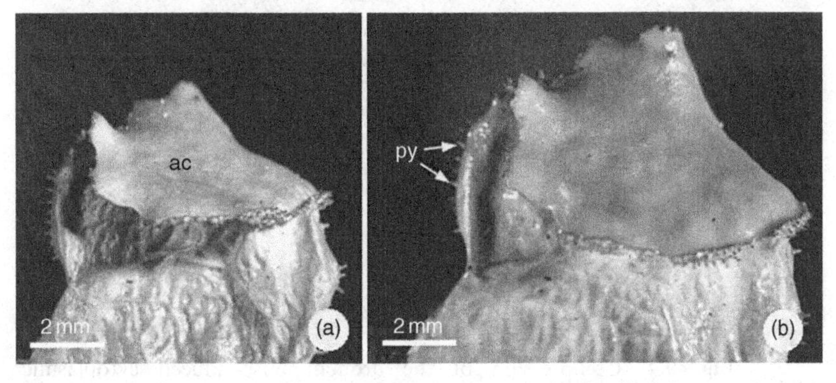

Fig. 9.3. *Cetraria islandica* ("Icelandic moss"): fertile lobe margin with large ascoma (ac) in the desiccated (a) and fully hydrated state (b). Microconidia-producing pycnidia (py) are contained in marginal projections. Dramatic shrinkage and swelling occur during drying and wetting cycles, owing to poikilohydric water relations.

protoplast was dissolved and replaced by water. Irreversibly cavitated xylem vessels lose their function as a water conduit from root to shoot (Comstock & Sperry, 2000). Drought-stress-induced cavitation of fungal cells is a distinctly more dramatic event: the living cell, not just a water column, is affected. Desiccated lichen thalli contain less than 20% water per unit dry mass. How can fungal cells survive anhydrobiosis and cavitation without suffering irreversible damage, where does the cavitation bubble come from and what happens to the cavitated symplast during rehydration?

As in xylem cavitation the exact origin of cavitation bubbles in drought-stressed fungal cells is often not known. It may originate from vaporization of water under negative pressure, from gassing out of dissolved molecules (e.g. CO_2) upon loss of water as a solvent, or from air seeding, in the worst case from outside via rupture of the peripheral boundary; in the living cell this would be a rupture in the plasma membrane. In his detailed light microscopic studies on drought-stress-induced structural changes in the foliose *Peltigera canina*, Goebel (1926) recognized gas-filled cortical cells, which he assumed to be dead and empty, functionally comparable to the

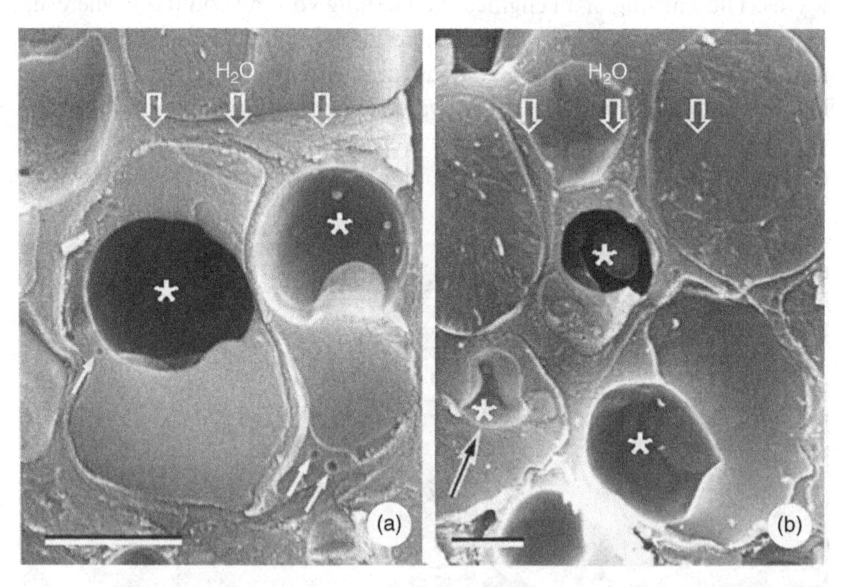

Fig. 9.4. Compression of the drought-stress-induced cytoplasmic cavitation bubble (asterisks) by the rehydrating protoplast, as seen in LTSEM micrographs of frozen-hydrated, freeze-fractured cortical cells of *Peltigera canina*. (a) Small arrows point to the gas-filled centres of freeze-fractured concentric bodies. (b) Large arrow points to a very irregularly compressed cavitation bubble. Scale bars, 5 μm.

velamen of roots of tropical, epiphytic orchids. In ultrastructural investigations the desiccated cortical cells are seen to be neither dead nor empty, but cavitated (Honegger & Hugelshofer, 2000). The fate of the symplastic cavitation bubble during rehydration was visualized with LTSEM techniques applied to frozen-hydrated specimens. The cortex of *Peltigera leucophlebia* and *P. membranacea* (neighbouring species of *P. aphthosa* and *P. canina*) is several cell layers thick. When water is added to the thallus surface the top cortical cells rehydrate within less than 30 seconds, but it takes more than a minute to fully rehydrate the whole cortex. Therefore a gradient from complete desiccation to full hydration can be studied among cortical cells of samples that have been cryofixed during rehydration. One usually excentrically located cavitation bubble is seen per desiccated fungal cell. As shown in ultrathin sections of freeze-substituted cells the cavitation bubble has a smooth lining, but is not membrane-bound (Honegger, 1995). During rehydration the bubble is compressed by the swelling cytoplasm and cell organelles and thus achieves irregular outlines (Fig. 9.4a–b).

Concentric bodies Cells of lichen-forming and of numerous desiccation-tolerant non-lichenized ascomycetes collected in the wild contain clusters of concentric bodies, round cell organelles (approximately 0.3 μm in diameter) of unknown origin and function. Clusters of concentric bodies are usually found near the cell periphery (Fig. 9.5a–b) (reviews: Honegger, 1993, 2001). Concentric bodies are not membrane-bound, but have a gas-filled centre surrounded by proteinaceous material, the composition and origin of the gas filling being unknown. Concentric bodies were postulated to be viruses (Ahmadjian, 1993), although they lack features characteristic of viral particles. It has been hypothesized that concentric bodies are the remains of cytoplasmic cavitation events, i.e. denatured proteins formerly

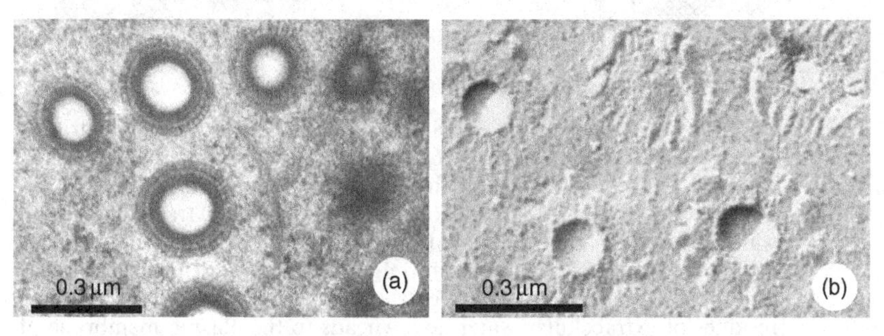

Fig. 9.5. Concentric bodies, as seen in transmission electron micrographs of an ultrathin section (a) or freeze–etch replica (b) of *Peltigera canina*. Proteinaceous material surrounds a gas-filled centre.

lining the cavitation bubble (Honegger, 1995). No concentric bodies have been found in axenic cultures of lichen-forming and non-lichenized ascomycetes that were kept on agarized, continuously moist media.

Ice nucleation Desiccated lichen-forming fungi and their photobionts survive cold temperature extremes, but what happens when fully hydrated thalli are subjected to a rapid temperature decline, as regularly experienced in polar or alpine ecosystems? Schroeter & Scheidegger (1995) simulated such situations under laboratory conditions and investigated the structural changes with LTSEM techniques applied to cryofixed specimens of the foliose-umbilicate *Umbilicaria aprina* from the Antarctic. At a cooling rate of 1 K min^{-1} down to −20 °C, prior to cryofixation, cellular

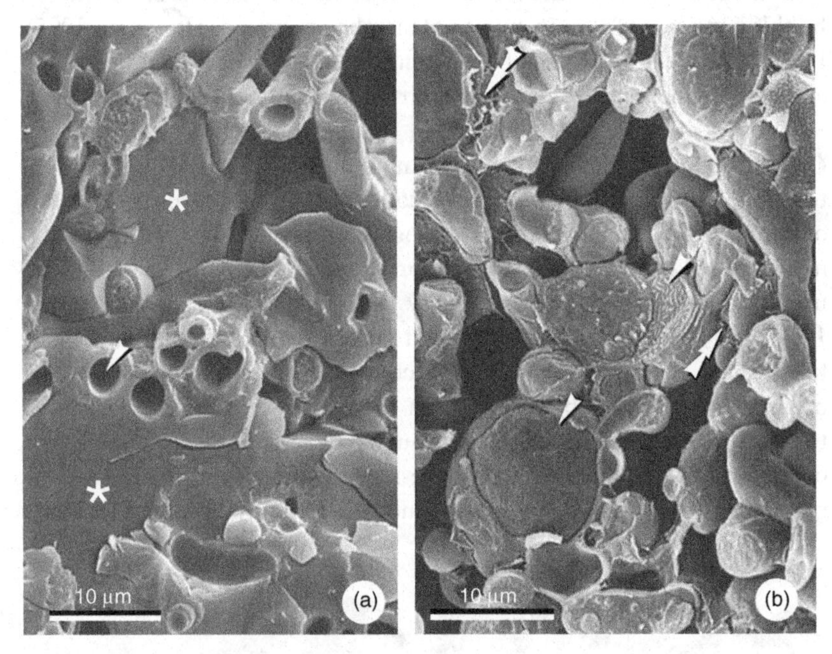

Fig. 9.6. Ice nucleation. LTSEM micrographs of fully hydrated, freeze–fractured specimens of *Umbilicaria aprina* from the Antarctic, which were cryoimmobilized either after cooling down to −20 °C (a) or after re-warming to +15 °C (b). (a) Asterisks refer to extracellular ice on wall surfaces of the desiccated, cavitated (arrowhead) fungal cells. (b) Upon thawing the fungal and algal cells achieved full turgescence after absorption of the melting extracellular ice. Arrows point to small residues of extracellular water, arrowheads to the plasma membrane of freeze–fractured *Trebouxia* cells. From Schroeter & Scheidegger (1995), with kind permission of the authors and Blackwell Publishing Ltd.

water was shown to diffuse out of the fully hydrated green algal and fungal cells and to crystallize on the hyphal surfaces in the intercellular, normally gas-filled space of the medullary and algal layers. Owing to substantial water loss the green algal cells collapsed, but the fungal cells cavitated (Fig. 9.6a). This astonishing feature refers to very powerful ice-nucleating sites at hyphal surfaces, facilitating avoidance of damaging ice-crystal formation within the cell. Upon warming this extracellular water reservoir was re-mobilized and absorbed by the fungal and algal cells, which regained turgescence (Fig. 9.6b) and metabolic activity (Schroeter & Scheidegger, 1995). Mycobiont-derived, protein-based ice nucleation activity at $-5\,^{\circ}C$ or warmer was found in symbiotic and aposymbiotic, axenically cultured phenotypes of lichen-forming ascomycetes, but not in the cultured photobionts (Kieft, 1988: Kieft & Ahmadjian, 1989; Kieft & Ruscetti, 1990).

Outlook

The molecular basis of the impressive desiccation- and stress-tolerance of lichen-forming fungi and their photoautotrophic partners is poorly investigated. In the natural habitat constitutively induced protective mechanisms for cellular membrane and enzyme systems are essential. Acyclic polyols from fungal and green algal partners, centrally important mobile carbohydrates in the symbiotic interaction, are assumed to act as compatible solutes under drought stress (Farrar, 1988; Honegger, 1991; Honegger *et al.*, 1992). A glutathione-based antioxidant system was found to be activated in thalli collected in the wild, but not in sterile cultured isolates of *Cladonia vulcani* and its green algal photobiont, *Trebouxia excentrica*, on agar media (Kranner *et al.*, 2005). A major problem in experimental studies with cultured lichen symbionts is the simulation of rapidly changing, multifactorial environmental parameters under sterile culturing conditions. Insights into the genetics of desiccation tolerance in lichen-forming fungi and their photobionts would be of considerable interest in applied sciences.

Acknowledgements

My sincere thanks are due to Burkhard Schroeter, Christoph Scheidegger and Blackwell Publishing Ltd for their permit to include Fig. 9.6(a, b), to Sybille Erni for designing Fig. 9.1c, and to the Swiss National Science Foundation for generous financial support (grant Nr. 31-103860/1).

References

Ahmadjian, V. (1993). *The Lichen Symbiosis.* New York: John Wiley.

Belnap, J. & Lange, O. L. (eds.) (2003). *Biological Soil Crusts: Structure, Function, and Management.* Berlin: Springer.

Brown, D. H., Rapsch, S., Beckett, A. & Ascaso, C. (1987). The effects of desiccation on cell shape in the lichen *Parmelia sulcata* Taylor. *New Phytologist* **105**, 295–9.

Comstock, J. P. & Sperry, J. S. (2000). Theoretical considerations of optimal conduit length for water transport in vascular plants. *New Phytologist* **148**, 195–218.

de Bary, A. (1866). *Morphologie und Physiologie der Pilze, Flechten und Myxomyceten.* Leipzig: W. Engelmann.

de Los Angeles Herrera-Campos, M., Lücking, R., Perez, R.-E., Campos, A., Colin, P. M. & Pena, A. B. (2004). The foliicolous lichen flora of Mexico. V. Biogeographical affinities, altitudinal preferences, and an updated checklist of 293 species. *Lichenologist* **36**, 309–27.

Englund, B. (1977). The physiology of the lichen *Peltigera aphthosa*, with special reference to the blue-green phycobiont (*Nostoc* sp.). *Physiologia Plantarum* **41**, 298–304.

Farrar, J. (1988). Physiological buffering. In *CRC Handbook of Lichenology*, ed. M. Galun, vol. 2, pp. 101–5. Boca Raton, FL: CRC Press.

Goebel, K. von (1926). Ein Beitrag zur Biologie der Flechten. *Annales du Jardin Botanique de Buitenzorg* **36**, 1–83.

Hawksworth, D. L. (1988). Effects of algae and lichen-forming fungi on tropical crops. In *Perspectives of Mycopathology*, ed. V. P. Agnihotri, A. K. Sarbhoy & D. Kumar, pp. 76–83. New Delhi: Malhotra Publishing House.

Honegger, R. (1982). Cytological aspects of the triple symbiosis in *Peltigera aphthosa. Journal of the Hattori Botanical Laboratory* **52**, 379–91.

Honegger, R. (1984). Cytological aspects of the mycobiont-phycobiont relationship in lichens. Haustorial types, phycobiont cell wall types, and the ultrastructure of the cell surface layers in some cultured and symbiotic myco- and phycobionts. *Lichenologist* **16**, 111–27.

Honegger, R. (1991). Functional aspects of the lichen symbiosis. *Annual Reviews of Plant Physiology and Plant Molecular Biology* **42**, 553–78.

Honegger, R. (1993). Developmental biology of lichens. *New Phytologist* **125**, 659–77.

Honegger, R. (1995). Experimental studies with foliose macrolichens: fungal responses to spatial disturbance at the organismic level and to spatial problems at the cellular level during drought stress events. *Canadian Journal of Botany* **73** (suppl. 1), 569–87.

Honegger, R. (1997). Metabolic interactions at the mycobiont-photobiont interface in lichens. In *The Mycota*, ed. K. Esser & P. A. Lemke, vol. V, *Plant Relationships*, ed. G. C. Carroll & P. Tudzynski, part A, pp. 209–21. Berlin: Springer.

Honegger, R. (1998). The lichen symbiosis – what is so spectacular about it? *Lichenologist* **30**, 193–212.

Honegger, R. (2001). The symbiotic phenotype of lichen-forming ascomycetes. In *The Mycota*, ed. K. Esser & P. A. Lemke, vol. IX, *Fungal Associations*, ed. B. Hock, pp. 165–88. Berlin: Springer.

Honegger, R. & Haisch, A. (2001). Immunocytochemical location of the (1→3), (1→4)-β-glucan lichenin in the lichen-forming ascomycete *Cetraria islandica* ("Icelandic moss"). *New Phytologist* **150**, 739–46.

Honegger, R. & Hugelshofer, G. (2000). Water relations in the *Peltigera aphthosa* group visualized with LTSEM techniques. *Bibliotheca Lichenologica* **75**, 113–26.

Honegger, R., Kutasi, V. & Ruffner, H. P. (1992). Polyol patterns in 11 species of aposymbiotically cultured lichen mycobionts. *Mycological Research* **95**, 905–14.

Honegger, R., Peter, M. & Scherrer, S. (1996). Drought stress-induced structural alterations at the mycobiont-photobiont interface in a range of foliose macrolichens. *Protoplasma* **190**, 221–32.

Huneck, S. & Yoshimura, I. (1996). *Identification of Lichen Substances*. Berlin: Springer.

Jürgens, N. & Niebel-Lohmann, A. (1995). Geobotanical observations on lichen fields of the Southern Namib desert. *Mitteilungen des Instituts für Allgemeine Botanik Hamburg* **25**, 135–56.

Kappen, L. (1988). Ecophysiological relationships in different climatic regions. In *CRC Handbook of Lichenology*, ed. M. Galun, vol. 2, pp. 37–100. Boca Raton, FL: CRC Press.

Kappen, L. (1993). Lichens in the Antarctic region. In *Antarctic Microbiology*, ed. E. I. Friedman, pp. 433–90. New York: Wiley-Liss.

Kappen, L., Lange, O. L., Schulze, E. D., Evenari, M. & Buschbom, U. (1979). Ecophysiological investigations on lichens in the Negev Desert. VI. Annual course of the photosynthetic production of *Ramalina maciformis* (Del.) Bory. *Flora* **168**, 85–108.

Kappen, L., Sommerkorn, M. & Schroeter, B. (1995). Carbon acquisition and water relations of lichens in polar regions – potentials and limitations. *Lichenologist* **27**, 531–45.

Kieft, T. L. (1988). Ice nucleation activity in lichens. *Applied and Environmental Microbiology* **54**, 1678–81.

Kieft, T. L. & Ahmadjian, V. (1989). Biological ice nucleation activity in lichen mycobionts and photobionts. *Lichenologist* **21**, 355–62.

Kieft, T. L. & Ruscetti, T. (1990). Characterization of biological ice nuclei from a lichen. *Journal of Bacteriology* **172**, 3519–23.

Kranner, I., Cram, W. J., Zorn, M., Wornick, S., Yoshimura, I., Stabentheiner, E. & Pfeifhofer, H. (2005). Antioxidants and photoprotection in a lichen as compared with its isolated symbionts. *Proceedings of the National Academy of Sciences of the USA* **102**, 3141–6.

Lange, O. L. (1990). Twenty-three years of growth measurements on the crustose lichen *Caloplaca aurantia* in the Negev Desert. *Israel Journal of Botany* **39**, 383–94.

Lange, O. L., Schulze, E. D. & Koch, W. (1970). Experimentell-ökologische Untersuchungen an Flechten der Negev-Wüste. II. CO_2-Gaswechsel und Wasserhaushalt von *Ramalina maciformis* (Del.) Bory am natürlichen Standort während der sommerlichen Trockenperiode. *Flora* **159**, 38–62.

Lange, O. L., Kilian, E. & Ziegler, H. (1986). Water vapour uptake and photosynthesis of lichens: performance differences in species with green and blue-green algae as phycobionts. *Oecologia* **71**, 104–10.

Lange, O. L., Green, T. G. A. & Ziegler, H. (1988). Water status-related photosynthesis and carbon isotope discrimination in species of the lichen genus *Pseudocyphellaria* with green and blue-green photobionts and in photosymbiodemes. *Oecologia* **75**, 394–411.

Lange, O. L., Meyer, A., Zellner, H., Ullmann, I. & Wessels, D. C. J. (1990). Eight days in the life of a desert lichen: water relations and photosynthesis of *Teleschistes capensis* in the coastal fog zone of the Namib desert. *Madoqua* **17**, 17–30.

Lange, O. L., Meyer, A., Ullmann, I. & Zellner, H. (1991). Mikroklima, Wassergehalt und Photosynthese von Flechten in der küstennahen Nebelzone der Namib-Wüste: Messungen während der herbstlichen Witterungsperiode. *Flora* **185**, 233–66.

Lange, O. L., Belnap, J. & Reichenberger, H. (1998). Photosynthesis of the cyanobacterial soil-crust lichen *Collema tenax* from arid lands in southern Utah, USA: role of water content on light and temperature responses of CO_2 exchange. *Functional Ecology* **12**, 195–202.

Lücking, R. (1995). Biodiversity and conservation of foliicolous lichens in Costa Rica. *Mitteilungen der Eidgenössischen Forschungsanstalt für Wald, Schnee und Landschaft* **70**, 63–92.

Lücking, R. (1997). The use of foliicolous lichens as bioindicators in the tropics, with special reference to the microclimate. *Abstracta Botanica* **21**, 99–116.

Lücking, R., Wirth, V., Ferraro, L. I. & Caceres, M. E. S. (2003). Foliicolous lichens from Valdivian temperate rain forest of Chile and Argentina: Evidence of an austral element, with the description of seven new taxa. *Global Ecology and Biogeography* **12**, 21–36.

Pérez, F. L. (1997a). Geoecology of erreatic lichens of *Xanthoparmelia vagans* in an equatorial Andean Paramo. *Plant Ecology* **129**, 11–28.

Pérez, F. L. (1997b). Geoecology of erratic globular lichens of *Catapyrenium lachneum* in the high Andean Paramo. *Flora* **192**, 241–59.

Sancho, L. G. & Pintado, A. (2004). Evidence of high annual growth rate for lichens in the maritime Antarctic. *Polar Biology* **27**, 312–19.

Sanders, W. B. (2002). In situ development of the foliicolous lichen *Phyllophiale* (Trichotheliaceae) from propagule germination to propagule production. *American Journal of Botany* **89**, 1741–6.

Scheidegger, C., Schroeter, B. & Frey, B. (1995). Structural and functional processes during water vapour uptake and desiccation in selected lichens with green algal photobionts. *Planta* **197**, 399–409.

Scherrer, S. & Honegger, R. (2003). Inter- and intraspecific variation of homologous hydrophobin (H1) gene sequences among *Xanthoria* spp. (lichen-forming ascomycetes). *New Phytologist* **157**, 375–89.

Scherrer, S., De Vries, O. M. H., Dudler, R., Wessels, J. G. H. & Honegger, R. (2000). Interfacial self-assembly of fungal hydrophobins of the lichen-forming ascomycetes *Xanthoria parietina* and *X. ectaneoides*. *Fungal Genetics and Biology* **30**, 81–93.

Scherrer, S., Haisch, A. & Honegger, R. (2002). Characterization and expression of *XPH1*, the hydrophobin gene of the lichen-forming ascomycete *Xanthoria parietina*. *New Phytologist* **154**, 175–84.

Schroeter, B. (1994). In situ photosynthetic differentiation of the green algal and the cyanobacterial photobiont in the crustose lichen *Placopsis contortuplicata*. *Oecologia* **98**, 212–20.

Schroeter, B. & Scheidegger, C. (1995). Water relations in lichens at subzero temperatures: structural changes and carbon dioxide exchange in the lichen *Umbilicaria aprina* from continental Antarctica. *New Phytologist* **131**, 273–85.

Sillett, S. C. & Antoine, M. E. (2004). Lichens and bryophytes in forest canopies. In *Forest Canopies*, ed. M. D. Lowman & H. B. Rinker, 2nd edn, pp. 151–74. Oxford: Elsevier Academic Press.

Sipman, H. J. M. (1994). Foliicolous lichens on plastic tape. *Lichenologist* **26**, 311–12.

Smith, R. I. Lewis (1995). Colonization by lichens and the development of lichen-dominated communities in the maritime Antarctic. *Lichenologist* **27**, 473–83.

Trembley, M. L. Ringli, C. & Honegger, R. (2002a). Hydrophobins DGH1, DGH2 and DGH3 in the lichen-forming basidiomycete *Dictyonema glabratum*. *Fungal Genetics and Biology* **35**, 247–59.

Trembley, M. L., Ringli, C. & Honegger, R. (2002b). Differential expression of hydrophobins DGH1, DGH2 and DGH3 and immunolocalization of DGH1 in strata of the lichenized basidiocarp of *Dictyonema glabratum*. *New Phytologist* **154**, 185–95.

Wessels, J. G. H. (1999). Fungi in their own right. *Fungal Genetics and Biology* **27**, 134–45.

Wösten, H. A. B. (2001). Hydrophobins: multipurpose proteins. *Annual Review of Microbiology* **55**, 625–46.

10

Development of the arbuscular mycorrhizal symbiosis: insights from genomics

JINYUAN LIU, MELINA LOPEZ-MEYER, IGNACIO MALDONADO-MENDOZA AND MARIA J. HARRISON

Boyce Thompson Institute for Plant Research, Ithaca, New York

Introduction

The majority of the vascular flowering plants have the ability to enter into symbiotic associations with a unique group of soil fungi, the arbuscular mycorrhizal (AM) fungi. The symbiosis develops in the roots of the plant, where the plant provides the fungus with a source of carbon and the fungus delivers mineral nutrients to the roots. In particular, the transfer of phosphorus from the AM fungus to the plant is widely documented, but there is evidence for zinc and nitrogen transport also (Hodge *et al.*, 2001). For both symbionts, significant quantities of nutrients may be exchanged. It is estimated that the plant allocates up to 20% of its photosynthate to the roots to support the fungal symbiont, and some studies suggest that in an AM symbiosis the plant receives all of its phosphorus via the fungus (Bago *et al.*, 2000; Smith *et al.*, 2003). Phosphorus is a relatively immobile nutrient and is often present at concentrations in the soil that are limiting for plant growth. Consequently, improvements in phosphorus supply resulting from the AM fungus can have a significant impact on plant health and subsequently on plant biodiversity and ecosystem productivity (van der Heijden *et al.*, 1998).

The AM symbiosis is an ancient association. Both molecular data and fossil evidence suggest that the AM fungi originated 460 MYA (Redeker *et al.*, 2000) at a time when bryophytes were the predominant plant form. There is no direct evidence that the early bryophytes formed interactions with the early AM fungi; however, some of the extant bryophytes can form associations with AM fungi and so it is speculated that there may have been interactions with their ancestors also (Schüßler, 2000). Fossils from

The first three authors contributed equally to this work.

Fungi in the Environment, ed. G. M. Gadd, S. C. Watkinson & P. S. Dyer. Published by Cambridge University Press.

the Devonian Rhynie Chert provide direct evidence that the ancestral AM fungi formed associations with the earliest land plants and it has been suggested that the AM fungi may have facilitated their colonization of land (Pirozynski & Malloch, 1975; Remy *et al.*, 1994). Consistent with this ancient origin, the ability to form an AM symbiosis is distributed widely throughout the plant kingdom and today includes angiosperms, gymnosperms, pteridophytes and some bryophytes (Smith & Read, 1997).

The AM fungi are all members of the Glomeromycota, a sister clade to the Ascomycota and Basidiomycota (Schüßler *et al.*, 2001). There is only one non-mycorrhizal member of the Glomeromycota, *Geosiphon pyriforme*, a fungus that forms an intriguing symbiosis with a cyanobacterium (Gehrig *et al.*, 1996). All of the other members of the Glomeromycota form AM associations with vascular plants. In addition, all AM fungi are obligate symbionts and are entirely dependent on their plant host for carbon. This dependency has made them difficult to study and it is not currently possible to culture AM fungi in the absence of a plant host. *In vitro* cultivation methods with plant roots have been developed and are extremely useful, but in comparison with other groups of fungi we still know relatively little about their biology (Bécard & Fortin, 1988; St-Arnaud *et al.*, 1996).

Development of the AM symbiosis is an intricate process. Although the symbiosis causes no dramatic alterations to the external appearance of the roots, there are significant alterations within the cortical cells; development of the association requires the coordinate differentiation of both symbionts to create the unique structures of the symbiosis (Gianinazzi-Pearson, 1996; Parniske, 2004; Harrison, 2005). Initially, AM fungal spores germinate and the hyphal germ tube makes contact with a root. The fungus then differentiates on the epidermis, forming an appressorium via which it penetrates the root. The formation of appressoria occurs only on the root surface, indicating the requirement for a specific signal(s) from the plant. Once inside the root, the pattern of growth varies slightly depending on the species of the plant and fungal symbionts involved. Most of the recent molecular and genomic studies have focused on three plant species, *Medicago truncatula*, *Lotus japonicus* and rice, in association with a range of fungi from the genus *Glomus*. In these particular associations, the AM fungus grows through the intercellular spaces of the root to the inner cortex where the fungus forms arbuscules within the cortical cells. Arbuscules are highly branched hyphae that are enveloped in a plant-derived membrane, the peri-arbuscular membrane. Nutrient exchange takes place across the interface between the arbuscule and the cortical cell; the arbuscule is of central importance in the symbiosis.

Table 10.1. Medicago truncatula *arbuscular mycorrhizal cDNA libraries*

EST sequences are available in the TIGR *Medicago truncatula* gene index (http://www.tigr.org/tigr-scripts/tgi/T_index.cgi?species=medicago)

Library name	TIGR cat no.	No. of ESTs	Description	References
MHAM	T1682	7351	ESTs from roots of *M. truncatula* colonized with *Glomus versiforme*	Liu *et al.*, 2003
MHAM2	#GFS	1679	Same as MHAM	J. Liu *et al.*, 2005 (unpublished)
MtBC	5520	8567	ESTs from roots of *M. truncatula* three weeks after inoculation with *G. intraradices*	Journet *et al.*, 2002
MTAMP	#ARE	3448	ESTs from *M. truncatula* mycorrhizal roots inoculated with *G. intraradices*	K. Manthey *et al.*, 2002 (unpublished)
MTGIM	#ARB	1686	ESTs from *M. truncatula* mycorrhizal roots inoculated with *G. intraradices* (Subtracted library)	Wulf *et al.*, 2003
M. truncatula mycorrhizal roots 3 weeks	#9CR	37	Suppression subtractive hybridization of *M. truncatula* and *G. mosseae*	Brechenmacher *et al.*, 2004
M. truncatula J5 roots	#G7D	29	Suppression subtractive hybridization of *M. truncatula* and *G. mosseae*	Weidmann *et al.*, 2004

The AM symbiosis has been studied for many years and the process of invasion of the roots, the patterns of growth within the root, and development of arbuscules have been well documented (Bonfante-Fasolo, 1984). Now, with the era of genomics, genome-wide insights into the molecular events that underlie the symbiosis are possible. Over the past few years, significant numbers of expressed sequence tags (EST) have been generated from mycorrhizal cDNA libraries and deposited in open-access databases. The majority of these were generated from *M. truncatula* in association with a range of AM fungi, including *Glomus versiforme*, *G. intraradices* and *G. mosseae* (Table 10.1). The tissues from which the libraries were made cover a range of developmental stages so the collections include genes expressed at all stages of the association. The majority of the ESTs in these collections represent *M. truncatula* genes but a few *Glomus* genes are also represented (Journet *et al.*, 2002). Although genomics tools are available, the AM symbiosis still presents a number of unique challenges. The fungal symbionts are not easy to manipulate and the infection process is asynchronous. Only a few appressoria form on the surface of the roots, so the number of responding cells is always a minor component of the whole root. Following invasion, most of the interaction occurs in cells deep within the root cortex. Even in highly colonized roots, the biomass of the fungal symbiont is relatively small and the number of cells in which arbuscules develop, as a percentage of the whole root system, is small. Despite these challenges, *in silico* and transcriptional profiling analyses have enabled the identification of genes that are differentially regulated at different stages of the symbiosis and new insights into the AM symbiosis have been obtained (Journet *et al.*, 2002; Liu *et al.*, 2003; Wulf *et al.*, 2003; Kuster *et al.*, 2004).

Gene expression during the early stages of the AM symbiosis

The earliest visible changes in the AM fungus occur as it approaches the plant root. In the vicinity of the root, the hyphae undergo extensive branching, a process that is thought to increase the chance of subsequent contact with the root (Bécard *et al.*, 2004). Recently, specific plant sesquiterpenes that induce this response have been identified (Akiyama *et al.*, 2005). There is also evidence of a diffusible fungal signal that is perceived by the plant. By analogy to Nod factor, the mycorrhizal fungal signal molecule has been named Myc factor (Kosuta *et al.*, 2003). The molecular nature of the fungal signal molecule is currently unknown.

Following contact between the symbionts, the earliest events in the symbiosis include the formation of appressoria on the root epidermis. Penetration of the hyphae occurs via the appressoria and the fungus grows into the cortex of the root. All of these events occur in a non-synchronous manner.

Recently, with the use of genomic approaches, several groups of genes induced in these early stages of the symbiosis have been reported. Many of the genes are predicted to play a role in defence responses. This finding is consistent with earlier studies in which the transient induction of some defence response genes had been noted (Harrison, 1999; Ruiz-Lozano *et al.*, 1999; Blilou *et al.*, 2000; Bonanomi *et al.*, 2001). Transcriptional profiling of the AM symbiosis formed between *M. truncatula* and *G. versiforme* identified clusters of genes whose expression is altered during development of symbiosis (Liu *et al.*, 2003). One group of genes, labelled as cluster 2, showed a transient increase in expression during the early stages of the association. Within this cluster, a significant number of genes show similarity to defence- and stress-related genes or putative defence-associated signal transduction components. Of these early induced genes, seven out of fourteen are represented by ESTs present in other mycorrhizal cDNA libraries, including *M. truncatula/G. intraradices* and *M. truncatula/G. mosseae* libraries. All of these genes are also present in cDNA libraries from non-mycorrhizal tissues, including nodulated roots and pathogen-infected roots (Table 10.2). This indicates that these genes are not expressed exclusively in the mycorrhizal symbiosis. It is possible that they play a role in the coordination of symbiotic and defence programmes of the plant. AM-specific genes (cluster 1) were also described in this study (Liu *et al.*, 2003); however, the induction of expression of these genes occurred later than that of those in cluster 2. Because development of the symbiosis is asynchronous, further analyses are needed to determine the precise timing of induction of these AM-specific genes. It is possible that they also play a role in the early stages of the symbiosis.

The analysis of sequences from Suppression Subtractive Hybridization (SSH) libraries developed from colonized and non-colonized roots has proven a useful strategy to identify genes specifically induced during the symbiosis. Brechenmacher *et al.* (2004) constructed an SSH library by using RNA from roots of *M. truncatula* 20 days post-inoculation (dpi) with *G. mosseae*. RNA from non-mycorrhizal roots was used to drive the subtraction. Analysis of ESTs from this library, coupled with expression analysis by reverse Northern hybridization, enabled the identification of 12 plant genes induced in the *M. truncatula/G. mosseae* symbiosis (Brechenmacher *et al.*, 2004). Although the RNA used to construct the SSH library came from fully developed mycorrhizal root material (20 dpi), additional expression analysis, using RNA from *M. truncatula* 5 days post-inoculation with *G. mosseae*, allowed the identification of genes induced in the early stages of the symbiosis (Table 10.2). At 5 days post-inoculation, fungal development was limited to appressoria at the root surface. Six out of the 12 mycorrhiza-induced genes were shown to be up-regulated at 5 dpi (Brechenmacher *et al.*, 2004) (Table 10.2). Three of these genes are represented by ESTs present in non-mycorrhizal libraries; ESTs corresponding to the other three genes are present only in mycorrhizal libraries. BLAST results suggest that some of these genes are likely to be involved in plant defence responses. Genes whose EST distributions suggest that they are mycorrhiza-specific could be from either the plant or the fungal symbiont. The BLAST hits may provide a clue as to the likely origins of the gene.

Table 10.2. *Frequency and distribution of early mycorrhiza-induced genes in the* M. trucatula *Gene Index (TIGR)*

dpi, Days post-inoculation.

Gene ID	Putative annotation	Induction time	Myc 1[a]	Myc 2[b]	Others[c]	References
TC100453	Receptor-like kinase Xa21-binding protein 3	8 dpi	5	2	43	Liu *et al.*, 2003 (cluster 2)
TC100808	Syringolide-induced protein B15-3-5	8 dpi	3	1	35	"
TC94517	Ubiquilin 4	8 dpi	3	0	19	"
TC106537	Chalcone synthase	8 dpi	1	3	111	"
TC94011	Phi-1 protein	8 dpi	1	0	24	"
TC106632	HMGR	8 dpi	2	0	14	"
TC100257	Hypothetical protein	8 dpi	2	0	6	"
TC94597	SRC2	8 dpi	7	6	32	"
TC106633	HMGR	8 dpi	1	1	37	"
TC101486	Hypothetical protein	8 dpi	2	0	13	"
TC100498	Protein kinase	8 dpi	2	1	103	"
TC100501	Protein kinase	8 dpi	2	0	1	"
TC106447	Unknown protein	8 dpi	1	0	8	"
TC107128	Arabinogalactan protein	8 dpi	3	6	26	"
TC106808	Glutamine synthetase	20 dpi	11	8	54	Brechenmacher *et al.*, 2004
TC100851	Nodulin 26-like major intrinsic protein	5 dpi	0	7	27	"
TC109466	PR10	20dpi	0	4	2	"
TC95018	Germin-like protein 1	5 dpi	0	22	0	"
TC107737	Probable wound induced protein	20 dpi	0	1	6	"
TC100720	Glutathione-S-transferase	20 dpi	0	46	0	"
TC102881	Hypothetical protein	5 dpi	0	7	0	"
TC94155	Extensin 3 precursor	5 dpi	0	2	26	"
TC94753	Unknown	5 dpi	0	34	0	"
TC96715	GTPase putative	5 dpi	0	1	6	"
TC101627	Transfactor-like protein	20 dpi	2	9	4	"
AJ311244	Unknown (singleton)	20 dpi	0	1	0	"
TC94467	Fiber annexin	Pre-contact	0	7	57	Weidmann *et al.*, 2004
TC101563	Map kinase	Pre-contact	0	3	6	"

TC107357	Nitrate reductase	Pre-contact	3	2	19	”
TC97792	U1 snRNP protein A	Pre-contact	0	1	3	”
TC100220	DEAD box RNA helicase	Pre-contact	8	2	54	”
TC94493	eIF4-like protein	Pre-contact	0	3	55	”
TC106541	40 S ribosomal protein S3	Pre-contact	3	7	65	”
TC100906	60 S ribosomal protein L7-3	Pre-contact	1	1	30	”
TC106429	60 S ribosomal protein L18a	Pre-contact	0	5	24	”
TC106485	EF-1 alpha	Pre-contact	49	37	976	”
TC101200	CLB1 protein	Pre-contact	0	5	17	”
TC106949	Unknown	5dpi	0	3	18	”
BQ148998	Acyl CoA:diacylglycerol acyltransferase (singleton)	5dpi	0	1	0	”
TC107487	HMGI/Y	5dpi	0	3	18	”
TC106952	Isopentenyl pyrophosphate isomerase	5dpi	4	1	49	”
TC94937	Putative F1F0-ATPase inhibitor protein	5dpi	0	8	17	”
TC107141	Putative PGPD14 protein (pollen germination related protein)	5dpi	3	4	26	”
TC94425	Methylene-tetra-hydrofolate reductase (MTHFR2)	5dpi	1	3	60	”
TC106690	Putative Hs1pro-1 homologue	5dpi	19	5	120	”
TC100672	Nitrate reductase precursor	5dpi	11	5	39	”
TC94311	Receptor-like protein kinase homologue RK20-1	5dpi	15	7	100	”
TC100948	ADR6 protein	5dpi	6	3	21	”
TC106910	GDP-mannose pyrophosphorylase	5dpi	3	2	22	”
TC94484	Expansin	5dpi	0	1	12	”
TC101619	Extensin (proline rich-protein precursor)	5dpi	1	3	11	”
AL377325	Cyclin B (singleton)	5dpi	0	1	0	”
TC94399	Hypothetical protein	5dpi	10	6	59	”
TC94253	Putative albumin 1 precursor	5dpi	5	10	5	”

[a] Myc 1, MHAM-MHAM2 (cDNA libraries *M. truncatula*/*G. versiforme*)
[b] Myc 2, MtBC, MTAMP, MTGIM (*M. truncatula*/*G. intraradices*), SSH libraries *M. truncatula*/*G. mosseae*
[c] Others, all other *M. truncatula* libraries in the *M. trucatula* Gene Index (TIGR) except the libraries included in Myc 1 and Myc 2

In a second study utilizing the SSH library approach, Weidmann *et al.* (2004) identified genes induced in roots prior to appressorium formation. In this work, a library was constructed from *M. truncatula* roots 5 days post-inoculation with *G. mosseae*. Screening of this library led to the identification of 29 clones that were up-regulated in *G. mosseae*-inoculated *M. truncatula* roots in early stages of the symbiosis (Table 10.2). Additional expression analysis of 11 of these genes showed that they were up-regulated in plant roots that were separated from the fungus by a nitrocellulose membrane. This suggests that these genes are induced by a fungal signal capable of passing through the nitrocellulose. The expression of these genes during the pre-contact phase suggests that they may be involved in the earliest events of the mycorrhizal association (Weidmann *et al.*, 2004). The identification of 'pre-contact' genes is consistent with previous observations in which the induction of *MtENOD11* was detected in plant roots separated from the fungus by a Cellophane membrane (Kosuta *et al.*, 2003).

Collectively, these studies have resulted in lists of genes whose expression is activated during the pre-appressoria and appressoria stages of the AM symbiosis (Liu *et al.*, 2003; Brechenmacher *et al.*, 2004; Weidmann *et al.*, 2004). Interestingly, there is almost no overlap in the mycorrhiza-induced genes reported in these different studies. This might be due to the fact that the species of fungus and the experimental systems differ. Alternatively, it might suggest that the transcriptional events involved in the symbiosis are complex and the description of players is still far from complete. There is one example of a gene, however, that was identified as a mycorrhiza-induced gene by two different studies (Wulf *et al.*, 2003; Brechenmacher *et al.*, 2004). The gene, for a member of a germin-like protein family, was demonstrated to be induced early in the AM symbiosis (Brechenmacher *et al.*, 2004). Additional characterization revealed that it is expressed only in mycorrhizal roots and that transcripts are present in cells containing arbuscules (Doll *et al.*, 2003). As expression was also noted prior to arbuscule formation, this suggests that it may be expressed in other cell types also. Germin-like proteins are encoded by large multi-gene families and represent a very heterogeneous group of proteins. Some of them display oxalate oxidase activity, whereas others have been identified as superoxide dismutases (Yamahara *et al.*, 1999; Carter & Thornburg, 2000).

In addition to studies focused on the identification of plant genes, similar approaches have also been taken to identify AM fungal genes involved in different stages of the symbiosis. By differential display

Table 10.3. *Genes from arbuscular mycorrhizal fungi induced during the mycorrhizal symbiosis*

Gene ID	Putative annotation	Developmental stage	References
AJ419663	No hit	Pre-symbiotic	Tamasloukht *et al.*, 2003
AJ419664	Adrenal gland protein	Pre-symbiotic	"
AJ419665	P450-monooxygenase	Pre-symbiotic	"
AJ419666	NCO4	Pre-symbiotic	"
AJ419667	24*S* mt rRNA	Pre-symbiotic	"
AJ419668	24*S* mt rRNA	Pre-symbiotic	"
AJ419669	mt ATP/ADP translocase	Pre-symbiotic	"
AJ419670	Pyruvate carboxylase	Pre-symbiotic	"
CF803240	MAP3k-like protein kinase	Early appressoria	Breuninger & Requena, 2004
CF803244	Leucine zipper-transmembrane protein 1	Early appressoria	"
CF803272	Ca-promoted Ras inactivator (CAPRI)	Early appressoria	"
CF803271	Fatty acid CoA ligase	Early appressoria	"
CF803245	Increased rDNA-silencing; Irs4p	Early appressoria	"
CF803243	KARP-binding protein	Early appressoria	"
CF803295	Caspase-8	Early appressoria	"
CF803298	Similar to avrRpt2-induced protein	Early appressoria	"
CF803304	Hypothetical protein	Early appressoria	"
TC73828	Peptidyl prolyl cis–trans isomerase	Throughout life cycle	Brechenmacher *et al.*, 2004
TC66256	Thioredoxin homologue	Throughout life cycle	"
TC62645	No hit	Throughout life cycle	"
TC61523	No hit	Appressoria formation	"
TC71889	No hit	Appressoria formation	"
AJ311241	No hit (singleton)	Arbuscules	"

analysis and SSH, seven fungal genes were identified as up-regulated in the pre-symbiotic stages. These genes are proposed to be involved in the activation of respiration in the fungus in response to root exudates (Tamasloukht *et al.*, 2003) (Table 10.3). In addition, genes induced in early stages of the symbiosis were discovered by sequencing and analysing a SSH library (Breuninger & Requena, 2004) (Table 10.3). Nine *G. mosseae* genes that are up-regulated in the fungus during the recognition and appressoria formation stage were identified in this study. Based on their BLAST hits, these genes were predicted to be involved in signalling, metabolism, translation, defence and stress. Six more genes from *G. mosseae* were discovered by SSH, expression profiling and EST sequencing (Brechenmacher *et al.*, 2004). Expression analysis indicated that two of these genes may be associated with appressorium formation and one with arbuscule formation. The other three genes were expressed throughout the life cycle of the fungus (Table 10.3). Further analysis of these genes may provide insights into the events occurring in the fungus during the different phases of its interaction with the plant.

Gene expression associated with arbuscule development

Arbuscules are complex structures that are unique to this symbiosis and a unifying feature of the arbuscular mycorrhizal symbiosis. In the 'Arum' type AM associations arbuscules develop in the inner cortical cells (Gallaud, 1905). Hyphal branches that arise from the intercellular hyphae penetrate the cortical cell wall and the arbuscule develops inside the cell by repeated dichotomous branching. In many cases, the branching is extensive and the arbuscule fills the cortical cell. As the arbuscule develops, the plant plasma membrane extends around the hyphal branches so that the arbuscule remains in an apoplastic compartment (Harrison, 1997). It takes a considerable amount of membrane to envelop the arbuscule and it is estimated that the plant plasma membrane increases in length four-fold (Alexander *et al.*, 1989). The region of the membrane that surrounds the fungus is referred to as the peri-arbuscular membrane. As this membrane continues to synthesize cell wall material, an extracellular matrix layer is deposited around the arbuscule. Immunocytolocalization and cytochemistry analyses have demonstrated that this matrix is composed of the same components as the wall of the cell including pectins, xyloglucans, non-esterified polygalacturonans, arabinogalactans and hydroxyproline-rich glycoproteins (HRGP); H^+-ATPase activity has also been demonstrated in this compartment (Gianinazzi-Pearson *et al.*, 1991; Balestrini *et al.*, 1994; Bonfante and Bianciotto, 1995; Gollotte *et al.*, 1995; Perfect *et al.*, 1998).

Arbuscule development is accompanied by significant alterations in the cortical cell. The cytoskeleton rearranges, and the microtubules orient towards, and cover, the developing arbuscule (Genre & Bonfante, 1997, 1998, 1999; Blancaflor *et al.*, 2001). It is presumed that this alteration facilitates the trafficking of materials to the developing membrane and matrix. In addition, the large vacuole, characteristic of the cortical cell, fragments to form smaller vacuoles. The nucleus migrates towards the developing arbuscule and there is an increase in the number of organelles

(Carling & Brown, 1982; Balestrini *et al.*, 1992; Bonfante & Bianciotto, 1995). The membrane interface between the arbuscule and the cortical cell is extensive and phosphate transfer to the plant occurs at this location. It has been suggested, but not demonstrated directly, that the fungus obtains carbon over this interface also (Bonfante-Fasolo, 1984; Gianinazzi-Pearson *et al.*, 1996; Harrison *et al.*, 2002). Surprisingly, arbuscules have a relatively short life and after approximately ten days (depending on the species involved) the arbuscule collapses and dies, after which the cortical cell returns to its original state (Alexander *et al.*, 1988, 1989). The molecular events involved in the formation of the arbuscules are beginning to be revealed and genomics approaches are providing new target genes that may play a role in the development, function and turnover of these specialized structures.

Different strategies have been adopted to identify plant and fungal genes expressed during arbuscule development. Initial approaches included differential screening (Tahiri-Aloui & Antoniw, 1996; Murphy *et al.*, 1997; van Buuren *et al.*, 1999a,b) and differential display (Martin-Laurent *et al.*, 1996, 1997). Recently, analysis of SSH libraries, transcriptional profiling with macro- and micro-arrays (Liu *et al.*, 2003; Wulf *et al.*, 2003; Brechenmacher *et al.*, 2004; Breuninger & Requena, 2004; Grunwald *et al.*, 2004; Kuster *et al.*, 2004) and DNA *in silico* analyses of the EST databases have enabled the identification of mycorrhiza-induced genes that are likely to be associated with arbuscule development (Journet *et al.*, 2002; Manthey *et al.*, 2004) (Table 10.4). In some cases, targeted searches for specific genes that are predicted to be associated with arbuscule development or functioning have been performed. A *M. truncatula* mycorrhiza-induced phosphate transporter, *MtPT4*, was identified by this type of approach. Analysis of phosphate transporter sequences from a wide array of organisms led to the identification of phosphate-transporter-specific sequence motifs. The sequence motifs were subsequently used to screen the EST databases and *MtPT4* was identified (Harrison *et al.*, 2002). Genomics approaches were also used to identify a mycorrhiza-induced phosphate transporter in rice, although here the strategy was different. Bioinformatics tools were used to survey the complete rice genome and this enabled identification of all likely rice phosphate transporter sequences. RT–PCR analyses then enabled the identification of *OsPT11*, a gene expressed only in mycorrhizal roots (Paszkowski *et al.*, 2002).

Even though a large number of genes have been identified as induced in response to development of the AM symbiosis, the spatial expression patterns of only a few of these have been characterized. In order to

Table 10.4. *Genes expressed in cells containing arbuscules*

Gene ID	Putative annotation	Reference
Genes localized by in situ *hybridization/immunolocalization*		
X58180	*PAL* phenylalanine ammonia lyase	Harrison & Dixon, 1994
L02901	*CHS* chalcone synthase	Harrison & Dixon, 1994
M11939	*PAL* phenylalanine ammonia lyase	Blee & Anderson, 1996
M19052	*CHT* chitinase	Blee & Anderson, 1996
MTU38651	*Mtst1* hexose transporter	Harrison, 1996
	HPRP hydroxyproline-rich glycoprotein	Balestrini *et al.*, 1997
AF106929	*AGP* arabinogalactan protein	van Buuren *et al.*, 1999b
AF167326	Chitinase class III	Bonanomi *et al.*, 2001
AJ132892	*Mtha1* plasma membrane H^+-ATPase	Krajinski *et al.*, 2002
AY184807	*MtGlp1* germin-like protein	Doll *et al.*, 2003
M60171	*Tubα3* alpha tubulin	Bonfante *et al.*, 1996
AJ250864/ AJ251304	*AOS* allene oxide synthase	Hause *et al.*, 2002
X66376	*JIP23* jasmonate-induced protein	Hause *et al.*, 2002
Y14214/ AF022873	*LePT1* phosphate transporter	Rosewarne *et al.*, 1999
U92438	Acid invertase	Blee & Anderson, 2002
A29484	Sucrose synthase nodulin 100	Blee & Anderson, 2002
AJ583446	*DXR* 1-deoxy-d-xylulose 5-phosphate reductoisomerase	Hans *et al.*, 2004
Genes localized by using promoter–reporter gene fusion activation		
J03679	GST (stress-related PRP-1 regulated)	Strittmatter *et al.*, 1996
M80492	H^+-ATPase *pma2*	Gianinazzi-Pearson *et al.*, 2000
X66737	H^+-ATPase *pma4*	Gianinazzi-Pearson *et al.*, 2000
AJ297721	*MtENOD11* repetitive proline-rich protein (RPRP)	Journet *et al.*, 2001
AJ318822	Potato phosphate transporter	Rausch *et al.*, 2001
AY116211	*MtPT4* phosphate transporter	Harrison *et al.*, 2002
AJ564166	*VfLb29 Vicia faba* leghaemoglobin	Vieweg *et al.*, 2004
AY619008	*MtXTH1* xyloglucan endo-*trans*-glycosylase/hydrolase	Maldonado-Mendoza *et al.*, 2005
Genes localized by promoter–reporter gene fusions derived from SSH or EST analysis		
AY134608	*MtGST1* glutathione-S-transferase	Wulf *et al.*, 2003
AY308955	*MtCel1* cellulase	Liu *et al.*, 2003
AY308958	*MtSCP1* serine carboxypeptidase-like	Liu *et al.*, 2003
AJ276900	*MtTi1* trypsin inhibitor	Grunwald *et al.*, 2004
AC137078	*MtTubb1* β-tubulin	Manthey *et al.*, 2004
AC146590	*MtAnn2* annexin	Manthey *et al.*, 2004

determine whether a gene shows a spatial expression that correlates with development of arbuscules it is necessary to carry out either *in situ* hybridization or analysis of a promoter reporter gene construct in transgenic roots. These approaches clearly involve a substantial amount of additional work. Although it is currently difficult to obtain promoter sequences in a high throughput manner, the availability of complete genome sequences will make this possible in the future. Another tool that facilitates the study of the spatial expression patterns of genes involved in the AM symbiosis is the use of plants containing transgenic roots obtained by *Agrobacterium rhizogenes*-mediated transformation (Boisson-Dernier *et al.*, 2001). The plants develop a transgenic root system; by using standard procedures, they can be inoculated with AM fungi and a normal AM symbiosis develops. This methodology, coupled with easy identification of promoter sequences, will allow the rapid characterization of the spatial expression patterns of mycorrhiza-induced genes and the identification of those that are expressed coincident with arbuscule formation.

Mycorrhiza-induced genes that have been shown to be expressed in arbuscule-containing cells are listed in Table 10.4. Their putative functions range through defence- and stress-related activities, carbon and phosphate transport, cytoskeleton structure, hormone responses, signalling and cell wall modification. The spatial expression patterns of these genes vary. Some are expressed exclusively in the arbuscule-containing cortical cells; others are expressed in cells with arbuscules and also in other cell types. One class is expressed systemically throughout the root system. The *M. truncatula* mycorrhiza-induced phosphate transporter, *MtPT4*, is one example of a gene showing arbuscule-specific expression. *MtPT4* is located in the peri-arbuscular membrane in root cortical cells (Harrison *et al.*, 2002) where the protein is thought to function in the uptake of phosphate across the peri-arbuscular membrane. *MtGST1*, with homology to a 2,4-D-inducible glutathione transferase of soybean, shows expression in arbuscule-containing cells and also adjacent cells in the root cortex (Wulf *et al.*, 2003). Finally, *Mt-XTH1*, a xyloglucan endo-*trans*-glycosylase/hydrolase gene, shows expression throughout the root in both cells with arbuscules and all other cell types. The expression pattern suggests that it responds to a systemic signal (Maldonado-Mendoza *et al.*, 2005).

These analyses have identified genes whose expression patterns correlate with arbuscule development; the next stage is to identify their roles. Recently, it has been shown that RNA interference can be used for reverse genetics in *M. truncatula* (Limpens *et al.*, 2004) and *Lotus japonicus* (Kumagai & Kouchi, 2003). Coupled with the root transformation system, this approach will enable analysis of the functions of the mycorrhiza-induced, arbuscule-associated genes.

Genes encoding cell-wall-associated proteins expressed during the AM symbiosis

Throughout the symbiosis, the AM fungus lives in the apoplastic spaces of the root. Starting from the initial contact between the fungal hypha with the root surface, through entry of the hypha into the cortex, and through to arbuscule formation, the fungus remains in constant contact with the cell wall. Even the arbuscules, which form within the cortical cells, remain in an apoplastic compartment. In this case, it is a new apoplastic compartment that forms as a result of envelopment of the arbuscule by the peri-arbuscular membrane. Here, it is possible that there is targeted secretion of both enzymes and cell wall components to the new peri-arbuscular compartment. Based on the apoplastic location of the fungus, it might be predicted that the plant cell wall and cell wall proteins will have critical functions in the symbiosis. The large EST data sets coupled with a variety of molecular, *in silico* and transcriptional profiling approaches have enabled the identification of cell wall proteins that show mycorrhiza-regulated expression and therefore might be important in the AM symbiosis (Table 10.5) (Tahiri-Aloui & Antoniw, 1996; Murphy *et al.*, 1997; Lapopin *et al.*, 1999; van Buuren *et al.*, 1999b; Journet *et al.*, 2002; Liu *et al.*, 2003; Wulf *et al.*, 2003; Kuster *et al.*, 2004; Manthey *et al.*, 2004). This includes known cell-wall-associated genes and novel genes predicted to play a role in the cell wall modification during the establishment of the AM symbiosis.

Numerous genes predicted to encode proline-rich proteins are induced in mycorrhizal roots, including ENOD2 (van Rhijn *et al.*, 1997), ENOD5 (Albrecht *et al.*, 1998), ENOD12 (Albrecht *et al.*, 1998; Journet *et al.*, 2001) and ENOD11 (Journet *et al.*, 2001). In addition there are newly identified mycorrhiza-induced proline-rich proteins represented by ESTs MtC00003, AJ308153 and BE999060 (Table 10.5). This large number of mycorrhiza-regulated proline-rich proteins strongly suggests that they have a role in the development of the AM symbiosis. Many of these genes, including ENOD 2, 5, 12 and 11, are also induced in legumes during their interaction with *Rhizobium*. The roles of these proteins are currently unknown but it has been suggested that they may alter the composition and the amount of cross-linking in the wall, making it more porous and possibly easier for the microsymbionts to penetrate the wall (Journet *et al.*, 2001; Chabaud *et al.*, 2002).

Table 10.5. *Genes encoding proteins involved in cell wall modification expressed in mycorrhizal roots*

Gene ID	Putative annotation	References
—	Cellulose	Bonfante-Fasolo *et al.*, 1990
—	Pectins	Bonfante-Fasolo *et al.*, 1990
—	β-1,4-glucans	Balestrini *et al.*, 1994
—	Hydroxyproline-rich glycoproteins	Balestrini *et al.*, 1996; 1997
P11728	*MsENOD2*	van Rhijn *et al.*, 1997
AAB23536	*PsENOD5*	Albrecht *et al.*, 1998
AF106929	Arabinogalactan protein	van Buuren *et al.*, 1999b
X68032	*MtENOD12*, repetitive proline-rich protein	Albrecht *et al.*, 1998; Journet *et al.*, 2001
AJ297721	*MtENOD11*, repetitive proline-rich protein	Journet *et al.*, 2001; Chabaud *et al.*, 2002
AF093507	Xyloglucan endo-*trans*-glycosylase/ hydrolase (*Mt-XTH1*)	Maldonado-Mendoza *et al.*, 2005
MtC00003	Repetitive proline-rich *MtPRP1*-like protein	Journet *et al.*, 2002
MtC00316	Arabinogalactan-protein precursor	Journet *et al.*, 2002; Kuster *et al.*, 2004
MtC00582	Endo-β-1,4-glucanase	Journet *et al.*, 2002
MtC30178.2	Extensin	Journet *et al.*, 2002
AY308955	*MtCEL1*	Liu *et al.*, 2003
AW587066	*Pyrus communis* arabinogalactan-protein	Liu *et al.*, 2003
AW586261	Putative ripening-related protein	Liu *et al.*, 2003
AW587040	Putative ripening-related protein	Liu *et al.*, 2003
AW584547	Expansin	Liu *et al.*, 2004
AJ621870	Expansin	Weidmann *et al.*, 2004
AJ621871	Extensin-like protein S3	Weidmann *et al.*, 2004
AJ308165	Extensin	Grunwald *et al.*, 2004
AJ308140	Extensin	Grunwald *et al.*, 2004
AJ308153	Proline-rich protein	Grunwald *et al.*, 2004
BE999060	*MtN8*, proline-rich protein	Manthey *et al.*, 2004

Five genes predicted to encode extensin-like glycoproteins have been reported as induced in mycorrhizal roots. These proteins are hydroxyproline-rich and are also suggested to cause loosening of plant cell walls by altering the extent of cross-linking of cell wall polymers (Gray-Mitsumune *et al.*, 2004). Hydroxyproline-rich glycoproteins have been detected at the peri-arbuscular interface in leek (Bonfante-Fasolo *et al.*, 1991) and at the coil and peri-arbuscular interfaces in maize cells (Balestrini *et al.*, 1997). It is possible that the newly identified extensions will be located at these interfaces also. Two of the extensin-like glycoproteins showed induction in *Pisum sativum* mycorrhizal associations in wild-type lines but not in a mycorrhizal mutant, suggesting that expression of these genes is associated with growth of the fungus in the cortex (Grunwald *et al.*, 2004). The other two genes, *AW584547* and *AJ621870*, predicted to encode extensin-like proteins were also identified as up-regulated in the AM roots (Liu *et al.*, 2004; Weidmann *et al.*, 2004).

Two arabinogalactan proteins (AGPs) showed induction in AM roots (van Buuren *et al.*, 1999b; Journet *et al.*, 2002; Kuster *et al.*, 2004). AGPs are hydroxyproline-rich glycoproteins that show extensive glycosylation. They are predicted to have adhesive properties, and are suggested to participate in a range of plant cellular processes including cell wall assembly. Genes predicted to encode AGPs are up-regulated in a number of plant microbe associations, including the interaction between *Gunnera* and *Nostoc* (Rasmussen *et al.*, 1996; Showalter, 2001).

Three endo-β-1,4-glucanase (EGase) genes were shown to be up-regulated during the AM symbiosis. A detailed analysis of the expression patterns of one of these genes, *MtCel1*, was carried out, including analysis of the spatial expression patterns. These analyses revealed that expression of *MtCel1* occurred specifically in cells that contain arbuscules (Liu *et al.*, 2003). *MtCel1* shows most similarity to genes of the E-type EGase sub-family II, which has been suggested to be involved in cell wall biosynthesis (Brummell *et al.*, 1997). *MtCel1* is predicted to encode a membrane-anchored protein and it is possible that it is anchored in the peri-arbuscular membrane. If this is the case, it could be involved in the assembly of the cellulose–hemicellulose matrix that is deposited in the peri-arbuscular compartment. The induction of genes encoding EGases has been noted in other plant–microbe interactions, including plant–nematode interactions where plant EGases are induced during giant cell and syncytium formation (Goellner *et al.*, 2001). In the latter example, the EGases are not membrane-anchored proteins and are predicted to be secreted to the wall, where they are suggested to play a role in the degradation of the matrix. An EST, MtC00582, predicted to represent a new mycorrhiza- induced EGase, shares similarity with two EGases from strawberry fruit that were suggested to play a role in cell wall expansion and softening during ripening (Llop-Tous *et al.*, 1999). The distribution of the ESTs representing the *MtC00582* gene suggests that it is expressed in a wide range of tissues. There are other similarities between cell-wall-associated genes involved in fruit ripening and the

AM symbiosis. Two genes (*AW586261* and *AW587040*) that are induced in mycorrhizal roots and encode proteins of unknown function share similarity to genes induced in grapes during fruit ripening (Davies & Robinson, 2000). Fruit ripening and the mycorrhizal symbiosis are clearly very different; however, both processes are accompanied by significant modifications to cell wall architecture and these recent gene expression data suggest that some of the underlying molecular events may be similar. The expression of some of these cell-wall-associated genes appears to be highly specific and occurs only in cortical cells containing arbuscules. However, there are also alterations to the cell walls that occur systemically throughout the mycorrhizal root system. Analysis of a mycorrhiza-induced xyloglucan endo-*trans*-glycosylase/hydrolase (Mt-XTH1) revealed that it is expressed in cells throughout the mycorrhizal root system (Maldonado-Mendoza *et al.*, 2005). XTHs are enzymes that catalyse the hydrolysis and transglycosylation of xyloglucan polymers in plant cell walls. The authors proposed that the Mt-XTH1 might be involved in systemic modifications to the cell wall structure to facilitate development and proliferation of the fungus in the apoplast (Maldonado-Mendoza *et al.*, 2005).

Over the past few years, many laboratories have focused on the identification of genes expressed in mycorrhizal roots. These activities have resulted in a collection of genes whose expression patterns suggest that they may be important for the AM symbiosis. A variety of additional analyses have linked expression of these genes to a specific stage or structure of the symbiosis. In the future, it will be important to identify those that are critical for the symbiosis and their roles in the interaction.

As we look to the future, the genome of one mycorrhizal host plant, rice, has been sequenced and the sequencing of two additional host species, *Medicago truncatula* and *Lotus japonicus*, is in progress. Furthermore, the genome of the AM fungus *G. intraradices* will be sequenced also. This will provide additional information and tools with which to analyse the symbiosis. In addition, it will allow genome-wide comparisons of mycorrhiza-associated genes from monocotyledonous and dicotyledonous hosts, and the comparison of AM fungal symbionts with other symbiotic and non-symbiotic fungi.

Acknowledgements

Financial support for this work was provided by The National Science Foundation, Plant Genome Program (DBI-0110206) and Atlantic Philanthropies Inc.

References

Akiyama, K., Matsuzaki, K.-I. & Hayashi, H. (2005). Plant sesquiterpenes induce hyphal branching in arbuscular mycorrhizal fungi. *Nature* **435**, 824–7.

Albrecht, C., Geurts, R., Lapeyrie, F. & Bisseling, T. (1998). Endomycorrhizae and rhizobial nod factors activate signal transduction pathways inducing *PsENOD5* and *PsENOD12* expression in which Sym8 is a common step. *Plant Journal* **15**, 605–14.

Alexander, T., Meier, R., Toth, R. & Weber, H. C. (1988). Dynamics of arbuscule development and degeneration in mycorrhizas of *Triticum aestivum* L. and *Avena sativa* L. with reference to *Zea mays* L. *New Phytologist* **110**, 363–70.

Alexander, T., Toth, R., Meier, R. and Weber, H. C. (1989). Dynamics of arbuscule development and degeneration in onion, bean and tomato with reference to vesicular-arbuscular mycorrhizae in grasses. *Canadian Journal of Botany* **67**, 2505–13.

Bago, B., Pfeffer, P. E. & Shachar-Hill, Y. (2000). Carbon metabolism and transport in arbuscular mycorrhizas. *Plant Physiology* **124**, 949–57.

Balestrini, R., Berta, G. & Bonfante, P. (1992). The plant nucleus in mycorrhizal roots: positional and structural modifications. *Biology of the Cell* **75**, 235–43.

Balestrini, R., Romera, C., Puigdomenech, P. & Bonfante, P. (1994). Location of a cell-wall hydroxyproline-rich glycoprotein, cellulose and β-1,3-glucans in apical and differentiated regions of maize mycorrhizal roots. *Planta* **195**, 201–9.

Balestrini, R., Hahn, M. G., Faccio, A., Mendgen, K. & Bonfante, P. (1996). Differential localization of carbohydrate epitopes in plant cell walls in the presence and absence of arbuscular mycorrhizal fungi. *Plant Physiology* **111**, 203–13.

Balestrini, R., José-Estanyol, M., Puigdoménech, P. & Bonfante, P. (1997). Hydroxyproline-rich glycoprotein mRNA accumulation in maize root cells colonized by an arbuscular mycorrhizal fungus as revealed by *in situ* hybridization. *Protoplasma* **198**, 36–42.

Bécard, G. & Fortin, J. A. (1988). Early events of vesicular-arbuscular mycorrhiza formation on Ri T-DNA transformed roots. *New Phytologist* **108**, 211–18.

Bécard, G., Kosuta, S., Tamasloukht, M., Sejalon-Delmas, N. & Roux, C. (2004). Partner communication in the arbuscular mycorrhizal interaction. *Canadian Journal of Botany* **82**, 1186–97.

Blancaflor, E. B., Zhao, L. M. & Harrison, M. J. (2001). Microtubule organization in root cells of *Medicago truncatula* during development of an arbuscular mycorrhizal symbiosis with *Glomus versiforme*. *Protoplasma* **217**, 154–65.

Blee, K. A. & Anderson, A. J. (1996). Defense-related transcript accumulation in *Phaseolus vulgaris* L. colonized by the arbuscular mycorrhizal fungus *Glomus intraradices*. *Plant Physiology* **110**, 675–88.

Blee, K. A. & Anderson, A. J. (2002). Transcripts for genes encoding soluble acid invertase and sucrose synthase accumulate in root tip and cortical cells containing mycorrhizal arbuscules. *Plant Molecular Biology* **50**, 197–211.

Blilou, I., Bueno, P., Ocampo, J. A. & García-Garrido, J. M. (2000). Induction of catalase and ascorbate peroxidase activities in tobacco roots inoculated with the arbuscular mycorrhizal *Glomus mosseae*. *Mycological Research* **106**, 722–5.

Boisson-Dernier, A., Chabaud, M., Garcia, F., Becard, G., Rosenberg, C., & Barker, D. G. (2001). *Agrobacterium* rhizogenes-transformed roots of *Medicago truncatula* for the study of nitrogen-fixing and endomycorrhizal symbiotic associations. *Molecular Plant-Microbe Interactions* **14**, 695–700.

Bonanomi, A., Wiemken, A., Boller, T. & Salzer, P. (2001). Local induction of a mycorrhiza-specific class III chitinase gene in cortical root cells of *Medicago truncatula* containing developing or mature arbuscules. *Plant Biology* **3**, 194–9.

Bonfante, P. & Bianciotto, V. (1995). Presymbiotic versus symbiotic phase in arbuscular endomycorrhizal fungi: morphology and cytology. In *Mycorrhiza: Structure, Function, Molecular Biology and Biotechnology*, ed. A. Varma & B. Hock, pp. 229–47. Berlin: Springer-Verlag.

Bonfante, P., Bergero, R., Uribe, X., Romera, C., Rigau, J. & Puigdomenech, P. (1996). Transcriptional activation of a maize alpha-tubulin gene in mycorrhizal maize and transgenic tobacco plants. *Plant Journal* **9**, 737–43.

Bonfante-Fasolo, P. (1984). Anatomy and morphology of VA mycorrhizae. In *VA Mycorrhizae*, ed. C. L. Powell & D. J. Bagyaraj, pp. 5–33. Boca Raton, FL: CRC Press.

Bonfante-Fasolo, P., Vian, B., Perotto, S., Faccio A. & Knox, J. P. (1990). Cellulose and pectin localization in roots of mycorrhizal *Allium porrum*: labelling continuity between host cell wall and interfacial material. *Planta* **180**, 537–47.

Bonfante-Fasolo, P., Tamagnone, L., Peretto, R.,Esquerré-Tugayé, M. T., Mazau, D., Mosiniak, M. & Vian, B. (1991). Immunocytochemical location of hydroxyproline rich glycoproteins at the interface between a mycorrhizal fungus and its host plants. *Protoplasma* **165**, 127–38.

Brechenmacher, L., Weidmann, S., van Tuinen, D., Chatagnier, O., Gianinazzi, S., Franken, P. & Gianinazzi-Pearson, V. (2004). Expression profiling of up-regulated plant and fungal genes in early and late stages of *Medicago truncatula-Glomus mosseae* interactions. *Mycorrhiza* **14**, 253–62.

Breuninger, M. & Requena, N. (2004). Recognition events in AM symbiosis: analysis of fungal gene expression at the early appressorium stage. *Fungal Genetics and Biology* **41**, 794–804.

Brummell, D. A., Catala, C., Lashbrook, C. C. & Bennett, A. B. (1997). A membrane-anchored E-type endo-1,4-β-glucanase is localized on Golgi and plasma membranes in higher plants. *Proceedings of the National Academy of Sciences of the USA* **94**, 4794–9.

Carling, D. E. & Brown, M. F. (1982). Anatomy and physiology of vesicular-arbuscular and nonmycorrhizal roots. *Phytopathology* **72**, 1108–14.

Carter, C. & Thornburg, R. W. (2000). Tobacco Nectarin I. Purification and characterization of a germin-like manganese superoxide dismutase implicated in the defence of floral reproductive tissues. *Journal of Biological Chemistry* **275**, 36 726–33.

Chabaud, M., Venard, C., Defaux-Petras, A., Becard, G. & Barker, D. G. (2002). Targeted inoculation of *Medicago truncatula* in vitro root cultures reveals MtENOD11 expression during early stages of infection by arbuscular mycorrhizal fungi. *New Phytologist* **156**, 265–73.

Davies, C. & Robinson, S. P. (2000). Differential screening indicates a dramatic change in mRNA profiles during grape berry ripening: cloning and characterization of cDNAs encoding putative cell wall and stress response proteins. *Plant Physiology* **122**, 803–12.

Doll, J., Hause, B., Demchenko, K., Pawlowski, K. & Krajinski, F. (2003). A member of the germin-like protein family is a highly conserved mycorrhiza-specific induced gene. *Plant Cell Physiology* **44**, 1208–14.

Gallaud, I. (1905). Etudes sur les mycorhizes endotrophes. *Revue Générale de Botanique* **17**, 5–48.

Gehrig, A., Schüßler, A. & Kluge, M. (1996). *Geosiphon pyriforme*, a fungus forming endocytobiosis with *Nostoc* (Cyanobacteria) is an ancestral member of the Glomales: Evidence by SSU rRNA analysis. *Journal of Molecular Evolution* **43**, 71–81.

Genre, A. & Bonfante, P. (1997). A mycorrhizal fungus changes microtubule orientation in tobacco root cells. *Protoplasma* **199**, 30–8.

Genre, A. & Bonfante, P. (1998). Actin versus tubulin configuration in arbuscule-containing cells from mycorrhizal tobacco roots. *New Phytologist* **140**, 745–52.

Genre, A. & Bonfante, P. (1999). Cytoskeleton-related proteins in tobacco mycorrhizal cells: γ-tubulin and clathrin localisation. *European Journal of Histochemistry* **43**, 105–11.

Gianinazzi-Pearson, V. (1996). Plant cell responses to arbuscular mycorrhiza fungi: getting to the roots of the symbiosis. *Plant Cell* **8**, 1871–83.

Gianinazzi-Pearson, V., Smith, S. E., Gianinazzi, S. & Smith, F. A. (1991). Enzymatic studies on the metabolism of vesicular-arbuscular mycorrhizas. *New Phytologist* **117**, 61–74.

Gianinazzi-Pearson, V., Dumas-Gaudot, E., Gollotte, A., Tahiri-Alaoui, A. & Gianinazzi, S. (1996). Cellular and molecular defence-related root responses to invasion by arbuscular mycorrhizal fungi. *New Phytologist* **133**, 45–57.

Gianinazzi-Pearson, V., Arnould, C., Oufattole, M., Arango, M. & Gianinazzi, S. (2000). Differential activation of H^+-ATPase genes by an arbuscular mycorrhizal fungus in root cells of transgenic tobacco. *Planta* **211**, 609–13.

Goellner, M., Wang, X. & Davis, E. L. (2001). Endo-β-1,4-glucanase expression in compatible plant-nematode interactions. *Plant Cell* **13**, 2241–55.

Gollotte, A., Gianinazzi-Pearson, V. & Gianinazzi, S. (1995). Immunodetection of infection thread glycoprotein and arabinogalactan protein in wild type *Pisum sativum* (L.) or an isogenic mycorrhiza-resistant mutant interacting with *Glomus mosseae*. *Symbiosis* **18**, 69–85.

Gray-Mitsumune, M., Mellerowicz, E. J., Abe, H., Schrader, J., Winzell, A., Sterky, F., Blomqvist, K., McQueen-Mason, S., Teeri, T. T. & Sundberg, B. (2004). Expansins abundant in secondary xylem belong to subgroup A of the α-Expansin gene family. *Plant Physiology* **135**, 1552–64.

Grunwald, U., Nyamsuren, O., Tamasloukht, M. B., Lapopin, L., Becker, A., Mann, P., Gianinazzi-Pearson, V., Krajinski, F. & Franken, P. (2004). Identification of mycorrhiza-regulated genes with arbuscule development-related expression profile. *Plant Molecular Biology* **55**, 553–66.

Hans, J., Hause, B., Strack, D. & Walter, M. H. (2004). Cloning, characterization, and immunolocalization of a mycorrhiza-inducible 1-deoxy-D-xylulose 5-phosphate reductoisomerase in arbuscule-containing cells of maize. *Plant Physiology* **134**, 614–24.

Harrison, M. J. (1996). A sugar transporter from *Medicago truncatula*: altered expression pattern in roots during vesicular-arbuscular (VA) mycorrhizal associations. *Plant Journal* **9**, 491–503.

Harrison, M. J. (1997). The arbuscular mycorrhizal symbiosis: an underground association. *Trends in Plant Sciences* **2**, 54–6.

Harrison, M. J. (1999). Molecular and cellular aspects of the arbuscular mycorrhizal symbiosis. *Annual Reviews of Plant Physiology and Plant Molecular Biology* **50**, 361–89.

Harrison, M. J. (2005). Signaling in the arbuscular mycorrhizal symbiosis. *Annual Reviews of Microbiology* **59**, 19–42.

Harrison, M. J. & Dixon, R. A. (1994). Spatial patterns of expression of flavonoid/isoflavonoid pathway genes during interactions between roots of *Medicago truncatula* and the mycorrhizal fungus *Glomus versiforme*. *Plant Journal* **6**, 9–20.

Harrison, M. J., Dewbre, G. R. & Liu, J. (2002). A phosphate transporter from *Medicago truncatula* involved in the acquisition of phosphate released by arbuscular mycorrhizal fungi. *Plant Cell* **14**, 2413–29.

Hause, B., Maier, W., Miersch, O., Kramell, R. & Strack, D. (2002). Induction of jasmonate biosynthesis in arbuscular mycorrhizal barley roots. *Plant Physiology* **130**, 1213–20.

Hodge, A., Campbell, C. D. & Fitter, A. H. (2001). An arbuscular mycorrhizal fungus accelerates decomposition and acquires nitrogen directly from organic material. *Nature* **413**, 297.

Journet, E. P., El-Gachtouli, N., Vernoud, V., de Billy, F., Pichon, M., Dedieu, A., Arnould, C., Morandi, D., Barker, D. G. & Gianinazzi-Pearson, V. (2001). *Medicago truncatula* ENOD11: a novel RPRP-encoding early nodulin gene expressed during mycorrhization in arbuscule-containing cells. *Molecular Plant-Microbe Interactions* **14**, 737–48.

Journet, E. P., van Tuinen, D., Gouzy, J., Crespeau, H., Carreau, V., Farmer, M. J., Niebel, A., Schiex, T., Jaillon, O., Chatagnier, O., Godiard, L., Micheli, F., Kahn, D., Gianinazzi-Pearson, V. & Gamas, P. (2002). Exploring root symbiotic programs in the model legume *Medicago truncatula* using EST analysis. *Nucleic Acids Research* **30**, 5579–92.

Kosuta, S., Chabaud, M., Lougnon, G., Gough, C., Denarie, J., Barker, D. G. & Becard, G. (2003). A diffusible factor from arbuscular mycorrhizal fungi induces symbiosis-specific MtENOD11 expression in roots of *Medicago truncatula*. *Plant Physiology* **131**, 952–62.

Krajinski, F., Hause, B., Gianinazzi-Pearson, V. & Franken, P. (2002). Mtha1, a plasma membrane H^+-ATPase gene from Medicago truncatula, shows arbuscule-specific induced expression in mycorrhizal tissue. *Plant Biology* **4**, 754–61.

Kumagai, H. & Kouchi, H. (2003). Gene silencing by expression of hairpin RNA in *Lotus japonicus* roots and root nodules. *Molecular Plant-Microbe Interactions* **16**, 663–8.

Kuster, H., Hohnjec, N., Krajinski, F., El Yahyaoui, F., Manthey, K., Gouzy, J., Dondrup, M., Meyer, F., Kalinowski, J., Brechenmacher, L., van Tuinen, D., Gianinazzi-Pearson, V., Puhler, A., Gamas, P. & Becker, A. (2004). Construction and validation of cDNA-based Mt6k-RIT macro- and microarrays to explore root endosymbioses in the model legume *Medicago truncatula*. *Journal of Biotechnology* **108**, 95–113.

Lapopin, L., Gianinazzi-Pearson, V. & Franken, P. (1999). Comparative differential RNA display analysis of arbuscular mycorrhiza in *Pisum sativum* wild type and a mutant defective in late stage development. *Plant Molecular Biology* **41**, 669–77.

Limpens, E., Ramos, J., Franken, C., Raz, V., Compaan, B., Franssen, H., Bisseling, T. & Geurts, R. (2004). RNA interference in *Agrobacterium rhizogenes*-transformed roots of *Arabidopsis* and *Medicago truncatula*. *Journal of Experimental Botany* **55**, 983–92.

Liu, J., Blaylock, L. & Harrison, M. J. (2004). cDNA arrays as tools to identify mycorrhiza-regulated genes: indentification of mycorrhiza-induced genes that encode or generate signaling molecules implicated in the control of root growth. *Canadian Journal of Botany* **82**, 1177–85.

Liu, J. Y., Blaylock, L. A., Endre, G., Cho, J., Town, C. D., Vanden Bosch, K. A. & Harrison, M. J. (2003). Transcript profiling coupled with spatial expression analyses reveals genes involved in distinct developmental stages of an arbuscular mycorrhizal symbiosis. *Plant Cell* **15**, 2106–23.

Llop-Tous, I., Dominguez-Puigjaner, E., Palomer, X. and Vendrell, M. (1999). Characterization of two divergent endo-beta -1,4-glucanase cDNA clones highly expressed in the nonclimacteric strawberry fruit. *Plant Physiology* **119**, 1415–22.

Maldonado-Mendoza, I. E., Dewbre, G. R., Blaylock, L. & Harrison, M. J. (2005). Expression of a xyloglucan endotransglucosylase/hydrolase gene, *Mt-XTH1*, from *Medicago truncatula* is induced systemically in mycorrhizal roots. *Gene* **345**, 191–7.

Manthey, K., Krajinski, F., Hohnjec, N., Firnhaber, C., Puhler, A., Perlick, A. M. & Kuster, H. (2004). Transcriptome profiling in root nodules and arbuscular mycorrhiza identifies a collection of novel genes induced during *Medicago truncatula* root endosymbioses. *Molecular Plant-Microbe Interactions* **17**, 1063–77.

Martin-Laurent, F., van Tuinen, D., Dumas-Gaudot, E., Gianinazzi-Pearson, V., Gianinazzi, S. & Franken, P. (1997). Differential display analysis of RNA accumulation in arbuscular mycorrhiza of pea and isolation of a novel symbiosis-regulated plant gene. *Molecular and General Genetics* **256**, 37–44.

Martin-Laurent, F. A., Franken, P., van Tuinen, D., Dumas-Gaudot, E., Schlichter, U., Antoniw, J. F., Gianinazzi-Pearson, V. & Gianinazzi, S. (1996). Differential display reverse transcription polymerase chain reaction: a new approach to detect symbiosis-related genes involved in arbuscular mycorrhiza. In *Mycorrhizas in Integrated Systems: from Genes to Plant Development*, ed. C. Azcón-Aguilar & J. M. Barea, pp. 195–8. Brussels, Belgium: European Commission.

Murphy, P. J., Langridge, P. & Smith, S. E. (1997). Cloning plant genes differentially expressed during colonization of roots of *Hordeum vulgare* by the vesicular-arbuscular mycorrhizal fungus *Glomus intraradices*. *New Phytologist* **135**, 291–301.

Parniske, M. (2004). Molecular genetics of the arbuscular mycorrhizal symbiosis. *Current Opinion in Plant Biology* **7**, 414–21.

Paszkowski, U., Kroken, S., Roux, C. and Briggs, S. P. (2002). Rice phosphate transporters include an evolutionarily divergent gene specifically activated in arbuscular mycorrhizal symbiosis. *Proceedings of the National Academy of Sciences of the USA* **99**, 13 324–9.

Perfect, S. E., O'Connell, R. J., Green, E. F., Doering-Saad, C. & Green, J. R. (1998). Expression cloning of a fungal proline-rich glycoprotein specific to the biotrophic interface formed in the *Colletotrichum*-bean interaction. *Plant Journal* **15**, 273–9.

Pirozynski, K. A. & Malloch, D. W. (1975). The origin of land plants: a matter of mycotrophism. *Biosystems* **6**, 153–64.

Rasmussen, U., Johansson, C., Renglin, A., Peterson, C. & Bergman, B. (1996). A molecular characterization of the *Gunnera-Nostoc* symbiosis: comparison with *Rhizobium*- and *Agrobacterium*-plant interactions. *New Phytologist* **133**, 391–8.

Rausch, C., Daram, P., Brunner, S., Jansa, J., Laloi, M., Leggewie, G. & Bucher, M. (2001). A phosphate transporter expressed in arbuscule-containing cells in potato. *Nature* **414**, 462–6.

Redeker, D., Kodner, R. & Graham, L. (2000). Glomalean fungi from the Ordovician. *Science* **289**, 1920–1.

Remy, W., Taylor, T. N., Hass, H. & Kerp, H. (1994). Four hundred-million-year-old vesicular arbuscular mycorrhizae. *Proceedings of the National Academy of Sciences of the USA* **91**, 11 841–3.

Rosewarne, G., Barker, S., Smith, S., Smith, F. & Schachtman, D. (1999). A *Lycopersicon esculentum* phosphate transporter (LePT1) involved in phosphorus uptake from a vesicular-arbuscular mycorrhizal fungus. *New Phytologist* **144**, 507–16.

Ruiz-Lozano, J. M., Roussel, H., Gianinazzi, S. & Gianinazzi-Pearson, V. (1999). Defense genes are differentially induced by a mycorrhizal fungus and *Rhizobium* sp. in wild-type and symbiosis-defective pea genotypes. *Molecular Plant-Microbe Interactions* **12**, 976–84.

Schüßler, A. (2000). *Glomus claroideum* forms an arbuscular mycorrhiza-like symbiosis with the hornwort *Anthoceros punctatus*. *Mycorrhiza* **10**, 15–21.

Schüßler, A., Schwarzott, D. & Walker, C. (2001). A new fungal phylum, the Glomeromycota: phylogeny and evolution. *Mycological Research* **105**, 1414–21.

Showalter, A. M. (2001). Arabinogalactan-proteins: structure, expression and function. *Cell Molecular Life Sciences* **58**, 1399–417.

Smith, S. E. & Read, D. J. (1997). *Mycorrhizal Symbiosis*. San Diego, CA: Academic Press.

Smith, S. E., Smith, F. A., and Jakobsen, I. (2003). Mycorrhizal fungi can dominate phosphate supply to plants irrespective of growth responses. *Plant Physiology* **133**, 16–20.

St-Arnaud, M., Hamel, C., Vimard, B., Caron, M. & Fortin, J. A. (1996). Enhanced hyphal growth and spore production of the arbuscular mycorrhizal fungus *Glomus intraradices* in an *in vitro* system in the absence of host roots. *Mycological Research* **100**, 328–32.

Strittmatter, G., Gheysen, G., Gianinazzi-Pearson, V., Hahn, K., Niebel, A., Rohde, W. & Tacke, E. (1996). Infections with various types of organisms stimulate transcription from a short promoter fragment of the potato *gst1* gene. *Molecular Plant-Microbe Interactions* **9**, 68–73.

Tahiri-Alaoui, A. & Antoniw, J. F. (1996). Cloning of genes associated with the colonization of tomato roots by the arbuscular mycorrhizal fungus *Glomus mosseae*. *Agronomie* **16**, 699–707.

Tamasloukht, M., Sejalon-Delmas, N., Kluever, A., Jauneau, A., Roux, C., Becard, G. & Franken, P. (2003). Root factors induce mitochondrial-related gene expression and fungal respiration during the developmental switch from asymbiosis to presymbiosis in the arbuscular mycorrhizal fungus *Gigaspora rosea*. *Plant Physiology* **131**, 1468–78.

van Buuren, M. L., Trieu, A. T., Blaylock, L. A. & Harrison, M. J. (1999a). Isolation of genes induced during arbuscular mycorrhizal associations using a combination of subtractive hybridization and differential screening. In *Proceedings of the American Phytopathological Society 1998*, ed. G. K. Podila and J. Douds, pp. 91–9. St. Paul, MN: APS Press.

van Buuren, M. L., Maldonado-Mendoza, I. E., Trieu, A. T., Blaylock, L. A. & Harrison, M. J. (1999b). Novel genes induced during an arbuscular mycorrhizal (AM) symbiosis between *M. truncatula* and *G. versiforme*. *Molecular Plant-Microbe Interactions* **12**, 171–81.

van der Heijden, M. G. A., Klironomos, J. N., Ursic, M., Moutoglis, P., Streitwolf-Engel, R., Boller, T., Wiemken, A. & Sanders, I. R. (1998). Mycorrhizal fungal diversity determines plant biodiversity, ecosystem variability and productivity. *Nature* **396**, 69–72.

van Rhijn, P., Fang, Y., Galili, S., Shaul, O., Atzmon, N., Wininger, S., Eshed, Y., Lum, M., Li, Y., To, V., Fujishige, N., Kapulnik, Y. & Hirsch, A. M. (1997). Expression of early nodulin genes in alfalfa mycorrhizae indicates that signal transduction pathways used in forming arbuscular mycorrhizae and *Rhizobium*-induced nodules may be conserved. *Proceedings of the National Academy of Sciences of the USA* **94**, 5467–72.

Vieweg, M. F., Fruhling, M., Quandt, H. J., Heim, U., Baumlein, H., Puhler, A., Kuster, H. & Perlick, A. M. (2004). The promoter of the *Vicia faba* L. leghemoglobin gene VfLb29 is specifically activated in the infected cells of root nodules and in the arbuscule-containing cells of mycorrhizal roots from different legume and non-legume plants. *Molecular Plant-Microbe Interactions* **17**, 62–9.

Weidmann, S., Sanchez, L., Descombin, J., Chatagnier, O., Gianinazzi, S. & Gianinazzi-Pearson, V. (2004). Fungal elicitation of signal transduction-related plant genes precedes mycorrhiza establishment and requires the Dmi3 gene in *Medicago truncatula*. *Molecular Plant-Microbe Interactions* **17**, 1385–93.

Wulf, A., Manthey, K., Doll, J., Perlick, A. M., Linke, B., Bekel, T., Meyer, F., Franken, P., Kuster, H. & Krajinski, F. (2003). Transcriptional changes in response to arbuscular mycorrhiza development in the model plant *Medicago truncatula*. *Molecular Plant-Microbe Interactions* **16**, 306–14.

Yamahara, T., Shiono, T., Suzuki, T., Tanaka, K., Takio, S., Sato, K., Yamazaki, S. & Satoh, T. (1999). Isolation of a germin-like protein with manganese superoxide dismutase activity from cells of a moss, *Barbula unguiculata*. *Journal of Biological Chemistry* **274**, 33274–8.

IV

Pathogenic interactions in the environment

11

Functional genomics of plant infection by the rice blast fungus *Magnaporthe grisea*

JOANNA M. JENKINSON, RICHARD A. WILSON, ZACHARY CARTWRIGHT, DARREN M. SOANES, MICHAEL J. KERSHAW, AND NICHOLAS J. TALBOT

School of Biosciences, University of Exeter

An introduction to the phytopathogenic fungus *Magnaporthe grisea*

Magnaporthe grisea is a heterothallic, phytopathogenic ascomycete capable of infecting over 50 species of grass (Ou, 1985). The most economically important of the species infected by *M. grisea* is rice (Ou, 1985; Rossman *et al.*, 1990), which is the staple diet of almost half the global human population. Rice blast disease, caused by *M. grisea*, is an extremely serious disease; despite modern advances, such as the development of fungicides and breeding of resistant rice cultivars, every year between 11% and 30% of the rice harvest is destroyed by this disease. A serious blast epidemic occurred in Bhutan in 1995 in which 1090 tonnes of rice was lost, with up to 100% crop losses for some farmers (Thinlay *et al.*, 2000). The American Centre for Disease Control and Prevention has also classified rice blast disease as a significant biological weapon that could be deployed in acts of agricultural bioterrorism (Schaad *et al.*, 2003).

Rice blast disease manifests itself as a number of different pathologies affecting stems, leaves and panicles of the rice plant (Talbot, 2003). Blast infections of stem nodes, for example, can cause the rice stem to rot before maturation of the seed and can result in complete loss of the rice crop (Ou, 1985). If leaves of rice seedlings are infected, a reduction in photosynthetic capacity can occur; growth is therefore impeded and seedlings often die. Rice blast symptoms appear on leaves as ellipsoid, brown necrotic lesions, which take 4–5 days to develop after initial infection (Talbot, 2003). The life cycle of *M. grisea* is initiated when three-celled, asexual spores called

Fungi in the Environment, ed. G. M. Gadd, S. C. Watkinson & P. S. Dyer. Published by Cambridge University Press.

conidia, which are dispersed by wind, dew or rain splash, attach to the hydrophobic rice leaf surface (Ou, 1985; Talbot, 1995). The conidial apex releases an adhesive, which attaches the spore tightly to the leaf surface (Hamer *et al.*, 1988); within an hour a germ tube 15–30 μm in length develops. Within four hours the tube 'hooks', changes direction, swells and flattens against the surface of the rice leaf (Bourett & Howard, 1990). This marks the beginning of differentiation into a specialized infection structure called the appressorium (Gilbert *et al.*, 1996; De Zwaan *et al.*, 1999). As the appressorium develops, an inner layer of melanin and an outer layer of chitin are deposited in the appressorium cell wall; these layers are required for the appressorium to be able to generate up to 8 MPa of pressure (Bourett & Howard, 1990; de Jong *et al.*, 1997). The appressorium produces a penetration peg in a small melanin-free region of the appressorium in contact with the leaf. The turgor within the appressorium enables the penetration peg to pierce the tough plant epidermis (Howard *et al.*, 1991). This is in contrast to the stomatal entrance strategy employed by many biotrophic fungi (for a review, see Voegele & Mendgen, 2003).

After the penetration peg has entered the plant cell, it differentiates into branched intracellular hyphae, which colonize the plant and eventually produce spores to continue the life cycle (Heath *et al.*, 1990). In the field, asexual reproduction is thought to predominate; this has been suggested by population-level studies of the fungus carried out in Europe, the United States, Cambodia and the Philippines, which showed typically clonal populations of *M. grisea* with relatively low genotypic diversity (Zeigler, 1998). However, recent studies in regions close to the Himalayas have shown that this may not always be the case (Zeigler, 1998). The high genotypic diversity of *M. grisea* in the centre of diversity of the fungus suggests that sexual reproduction can occur in the field, and it may be possible that there are higher levels of genetic recombination and gene flow in certain populations of rice blast (Rathour *et al.*, 2004). Sexual reproduction requires fertile and opposite mating types, conditioned by two genes in *M. grisea*, *MAT1–1* and *MAT1–2*. Opposite mating types form sexual fruiting bodies, called perithecia, when grown together (Talbot, 2003). Perithecia produce an abundance of ascospores, which consist of four pairs of ascospores (Valent & Chumley, 1991). Ascospores can initiate a disease cycle just as asexually generated conidia can, and develop appressoria in the same way (Bourett & Howard, 1990).

There is evidence to show that *M. grisea* is able to infect roots, at least under laboratory conditions (Sesma & Osbourn, 2004). The last study furthermore provided evidence that *M. grisea* may have had an ancestral relationship to soil-borne pathogenic fungi. Up to 10% of rice plants infected via root inoculation showed fungal lesions on leaves, indicating systemic disease (Sesma & Osbourn, 2004). The appressorium, which is required for the fungus to penetrate the plant leaf epidermis, is not developed during root infection. The structures employed to penetrate the roots are similar to the hyphopodia of root-infecting fungi such as *Gaeumannomyces graminis*. Interestingly, gene-for-gene resistance, typically associated with leaf and node blast infections, also operates during root infection (Sesma & Osbourn, 2004).

Originally, the differentiation between fungal lifestyles in terms of necrotrophy and biotrophy was based on the method by which they derive their nutrition. Necrotrophic fungi kill the host plant cells and utilize the dead material as a nutritional source. Biotrophs, in contrast, derive their nutrition from living host tissue (Oliver & Ipcho, 2004). However, a number of fungal species have a period of biotrophic growth before changing their metabolism to necrotrophic growth during disease symptom development. Because of this, a third class of pathogenic fungi, the hemibiotrophs, can be defined (Oliver & Ipcho, 2004). The foundation of this class occurred as knowledge about fungal–plant interactions, phylogeny, mechanisms of infection and operation of defence pathways increased. For example, it is now becoming increasingly apparent that biotrophs and necrotrophs trigger different host defence pathways. Biotrophs appear to induce salicylate-dependent defence pathways (Reuber *et al.*, 1998) whereas necrotrophs induce jasmonate and ethylene-dependent defence pathways (Thomma *et al.*, 2001), although there is interaction between the two pathways (Schenk *et al.*, 2000). The differentiated response to fungal pathogens is, however, making the appropriateness of current definitions less clear (Oliver & Ipcho, 2004). Some fungal species, for example, do not fit clearly into any of these classes, with *M. grisea* being a classic example that is frequently classified as both a necrotroph and a hemibiotroph (Mendgen & Hahn, 2002; Hammond-Kosack & Parker, 2003; Asiegbu *et al.*, 2004), which is not an uncommon phenomenon (Oliver & Ipcho, 2004). *M. grisea* is still classified as a necrotroph by many observers because of the physical mechanism by which *M. grisea* enters the plant. Others, however, classify *M. grisea* as a hemibiotroph because *M. grisea* does not induce overt disease symptoms until 3 days after infection, and host cells are not damaged during the initial infection period (Talbot, 2003).

An introduction to functional genomics

Functional genomics is a term that encompasses a large number of genome-level analytical techniques including expression profiling, high-throughput genetic modification and gene mapping. The numbers of genomes that have been, or are in the process of being, sequenced, and the number of expressed sequence tag (EST) collections available to researchers, are the most obvious results of genomic technology. However, this wealth of information is now being exploited and interpreted by using genome-wide forms of functional genetic analysis such as high-throughput targeted mutation and post-genomic technologies, such as proteomics and transcriptional profiling, that are now giving new insights into the importance of patterns of protein abundance and networks of gene expression. Genomic technologies themselves will ultimately facilitate a systems biology approach that allows the study of organisms in terms of interacting components and integrated systems (Aderem, 2005).

Currently (March 2006) the complete published genomes of 355 organisms are listed on the GOLD online database (www.genomesonline.org) with a further 984 ongoing prokaryotic genome projects and 590 ongoing eukaryotic genome projects. Of the published 355 genomes, 26 are Archaea, 41 are eukaryotic and 207 are bacterial genomes. On initial inspection it appears that the fungal research community is lagging behind the bacterial research community but this difference simply reflects the relative sizes of the genomes and associated complexity of bacteria and fungi. In fact, of the 41 published, completed eukaryotic genomes, 13 are from fungal species. These are *Ashbya (Eremothecium) gossypii, Aspergillus fumigatus, Aspergillus oryzae, Cryptococcus neoformans, Candida glabrata, Debaryomyces hansenii* var. *hansenii, Encephalitozoon cuniculi, Kluveromyces lactis, Neurospora crassa, Phanerochaete chrysosporium, Saccharomyces cerevisiae, Schizosaccharomyces pombe* and *Yarrowia lipolytica* (www.genomesonline.org). However, there are an incredible 158 further large-scale fungal sequencing projects currently ongoing. These include 13 EST, 19 random sequence tag (RST) and 136 full genome projects (www.genomesonline.org).

The Broad Institute is currently sequencing 25 fungal genomes as part of the Fungal Genomes Initiative (http://www.broad.mit.edu/annotation/fungi/fgi/status.html). Having already completed *N. crassa*, a further 22 assemblies have been released, including the *M. grisea* genome, which is currently being finished, the first draft having been released in 2002 and the paper detailing the project appearing recently (Dean *et al*., 2005). Clearly, the number and quality of fungal genomes available for comparison is ever increasing.

The first eukaryotic genome to be sequenced was that of the budding yeast *Saccharomyces cerevisiae* in 1996 (Goffeau *et al.*, 1996). No other fungal genome sequences were published until that of the fission yeast *Schizosaccharomyces pombe* in 2002 (Wood *et al.*, 2002). Much of the analysis of the *S. cerevisiae* genome over the past nine years is available in databases such as the Comprehensive MIPS Yeast Genome Database (http://mips.gsf.de/proj/yeast/CYGD/db) and Saccharomyces Genome Database (www.yeastgenome.org) at Stanford University, which not only detail predicted open reading frames but incorporate information available on deletion mutant phenotypes and microarray and proteomic analyses. This offers an exciting glimpse into what could be achieved for other filamentous fungi, although this will require improvements in current technology for generation of whole-genome gene knockout collections and a high level of collaboration.

The initiation of such a relational database for *M. grisea* is being carried out at the MGOS website (www.mgosdb.org) as part of the NSF-funded genome analysis of pathogen–host recognition and subsequent responses in the rice blast patho-system (Dean *et al.*, 2005). This project has generated over 50 000 insertional mutants, a number of which have been assayed for growth rate, conidiation, pigmentation, auxotrophy and pathogenicity. The proportions of mutants showing aberration in these categories can be found at www.mgosdb.org/cgibin/mutant/mutant_stats.cgi; for example, currently (March 2006) of the 57 223 mutants assayed 3.18% show a non-pathogenic phenotype. Other data that can be retrieved from this database include: searchable serial analysis of gene expression (SAGE) libraries from 24 hour and 96 hour post-infections of rice with *M. grisea*; and a *M. grisea* SAGE library for 3 days of growth on minimal medium.

Functional genomics requires high-throughput analyses. A recent review of the techniques involved such as microarrays, targeted gene deletion, gene silencing and two-hybrid analyses has been published by Veneault-Fourrey & Talbot (2005). As a result, this chapter concentrates on the initial analysis of the *M. grisea* genome, from which a number of interesting conclusions can already be drawn, and then we focus on the published analysis of pathogenicity determinants from *M. grisea*.

The *M. grisea* genome sequence

In the past few years, draft genome sequences from a number of fungi have become available, including the phytopathogens *Magnaporthe grisea* (Dean *et al.*, 2005), *Ustilago maydis* (www.broad.mit.edu/annotation/fungi/ustilago_maydis/) and *Fusarium graminearum* (www.broad.mit.edu/

annotation/fungi/fusarium/). The availability of genome sequences from a range of pathogenic and non-pathogenic fungi enables us to perform the kind of comparative functional genomics that up to now could only be performed with yeast species (Dujon *et al.*, 2004). The sequence coverage of the *M. grisea* genome is greater than seven-fold and the number of predicted genes is 11 109 (Dean *et al.*, 2005). This is comparable to the number of predicted genes in the genome of *Neurospora crassa* (10 082) (Galagan *et al.*, 2003) but nearly double that of *S. cerevisiae* (6591) (Goffeau *et al.*, 1996). The apparent greater complexity of the genomes of the filamentous ascomycetes when compared with that of the unicellular yeasts may reflect the multicellular nature, and more diverse life histories and ecologies, of filamentous species.

Comparison of gene inventories between both pathogenic and non-pathogenic species of fungi enables us to ask the question: 'What makes a pathogen different from a non-pathogen?' In fact, there may be more than one answer to this question because pathogens are present in a wide range of taxonomic groups, often closely related to non-pathogens, and therefore pathogenicity is likely to have evolved many times within the kingdom Fungi. It can be speculated that there are three possible mechanisms that account for the evolution of pathogenic species (Tunlid & Talbot, 2002). First, the genomes of pathogens may have acquired novel genes that enable them to infect and colonize plants. These genes may have been acquired by horizontal transfer, which is defined as the stable transfer of genetic material between individuals not directly attributable to vertical (i.e. meiotic or mitotic) processes. Horizontal gene transfer has been demonstrated in bacteria, but there are only a few possible examples where this may have occurred in fungi (Rosewich & Kistler, 2000). Alternatively, they may have evolved by the duplication of ancestral genes and subsequent divergence of function due to the accumulation of mutations. Gene duplication is thought to have a primary role in the innovation of new genes (Ohno, 1970). If this is the case then certain gene families may contain more members in the genomes of pathogenic species compared with non-pathogens. For example, there are 122 predicted cytochrome P450 encoding genes in the genome of the plant *M. grisea* but only 37 in the genome of the closely related saprotroph *N. crassa* (Dean *et al.*, 2005). Cytochrome P450s have been shown to be involved in a number of processes essential for pathogenicity, including the biosynthesis of toxins and the detoxification of anti-fungal compounds (van den Brink *et al.*, 1998).

There are also nine putative cutinase-encoding genes in the genome of *M. grisea* but none in the genome of *N. crassa*. Cutinases are secreted methyl esterases that break down cutin, a waxy polymer that covers the surface of plant leaves. Based on the deletion of a cutinase-encoding gene, *CUT1*, in *M. grisea* it had been suggested that cutinases are not essential for the infection of the rice-host by this pathogen (Sweigard *et al.*, 1992). This hypothesis has been brought into question, however, by the discovery of another eight cutinase-encoding genes in the *M. grisea* genome, some of which are expressed in appressoria. The activities of these enzymes are therefore likely to have compensated for the loss of *CUT1* (Dean *et al.*, 2005).

Second, there may be genes in pathogens that are also present in non-pathogens but evolve a different role in pathogenic species. In this case, the difference between the pathogen and the non-pathogen will reside in the regulation of the gene. There are many examples of this, mainly concerning genes involved in signal transduction, such as mitogen-activated protein kinases (Xu & Hamer, 1996), adenylate cyclases (Choi & Dean, 1997), G-proteins (Liu & Dean, 1997) and cyclic AMP-dependent protein kinases (Mitchell & Dean, 1995).

There are also examples of metabolites that have different roles in pathogenic and non-pathogenic fungi. For example, *S. cerevisiae* accumulates glycerol in response to hyperosmotic shock in order to balance cellular osmotic pressure with the external environment (O'Rourke *et al.*, 2002). In contrast, *M. grisea* accumulates arabitol as its major compatible solute in response to osmotic shock (Dixon *et al.*, 1999). Accumulation of glycerol generates enormous turgor pressure in the appressorium that enables it to rupture the plant cuticle mechanically (de Jong *et al.*, 1997). Homologues of the genes encoding glycerol-synthesizing enzymes in yeast are also present in the genome of *M. grisea* but presumably their activity is controlled by different environmental cues (Dean *et al.*, 2005).

Third, it may be the case that pathogenicity is associated with gene loss. There are few documented examples of this strategy in eukaryotes, but analysis of the *Mycobacterium leprae* genome in comparison with that of *M. tuberculosis* provided evidence for such a mechanism in prokaryotic pathogens (Harrison & Gerstein, 2002).

Analysis of the EST collections of *M. grisea*

The functional analysis of genes involved in plant pathogenesis requires the combination of several investigative procedures including confirmation of gene expression, annotation of gene sequence and gene organization, the spatial and temporal pattern of gene expression, and abundance of transcript (Soanes *et al.*, 2002b). Expressed sequence tags

(ESTs) are single-pass, partial sequences of either 3′ or 5′ ends of complementary DNA (cDNA) clones and represent a rapid way of generating sequence information (Skinner *et al.*, 2001). ESTs are therefore directly derived from messenger RNA, and therefore confirm the expression of a given gene. In addition, comparison of EST-derived sequences with those available from genomic DNA allows intron positioning and 5′ and 3′ UTR definition. EST data sets are available in the public domain for a number of phytopathogenic fungi, many of which have no published genomic sequence. The published EST data are generally poorly annotated and in a flat-file format. The COGEME (Consortium for the Genomics of Microbial Eukaryotes) EST database was constructed to act as a resource for the phytopathogen community (Soanes *et al.*, 2002b). EST sequences were clustered into unisequences (unique sequences) to reduce redundancy and improve sequence quality. Unisequences were then annotated based on homology to known gene sequences and classified by function based on these annotations. Figure 11.1 shows the proportion of unisequences from *M. grisea* assigned to each functional category. The function of 70% of the *M. grisea* unisequences is still unknown, either because they have no

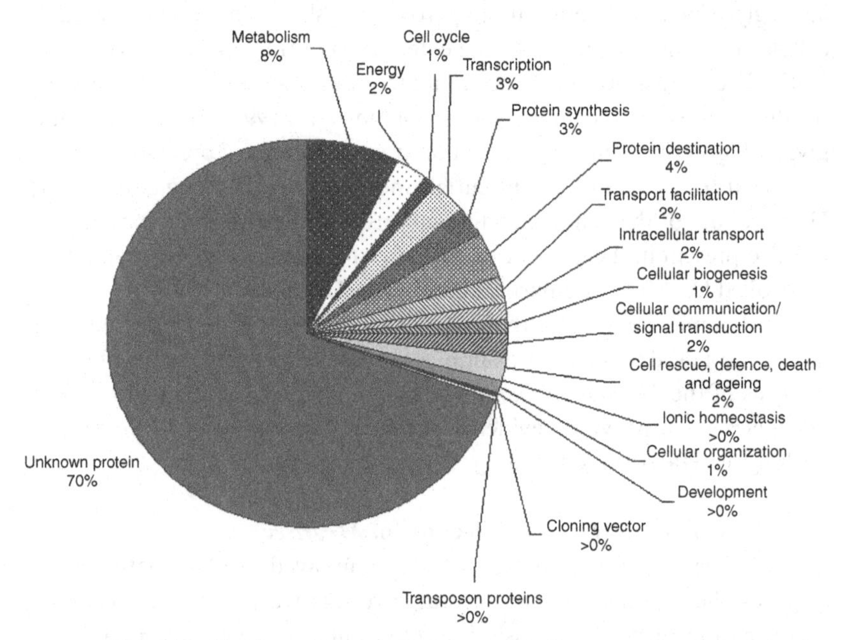

Fig. 11.1. Pie chart showing the known functions of unisequences in the COGEME database of EST sequences identified in *Magnaporthe grisea*. See text for details.

homologues among sequenced genes or because they have homology only to genes the function of which is unknown. The COGEME database is accessible via an easy-to-use web-based front-end (http://cogeme.ex.ac.uk/) (Soanes *et al.*, 2002b). As of March 2006 the database contains 59 765 unisequences from 18 species of fungal and oomycete phytopathogen and three non-pathogenic fungi.

If we assume that ESTs are randomly sequenced from non-normalized libraries, then the number of ESTs representing each gene is directly proportional to the mRNA abundance in the tissue from which the library was constructed. Comparison of EST sequence frequency across a number of different libraries can build up a transcript profile of that gene. Alternatively, the comparison of ESTs sequenced from two cDNA libraries can identify which genes are differentially regulated between the two conditions or tissue types from which the libraries were constructed. A statistical analysis method has been developed to identify whether the difference in EST sequence abundance between two libraries is significant (Audic & Claverie, 1997). An analysis of EST transcript abundance for 958 *M. grisea* genes across eight different cDNA libraries is available at the COGEME website (http://cogeme.ex.ac.uk/transcript.html).

Ebbole *et al.* (2004) have also generated a comprehensive study of 28 682 ESTs produced from nine cDNA libraries representing several growth conditions, cell types and *M. grisea* strains. Moderate sequencing of several diverse libraries was expected to reveal more unique gene sequences than deep sequencing of a single library, and researchers hoped to identify patterns of gene expression related to early events in the plant–fungal interaction, such as appressorium formation and turgor generation. Therefore, ESTs representing *M. grisea* developmental stages were derived from conidial, mycelial, appressorial and perithecial cDNA libraries. ESTs representing different growth conditions were derived from cDNA libraries of cultures grown on complete medium, minimal medium, nitrogen-starvation medium and rice cell wall medium. Given the importance of the *PMK1* gene for appressorium formation and invasive growth (Xu & Hamer, 1996), a non-pathogenic *pmk1* mutant strain was also included as a source of cDNA clones, 95% of the ESTs aligned to sequences found in the draft of the *M. grisea* genome. The combined ESTs for all libraries clustered into 3050 multisequence contigs, with 5127 singleton sequences, to yield an estimated 8177 unique gene sequences representing more than half the gene content of *M. grisea*. Statistical analysis of these EST sequences enabled characterization of the uniqueness of each library and the distribution of ESTs across multiple libraries. In the first instance, approximately 31%–50% of EST sequences were found to be library-specific; the remaining sequences were found to be present in at least one more library, with the majority found in two cDNA libraries and relatively few multisequences found in all nine cDNA libraries. Therefore, a large fraction of sequences were restricted to only one or two cDNA libraries, indicating real differences in gene expression patterns between cDNA libraries. To reveal similarities and differences between the cDNA libraries, the ESTs in each library were assigned to functional categories based on gene ontology terms (Ashburner *et al.*, 2000). Each library had a similar distribution of ESTs to each category, with the most abundant categories in each library representing proteins for cellular activities related to metabolism or physiological processes. However, the conidial library contained a larger fraction of ESTs representing genes for cell growth, and the ESTs derived from cultures grown on rice cell wall medium contained a large proportion (4.7%) of genes involved in plant cell wall degradation.

Analysis of the number of ESTs clustered in multisequence contigs was also used to estimate transcript abundance and allow a broad examination of gene expression patterns. Hierarchical clustering was used to group these genes according to expression patterns in the different libraries. The most abundantly represented gene, making up 2.2% of EST sequences from all cDNA libraries, was found to be the *M. grisea* homologue of *UV-1*, an appressorium and UV-inducible gene from *Bipolaris oryzae* (Kihara *et al.*, 2001). It was most abundantly expressed in the mating, appressorial, conidial and *pmk1* mutant cDNA libraries, but its function under these conditions is not known. Two other genes had a similar expression pattern to *UV-1*. These were a homologue of a protein that interacts with adenylate cyclase (Kulkarni & Dean, 2004) and a gene with similarity to Δ24-sterol methyltransferase. Considering that cAMP signalling is important for the induction of appressorium formation, the adenylate-cyclase-interacting protein might play a role in these developmental processes. The *MPG1* gene (Talbot *et al.*, 1993) was also found to be abundantly expressed and, although found in all libraries, was most prevalent in appressorium, mating, rice cell wall and conidial cDNA libraries. Expression of *UV-1*, the adenylate-cyclase-interacting protein, the Δ24-sterol methyltransferase and *MPG1* was also detected in the *pmk1* library. In contrast, ESTs of two virulence factors, *MAS1/GAS2* and *MAS3/GAS1* (Xue *et al.*, 2002), were abundant in the appressorium library but were not present in the *pmk1* library, suggesting that the expression of these genes is dependent on the *PMK1* MAP kinase signalling pathway

(Xu & Hamer, 1996). Therefore, comparison of gene expression in the *pmk1* library with that in other libraries could identify additional virulence factors that act early in the infection process.

By analysing the sequence of ESTs unique to the appressorium library, Ebbole *et al.* (2004) identified genes likely to contribute to early stages of the infection process. These included *MAS1* and *MAS3*, described above, and also a homologue of a low-affinity copper transporter and a gene with no known homologue. In a *M. grisea*-infected rice cDNA library (Kim *et al.*, 2001), the 511 cDNA clones sequenced generated 296 non-redundant ESTs. Of this group, 72 clones, representing 39 non-redundant ESTs, were identified as being fungal in origin; the remaining 221 clones, forming 106 non-redundant ESTs, were from rice. Approximately half of all ESTs (57.3%) were matched to known protein coding sequences in the NCBI database and allocated putative functions. These included 96 clones (32.8%) with homology to ubiquitous metabolic pathway enzymes, 54 clones (18.4%) having homology to components of the transcriptional or translational apparatus, and 86 clones (29.4%) having homology to cell defence or pathogenicity proteins. Most of the 72 fungal gene matches were categorized as being involved in gene expression and protein synthesis, suggesting (not surprisingly) active protein synthesis during invasive growth *in planta*. However, 19 fungal-derived clones, representing five non-redundant ESTs, were functionally assigned to cell defence or pathogenicity activities. These included the pathogenicity gene *MPG1* and a second hydrophobin gene homologue *MHP1*. Of the remainder, several cDNA clones were matched to various stress-inducible proteins, including *PEG1*, a homologue of the *Neurospora crassa* blue-light-inducible gene 3 (*BLI-3*) (Eberle & Russo, 1994), *PEG2*, a homologue of a nitrogen-starvation-induced gene of *Colletotrichum gloeosporioides*, and *PEG3*, a homologue of glucose-repressible gene 1 (*GRG1*) of *N. crassa* (McNally & Free, 1998). Therefore, analysis of ESTs generated from the infection of rice leaves by *M. grisea* has proved successful in identifying a small number of genes with homology to genes encoding pathogenicity determinants and cell defences from both interacting organisms, which are likely to be involved directly in the plant–pathogen interaction.

SAGE analysis of *M. grisea*

Serial analysis of gene expression (SAGE) (Velculescu *et al.*, 1995) was developed as a high-throughput method of gene expression analysis with the potential to address some of the drawbacks of EST analysis. It is cost-effective, fast and has the power to determine the abundance of every transcript in a population of cells. Moreover, whereas ESTs are typically generated from the cDNA libraries of single organisms, SAGE is capable of simultaneously monitoring the gene expression patterns of two interacting organisms, such as plant host and fungal pathogen. A decription of SAGE methodology can be found in Velculescu *et al.* (1995) with Robust-LongSAGE detailed in Gowda *et al.* (2004) and the superSAGE technique detailed by Matsumura *et al.* (2003, 2005). The difference between the techniques is the length of the tag generated. Longer tags more faithfully and unequivocally identify the gene of origin, which is particularly important for organisms with complex genomes.

In *M. grisea*, SAGE and its improved protocols have been used to compare the gene expression profiles of two differentially treated samples of *M. grisea*, and the rice–*M. grisea* interaction. Irie *et al.* (2003) used mRNA isolated from *M. grisea* cells that had been treated with cAMP, an initiator of appressorium formation, to identify genes that were altered in expression during appressorium formation. The importance of cAMP as a signalling molecule has already been implicated during the discussion of the MAPK-dependent pathways. A total of 5387 and 3938 tags were extracted from treated and untreated SAGE libraries, respectively, representing 2889 and 2342 unique genes. Using available EST and genomic sequencing data, the authors assigned putative gene names to 28 cAMP-induced and 9 cAMP-repressed genes. Most genes induced by cAMP were classified as being involved in sugar metabolism; other housekeeping genes and genes with unknown functions were also found. Some genes of interest that were up-regulated by cAMP included a polytetrahydroxynaphthalene reductase involved in melanin biosynthesis, the *MPG1* hydrophobin gene, the *MAS1* virulence factor and *MAC1*, encoding adenylate cyclase, which is necessary for appressorium formation. Use of the *pmk1* EST library had previously shown that *MAS1* expression was dependent on *PMK1* but *MPG1* was not, and yet the expression of both genes is modulated by cAMP. This reveals the complexity of signalling interactions within the fungus and the range of information that is already being revealed by the few assays that have been undertaken.

Matsumura *et al.* (2003) used the improved SAGE method superSAGE to analyse the gene expression profiles of both rice and *M. grisea* during the later stages of the plant–fungal interaction. The tags were counted and queried against the rice and *M. grisea* genome and EST sequences to allocate putative functions to genes expressed during the interaction. Although most tags were identified as being rice plant in origin, the most abundantly expressed encoding photosynthetic proteins (74 tags), representing 0.6% of the transcripts analysed, were identified as being from *M. grisea*. Two tags each represented a nucleoside-diphosphate kinase and 60*S* ribosomal protein. Incredibly, half of these (38 tags) were *MPG1*, which encodes a well-characterized hydrophobin. The expression of these two genes, and *MPG1*, during the plant–pathogen interaction was confirmed by RT–PCR. This indicates the very high expression of hydrophobins during plant infection.

Subtractive hybridization analysis to assess gene expression in *M. grisea* pathogenesis

Subtractive hybridization analysis is another technique that provides an assessment of patterns of gene expression. An advantage of this technique lies in the fact that gene expression is defined at a specific developmental stage. For example, Lu *et al.* (2005) have made a subtractive appressorium cDNA library at 24 hours following subtraction with conidium, aerial mycelia and a substrate mycelial mRNA. This suppression-subtractive hybridization approach allowed identification of 71 ESTs that are found only in mature appressoria (Lu *et al.*, 2005). As well as identifying genes known to be involved in appressorium differentiation and penetration that had already been characterized, such as *MAS1*, *MAS2* and *PTH11*, a number of the appressorium-specific genes were identified that are predicted to be involved in lipid metabolism within peroxisomes and detoxification, based on their homology to known sequences (Lu *et al.*, 2005). The importance of lipid metabolism in appressorium physiology, and sterol biosynthesis or modification, was also highlighted (Lu *et al.*, 2005); the results of the study were broadly consistent with the EST analysis performed by Ebbole *et al.* (2004). Previous subtractive analyses were rather more limited owing to lack of a genome sequence and therefore annotated gene sequences. Kamakura *et al.* (1999) used subtraction of an early-stage appressorium library (2.5 hours) and found 158 distinct clones expressed in the germ tubes, including a gene involved in melanin biosynthesis and a homologue of an avirulence gene (Kamakura *et al.*, 1999). Rauyaree *et al.* (2001) used differential hybridization of a cDNA library from a 48 hour infection of rice leaves. As with the SAGE analysis this approach meant that many of the genes identified as expressed were of plant origin when other time points of infection were used for subtraction. Fungal genes identified, however, included pathogenicity determinants such as cAMP-dependent kinase and *MPG1* (Rauyaree *et al.*, 2001).

Proteomics of *M. grisea*

Proteomic analysis of the rice blast fungus is in its infancy, but has proceeded following release of the genome sequence of the fungus (Dean *et al.*, 2005). Kim *et al.* (2004) used two-dimensional gel electrophoresis to detect changes in the levels of large numbers of proteins during fungal development, and to identify proteins that are more abundant in *M. grisea* during appressorium formation. Proteins were extracted at 4, 8 and 12 hours post-germination from conidia that had germinated on hydrophilic plates but were unable to form appressoria and from germinating conidia and appressoria that had formed on leaf-wax-coated, hydrophobic glass plates. After visualization of the proteins by two-dimensional polyacrylamide gel electrophoresis, protein spots labelled with [^{35}S]methionine or silver staining that were differentially expressed in appressoria compared with conidia were identified by mass spectrometry. Five proteins were found to be more abundant during appressorium formation and hence in the early stages of the plant–fungal interaction. These included scytalone dehydratase, a key enzyme of the melanin biosynthetic pathway and required for appressorial maturation, which was detectable just 4 hours after germination; a protein expressed at 12 hours with some identity to a serine carboxypeptidase Y (CPY) and two proteins expressed at 8 hours which share some identity with 20*S* proteosome subunits of *N. crassa* and are likely to be involved in protein turnover in the cell; and a protein with no match in the database. Although one of these proteins, scytalone dehydratase, had already been implicated in appressorium formation from previous studies (see Talbot, 2003), the work of Kim *et al.* (2004) demonstrates the potential of proteomics to identify global patterns of protein abundance under different developmental conditions. Furthermore, the study also identified the potential that protein turnover via the proteasome pathway may be an important consequence, or component, of appressorium development in *M. grisea*.

The MAPK signalling pathways of *M. grisea*

One area that has been relatively well studied in the pre-genomics era in filamentous fungi is the Mitogen Activated Protein Kinase (MAPK) signalling pathways (for review see Xu, 2000). Their importance is demonstrated by the mutant phenotypes that will be described but is also confirmed by the EST analysis and the creation of a *pmk1* EST library. These pathways offer the most immediate hope for modelling and prediction of protein–protein interactions in *M. grisea*; a relatively good model already exists from more advanced studies in budding yeast. This is partly due to the use of the functional genomics carried out in *S. cerevisiae*, which identified probable upstream and downstream targets of *M. grisea* MAPK that have since been studied experimentally. Important differences, however, clearly exist in the operation of MAPK signalling pathways between the two ascomycetes.

Signalling pathways inside cells allow the fungus to react to environmental cues, and often result in the increased (or decreased) expression of genes involved in infection-related development. MAPK pathways are found in a number of eukaryotic organisms from yeast to humans (Xu, 2000) and are highly conserved. The final kinase in the pathway is the serine/threonine kinase MAPK. This is activated by a serine/threonine/tyrosine kinase, MAPKK, which in turn is activated by a serine/threonine kinase MAPKKK. Activation of each element is brought about by a phosphorylation reaction. There are usually three protein kinase units that make up this pathway, although it has recently come to light that dual-specificity MAPK complexes exist with different MAPKs activated. As a result, a scaffold protein acts as a tether and gives specificity to the particular MAPK pathway (Park *et al.*, 2003). In *M. grisea* there are currently three well-characterized MAPKs: the MAPK Mps1 (MAP kinase for penetration and sporulation), Pmk1 (pathogenicity MAP kinase) and Osm1 (osmoregulation MAP kinase) (Xu, 2000).

Mps1 is the homologue of Slt2 in *S. cerevisiae*. Slt2 positively regulates growth and cell proliferation and responds to a variety of different signals such as low osmolarity, high temperature, mating pheromones and nutrient limitations (Zarzov *et al.*, 1996). Δ*slt2* Mutants are sensitive to high temperatures and cell-wall-degrading enzymes (Lee *et al.*, 1993). It has been shown that Mps1 complements an autolysis phenotype of Slt2 mutants. Mutants deficient in Mps1 are non-pathogenic owing to an inability to penetrate the host cuticle, but are still able to infect wounded plants. Mps1 is believed to be involved in appressorium maturation and penetration peg formation, as generation of normal turgor pressure appears to be unaffected in *Δmps1* mutants (Z. Cartwright, Holcombe & N. J. Talbot, unpublished observations). Other phenotypes include severely reduced conidiation, reduced aerial hyphae development and an autolytic ('easily wettable') phenotype. *Δmps1* mutants also have an increased sensitivity to cell-wall-degrading enzymes (Xu *et al.*, 1998).

Preliminary studies suggest that a Mps1:GFP fusion localizes to the nuclei of appressoria after approximately 4 hours, with a peak between 8 and 12 hours, when germinated on hydrophobic glass coverslips. This coincides with appressorium maturation, including changes in cell-wall morphology and penetration peg formation (Z. Cartwright & N. J. Talbot, unpublished observations). Both upstream and downstream elements of the Mps1 pathway, homologous to those of the Slt2 pathway, have been isolated in *M. grisea*, the putative upstream MAPKKK and MAPKK, MgBck1 and MgMmk2 being homologues of Bck1 and Mmk2 in *S. cerevisiae*. Preliminary studies suggest that mutants deficient in these proteins have a phenotype similar to that of *Δmps1* (Zhao *et al.*, 2005). Homologues of the downstream targets of Slt2 have also been identified in yeast. Recent studies show that both Swi4 and Swi6 interact with Slt2 (Madden *et al.*, 1997). Swi4 and Swi6 form a heteromeric dimer called SBF (SCB (Swi4/6 dependent cell cycle box) binding factor). Together, they bind to the repeated upstream regulatory sequence CACGAA (Nasmyth & Dirick, 1991; Baetz *et al.*, 2001). SBF is essential for the expression of G1-specific genes along with various cell wall biosynthetic genes, although it has been proposed that there are up to 200 putative targets of SBF (Iyer *et al.*, 2001) with diverse roles in morphogenesis, cell cycle function and cell wall biosynthesis. In *M. grisea*, the Swi4 and Swi6 homologues MgSwi4 and MgSwi6 have been identified. Preliminary data suggests that they interact with each other (Z. Cartwright & N. J. Talbot, unpublished), although putative roles for the MgSBF in infection-related development of *M. grisea* have yet to be determined.

The reduction in conidiation and the autolytic phenotype of *mps1* mutants is also shown by mutants of the hydrophobin *MPG1*, believed to be involved in the production of aerial structures and spores (Talbot *et al.*, 1993). Northern blot analysis has revealed a large down-regulation of *MPG1* expression when grown under standard conditions or under salt stress in *Δmps1* mutants. This may indicate that Mpg1 is a possible downstream target of the MPS1 pathway and controlled by the SBF.

Pmk1 is a MAPK that is required for both appressorium formation and invasive growth (Xu & Hamer, 1996). It is also responsible for the mass transfer of storage carbohydrate and lipid reserves to the appressorium (Thines *et al.*, 2000). Pmk1 is homologous to Fus3/Kss1 in *S. cerevisiae*, which is known to be involved in the pheromone response and filamentation (including invasive growth) pathways (Gustin *et al.*, 1998). *Δpmk1* Mutants are unable to form appressoria and are therefore unable to infect wounded rice tissue. They do, however, undergo germ-tube swelling and hooking, and germ tube tips appear highly deformed (Xu & Hamer, 1996). The addition of exogenous cAMP fails to rescue this phenotype, but does increase the rates of germ-tube swelling and hooking, suggesting that Pmk1 may act downstream of a cAMP-dependent signal (Xu & Hamer,

1996). Recent studies have shown that GFP-tagged Pmk1 localizes to the nuclei of appressoria during maturation, 12–24 hours after spore germination (Bruno *et al.*, 2004). High *PMK1:GFP* expression was also observed in developing conidia (Bruno *et al.*, 2004). Weak Pmk1GFP expression is visible in conidia and germ tubes and in vegetative and infectious hyphae. The upstream MAPKK and MAPKKK in the Pmk1 pathway have also been identified and characterized in *M. grisea.* Deletion mutants of the MAPKK Mst7 and MAPKKK Mst11, which are homologous to *S. cerevisiae* Ste7 and Ste11, respectively, have exhibited many phenotypes similar to *Δpmk1* (Zhao *et al.*, 2005); this similarity further suggests their upstream function. The *pmk1* EST library is also revealing genes whose expression is or is not dependent on *PMK1.*

Osm1 is a functional homologue of the Hog1 MAPK in *S. cerevisiae* (Dixon *et al.*, 1999). Hog1 is thought to regulate cellular turgor in yeast grown in hyperosmotic conditions by maintaining an osmotic gradient across the plasma membrane (Nevoigt & Stahl, 1997). It is also responsible for regulation of cytoskeletal re-organization during exposure to high solute concentrations (Brewster & Gustin, 1994). Stimulation of Hog1 leads to production of glycerol, via action of the glycerol-3-phosphate dehydrogenase-encoding gene *GPD1*, which results in maintenance of cellular turgor even after exposure to high osmotic stress (Dixon *et al.*, 1999). In *M. grisea, Δosm1* mutants are fully pathogenic and grow normally on standard media but levels of conidiation were approximately 10 times lower than in the wild-type strain and mutants were sensitive to hyperosmotic stress. During infection-related development, *Δosm1* mutants formed multiple appressoria when germinated under chronic osmotic stress, suggesting a possible role in negatively regulating appressorium development during severe osmotic stress. It has been demonstrated that *OSM1* may control accumulation of arabitol, the main compatible solute during response to hyperosmotic stress (Dixon *et al.*, 1999). This suggests that a different, more specific pathway has evolved in *M. grisea* to be independent of the conserved eukaryotic mechanisms responsible for regulation of turgor in cells (Dixon *et al.*, 1999). Homologues to the Hog1 upstream MAPKK and MAPKKK elements, SSK2 and PBS2, respectively, have been isolated in *M. grisea* (Zhao *et al.*, 2005). MgSSK2 and MgPBS2 knockout mutants have been shown to exhibit a phenotype similar to that of *Δosm1.*

Until recently, little was known about the degree of conservation between homologous MAPK pathways in fungi. With the release of many fungal genomes and EST databases, as detailed above, there is an opportunity to explore the level of functional conservation. There are some differences between the *S. cerevisiae* and the *M. grisea* pathways. In *S. cerevisiae*, the membrane-bound p21 activated kinase (PAK) kinase *STE20* is essential for the activation of Ste11 during mating and filamentous growth (Bardwell, 2004). In *M. grisea* this appears not to be the case, because the Ste20 homologue Mst20 appears to be dispensable for the formation of mature appressoria and therefore may not be involved in Pmk1 signalling (Li *et al.*, 2004). It has been demonstrated that Mst11 interacts with Ras1 and Ras2 homologues in *M. grisea* (Zhao *et al.*, 2005). This indicates a different method of upstream activation of the Pmk1 cascade. It will be interesting to see whether Mst20 is involved in the Osm1 pathway; this demonstrates that caution will be necessary in assigning the function of genes to a genome or EST collection.

One exciting prospect in the study of MAPK signalling is the probability of encountering cross-talk between two or more pathways and defining how specificity in signalling is achieved during plant infection. There is already some evidence that cross-talk does occur between MAPK signalling pathways in *M. grisea*. Studies of the levels of phosphorylation of both Mps1 and Pmk1 showed that Mps1 phosphorylation increased in vegetative hyphae in a *Δpmk1* mutant background, although Pmk1 phosphorylation did not change in an *Δmps1* mutant compared with that of the wild-type strain (Zhao *et al.*, 2005). This result would indicate that Mps1 phosphorylation may be up-regulated in an attempt to compensate for the cell-wall defects in vegetative hyphae associated with a Pmk1 mutation. During appressorium formation in a wild-type strain, however, Mps1 phosphorylation was repressed, suggesting that Pmk1 may negatively regulate Mps1 phosphorylation. Osm1 is also believed to prevent cross-talk between the hyperosmotic stress pathway and the *PMK1* pathway (Dixon *et al.*, 1999). Hog1 in *S. cerevisiae* prevents mis-communication between the HOG1 pathway and the pheromone-response pathway, to prevent mating occurring during hyperosmotic conditions (Hall *et al.*, 1996; O'Rourke & Herskowitz, 1998). It has been suggested that Osm1 normally acts to prevent activation of the Pmk1 pathway during hyperosmotic stress, and thus suppresses appressorium formation. In the absence of Osm1, Pmk1 may be phosphorylated at higher levels than usual during osmotic stress, resulting in multiple appressoria being formed from a single conidium (Dixon *et al.*, 1999). This analysis reveals the complexity of the interactions between these signalling pathways and the similarity and differences between *M. grisea* and *S. cerevisiae* and shows that despite a relatively well-defined model far more elucidation of these key signalling pathways is necessary.

Fungal hydrophobins involved in the pathogenicity of *M. grisea*

Hydrophobins are small, hydrophobic proteins that are produced by filamentous fungi during hyphal growth and development. They play important roles in many developmental processes carried out by fungi, including the generation of aerial hyphae, spore production and fruiting body formation. They are also involved in the virulence of some pathogenic fungi (Wessels, 1997; Kershaw & Talbot, 1998; Wösten, 2001).

Hydrophobins have limited sequence similarity and are only defined by the presence of eight cysteine residues, which are spaced in a characteristic manner. They have been classified into groups, Class I and II, based primarily on the distinct spacing of the cysteine residues but also on the grouping of the hydrophilic and hydrophobic residues along the primary sequence. The differences in structure are consistent with their different biochemical characteristics (Wessels, 1994).

The involvement of hydrophobins in many aspects of the fungal lifestyle is attributed to the ability of these proteins to undergo a process called interfacial self-assembly, a process first discovered by using purified SC3 hydrophobin from the basidiomycete *Schizophyllum commune* (Wösten *et al.*, 1993). When secreted into the aqueous environment that surrounds a fungal cell the protein is in a monomeric form, but at an interface with air or a hydrophobic surface the hydrophobins spontaneously polymerize into an amphipathic membrane-like form, thereby changing the nature of surfaces (Wösten *et al.*, 1993). In the development of aerial hyphae, hydrophobin secretion and self-assembly decreases the surface tension, allowing the hypha to emerge into the air (Wösten *et al.*, 1994). Hydrophobins will also self-assemble upon encountering a hydrophobic surface, thus providing a hydrophilic surface for the hyphal wall to adhere to, normally by using accessory adhesive molecules such as mucilages (Wösten *et al.*, 1994). As well as helping fungi to overcome the barriers encountered at natural interfaces, hydrophobins also form a protective coating on the surface of hyphae, spores and fruiting bodies, preventing dehydration and increasing hydrophobicity (Beever & Dempsey, 1978; Stringer *et al.*, 1991; Bell-Pederson *et al.*, 1992; Wösten *et al.*, 1994; Lugones *et al.*, 1996). The hydrophobic side of these layers can be visualized as rodlets by using transmission electron microscopy (Wösten *et al.*, 1994; Talbot *et al.*, 1996).

The *M. grisea* class I hydrophobin-encoding gene *MPG1* was originally identified as a differentially expressed gene during infection of rice by the blast fungus *M. grisea* (Talbot *et al.*, 1993). This was confirmed by the superSAGE analysis by Matsumura *et al.* (2003) and suggested that *MPG1* is the most highly expressed fungal gene during rice plant infection. *Δmpg1* Mutants are impaired in their ability to cause disease symptoms, and this correlates with a reduced ability to form appressoria. The self-assembly of the *MPG1* hydrophobin on the rice leaf surface is thought to act as a conformational cue for appressorium development (Talbot *et al.*, 1996). Appressorium differentiation in *M. grisea* is known to be regulated, at least in part, by the cyclic AMP-dependent signalling pathway, and the fact that application of cAMP to *mpg1* mutants restores appressorium formation is consistent with the action of Mpg1 prior to the developmental switch (Talbot *et al.*, 1996). It has also been shown that the *MPG1* gene appears to be positively regulated by the cAMP-dependent protein kinase A pathway (Soanes *et al.*, 2002a). *MPG1* promoter–GFP fusions established that *MPG1* expression occurs during germ-tube elongation and appressorium formation prior to penetration (Kershaw *et al.*, 1998; Soanes *et al.*, 2002a).

MPG1 also encodes a spore-wall rodlet protein similar in function to the rodlet layer encoded by *rodA* in *Aspergillus nidulans* and *EAS* in *Neurospora crassa* (Stringer *et al.*, 1991; Bell-Pederson *et al.*, 1992). Disruption of these spore rodlet proteins in *A. nidulans* and *N. crassa* results in a reduction of surface hydrophobicity but not in conidiogenesis. The *Δmpg1* mutant, however, does not sporulate well or form infection structures efficiently, producing approximately ten-fold fewer conidia than wild-type strains of *M. grisea*, with conidiophores lacking a complete whorl of spores (Talbot *et al.*, 1996). It has also been shown that aberrant expression of the *MPG1* hydrophobin in the morphological mutant *acropetal* results in chains of conidia forming in place of the normal whorls of sympodially arrayed spores (Lau & Hamer, 1998). The Mpg1 hydrophobin therefore has distinct functions during both conidiogenesis and appressorium development, and this emphasizes the fact that hydrophobins are able to act not only as cell-wall coat proteins but also as morphological determinants. The variety of morphogenetic roles of hydrophobins also demonstrates well the limitations of functional assignment on the basis of sequence homology.

The significance of intramolecular disulphide bridges in the Mpg1 hydrophobin has recently been examined *in vivo*. A series of *MPG1* alleles were generated in which each of the eight cysteine-encoding codons had been substituted with an alanine codon to prevent disulphide bridge formation. These alleles were expressed in an *Δmpg1* mutant. The expression of cysteine–alanine substitution alleles was shown to have deleterious consequences for the cell, with severe defects in development of hyphae and spores. Immunolocalization V5 epitope-tagged Mpgl variants revealed that alleles lacking a cysteine residue retain the capacity to self-assemble but are not secreted to the cell surface. This provided the first genetic evidence that disulphide bridges in a hydrophobin are dispensable for aggregation, but essential for secretion (Kershaw *et al.*, 2005).

Hydrophobins are normally differentially expressed and appear to be under complex transcriptional regulation to allow expression at various times during the life cycle of a fungus. The diverse functions of hydrophobins (Talbot, 1997) may simply reflect the developmental stage at which they are produced. The analysis of complete genome sequences and EST sets from diverse fungal species has revealed the presence of multiple hydrophobins in most fungi (Elliot & Talbot, 2004). *M. grisea* has six putative hydrophobin genes, three with homology to Class II hydrophobins and three others including the class I hydrophobin MPG1. All the EST and SAGE analyses have highlighted the importance and high expression of *MPG1* and an additional class II hydrophobin (Kim *et al.*, 2001). This hydrophobin is so highly expressed under many different conditions that this necessitates the development of techniques to identify genes that have a lower expression that is stage-specific, such as subtractive hybridization.

Conclusions and future prospects

Identifying the molecular mechanisms that underly the rice – *M. grisea* pathogenic interaction will be key to devising new control strategies for this devastating disease. Although genome sequencing has been successfully applied to both plant hosts and fungal pathogens, allowing the large-scale discovery of suites of genes with putative involvement in the disease, the identification and characterization of specific genes contributing to pathogenicity will require additional, functional, analysis beyond genome mining and comparative studies. Consequently, several high-throughput approaches have been initiated to identify genes directly involved in *M. grisea* appressorium formation and pathogenicity. These include the generation of expressed sequence tags (Ebbole *et al.*, 2004), the comprehensive use of the SAGE technique (Matsumura *et al.*, 2005), subtractive hybridization (Lu *et al.*, 2005) and proteomics (Kim *et al.*, 2004).

The current challenge is the full utilization of the information provided by these techniques. The importance of further and more detailed annotation of the genome sequence of *M. grisea* and EST data sets is apparent, and will be critical to its effective usage. Even when functional categories have been putatively assigned based on homology searches, a large proportion of genes remain unidentified, as demonstrated by Fig. 11.1. Functional analysis, by high-throughput gene replacement of these unknown genes, will be necessary to determine their precise roles in pathogenesis. The need for a comprehensive set of mutants to identify new genes critical to pathogenicity is beginning to be addressed by the ambitious MGOS project. It is hoped that the current technological bottleneck in generating targeted gene replacements at high throughput will soon be overcome; novel methods are beginning to be made available (see Gold *et al.*, 2001; Catlett *et al.*, 2003). When coupled with large-scale generation of insertional mutant collections, the opportunity for a rapid increase in our understanding of rice blast disease is very great. Systems-level analysis of rice blast disease is still some way in the future (Veneault-Fourrey & Talbot, 2005) but the rapid application of genomic techniques will allow greater insight into the plant infection process and expression of disease symptoms than has hitherto been possible.

References

Aderem, A. (2005) Systems biology: its practice and challenges. *Cell* **121**, 511–13.

Ashburner, M., Ball, C. A., Blake, J. A., Botstein, D., Butler, H., Cherry, J. M., Davis, A. P., Dolinski, K., Dwight, S. S., Eppig, J. T., Harris, M. A., Hill, D. P., Issel-Tarver, L., Kasarskis, A., Lewis, S., Matese, J. C., Richardson, J. E., Ringwald, M., Rubin, G. M. & Sherlock, G. (2000). Gene ontology: tool for the unification of biology. *Nature Genetics* **25**, 25–9.

Asiegbu, F. O., Choi, W., Jeong, J. S. & Dean, R. A. (2004). Cloning, sequencing and functional analysis of *Magnaporthe grisea MVP1* gene a *hex-1* homolog encoding a putative 'woronin body' protein. *FEMS Microbiology Letters* **230**, 85–90.

Audic, S. & Claverie, J. M. (1997). The significance of digital gene expression profiles. *Genome Research* **7**, 986–95.

Baetz, K., Jason, M., Haynes, J., Chang, M. & Andrews, B. (2001). Transcriptional coregulation by the cell integrity mitogen-activated protein kinase Slt2 and the cell cycle regulator Swi4. *Molecular and Cellular Biology* **21**, 6515–28.

Bardwell, L. (2004). A walk-through of the yeast mating pheromone response pathway. *Peptides* **25**, 1465–76.

Beever, R. E. & Dempsey, G. P. (1978). Function of rodlets on the surface of fungal spores. *Nature* **272**, 608–10.

Bell-Pederson, D., Dunlap, J. C. & Loros, J. J. (1992). The *Neurospora* circadian clock-controlled gene, *ccg-2*, is allelic to *eas* and encodes a fungal hydrophobin required for formation of the conidial rodlet layer. *Genes and Development* **6**, 2382–94.

Bourett, T. M. & Howard, R. J. (1990). In vitro development of penetration structures in the rice blast fungus *Magnaporthe grisea*. *Canadian Journal of Botany* **68**, 329–42.

Brewster, J. L. & Gustin, M. C. (1994). Positioning of cell growth and division after osmotic stress requires a MAP kinase pathway. *Yeast* **10**, 425–39.

Bruno, K. S., Tenjo, F., Li, L., Hamer, J. E. & Xu, J. R. (2004). Cellular localization and role of kinase activity of PMK1 in *Magnaporthe grisea*. *Eukaryotic Cell* **3**, 1525–32.

Catlett, N. L., Lee, B. N., Yoder, O. C. & Turgeon, B. G. (2003) Split-marker recombination for efficient target deletion of fungal genes. *Fungal Genetics Newsletter* **50**, 9–11.

Choi, W. & Dean, R. A. (1997). The adenylate cyclase gene *MAC1* of *Magnaporthe grisea* controls appressorium formation and other aspects of growth and development. *Plant Cell* **9**, 1973–83.

de Jong, J., McCormack, B. J., Smirnoff, N. & Talbot, N. J. (1997). Glycerol generates turgor in rice blast. *Nature* **389**, 244–5.

De Zwaan, T. M., Carrol, A. M., Valent, B. & Sweigard, J. A. (1999). *Magnaporthe grisea* Pth11p is a novel plasma membrane protein that mediates appressorium differentiation in response to inductive substrate cues. *Plant Cell* **11**, 2013–30.

Dean, R. A., Talbot, N. J., Ebbole, D. J., Farman, M. L., Mitchell, T. K., Orbach, M. J., Thon, M., Kulkarni, R., Xu, J.-R., Pan, H., Read, N. D., Lee, Y.-H., Carbone, I., Brown, D., Oh, Y. Y., Donofrio, N., Jeong, J. S., Soanes, D. M., Djonovic, S., Kolomiets, E., Rehmeyer, C., Li, W., Harding, M., Kim, S., Lebrun, M.-H., Bohnert, H., Coughlan, S., Butler, J., Calvo, S., Ma, L.-J., Nicol, R., Purcell, S., Nusbaum, C., Galagan, J. E. & Birren, B. W. (2005). The genome sequence of the rice blast fungus *Magnaporthe grisea*. *Nature* **434**, 980–6.

Dixon, K. P., Xu, J. R., Smirnoff, N. & Talbot, N. J. (1999). Independent signalling pathways regulate cellular turgor during hyperosmotic stress and appressorium-mediated plant infection by *Magnaporthe grisea*. *Plant Cell* **11**, 2045–58.

Dujon, B., Sherman, D., Fischer, G., Durrens, P., Casaregola, S., Lafontaine, I., de Montigny, J., Marck, C., Neuveglise, C., Talla, E., Goffard, N., Frangeul, L., Aigle, M., Anthouard, V., Babour, A., Barbe, V., Barnay, S., Blanchin, S., Beckerich, J. M., Beyne, E., Bleykasten, C., Boisrame, A., Boyer, J., Cattolico, L., Confanioleri, F., de Daruvar, A., Despons, L., Fabre, E., Fairhead, C., Ferry-Dumazet, H., Groppi, A., Hantraye, F., Hennequin, C., Jauniaux, N., Joyet, P., Kachouri, R., Kerrest, A., Koszul, R., Lemaire, M., Lesur, I., Ma, L., Muller, H., Nicaud, J. M., Nikoloski, M., Oztas, S., Ozier-Kalogeropoulos, O., Pellenz, S., Potier, S., Richard, G. F., Straub, M. L., Suleau, A., Swennen, D., Tekaia, F., Wesolowski-Louvel, M., Westhof, E., Wirth, B., Zeniou-Meyer, M., Zivanovic, I., Bolotin-Fukuhara, M., Thierry, A., Bouchier, C., Caudron, B., Scarpelli, C., Gaillardin, C., Weissenbach, J., Wincker, P. & Souciet, J. L. (2004). Genome evolution in yeasts. *Nature* **430**, 35–44.

Ebbole, D. J., Jin, Y., Thon, M. *et al.* (2004). Gene discovery and gene expression in the rice blast fungus, *Magnaporthe grisea*: analysis of expressed sequence tags. *Molecular Plant-Microbe Interactions* **17**, 1337–47.

Eberle, J. & Russo, V. E. (1994). *Neurospora crassa* blue light-inducible gene bli-3. *Biochemistry and Molecular Biology International* **34**, 737–44.

Elliot, M. A. & Talbot, N. J. (2004). Building filaments in the air: aerial morphogenesis in bacteria and fungi. *Current Opinions in Biology* **7**, 594–601.

Galagan, J. E., Calvo, S. E., Borkovich, K. A., Selker, E. U., Read, N. D., Jaffe, D., FitzHugh, W., Ma, L. J., Smirnov, S., Purcell, S., Rehman, B., Elkins, T., Engels, R., Wang, S. G., Nielsen, C. B., Butler, J., Endrizzi, M., Qui, D. Y., Ianakiev, P., Pedersen, D. B., Nelson, M. A., Werner-Washburne, M., Selitrennikoff, C. P., Kinsey, J. A., Braun, E. L., Zelter, A., Schulte, U., Kothe, G. O., Jedd, G., Mewes, W., Staben, C., Marcotte, E., Greenberg, D., Roy, A., Foley, K., Naylor, J., Stabge-Thomann, N., Barrett, R., Gnerre, S., Kamal, M., Kamvysselis, M., Mauceli, E., Bielke, C., Rudd, S., Frishman, D., Krystofova, S., Rasmussen, C., Metzenberg, R. L., Perkins, D. D., Kroken, S., Cogoni, C., Macino, G., Catcheside, D., Li, W. X., Pratt, R. J., Osmani, S. A., DeSouza, C. P. C., Glass, L., Orbach, M. J., Berglund, J. A., Voelker, R., Yarden, O., Plamann, M., Seller, S., Dunlap, J.,

Radford, A., Aramayo, R., Natvig, D. O., Alex, L. A., Mannhaupt, G., Ebbole, D. J., Freitag, M., Paulsen, I., Sachs, M. S., Lander, E. S., Nusbaum, C. & Birren, B. (2003). The genome sequence of the filamentous fungus *Neurospora crassa*. *Nature* **422**, 859–68.

Gilbert, R. D., Johnson, A. M. & Dean, R. A. (1996). Chemical signals responsible for appressorium formation in the rice blast fungus *Magnaporthe grisea*. *Physiological and Molecular Plant Pathology* **48**, 335–46.

Goffeau, A., Barrell, B. G., Bussey, H., Davis, R. W., Dujon, B., Feldmann, H., Galibert, F., Hoheisel, J. D., Jacq, C., Johnston, M., Louis, E. J., Mewes, H. W., Murakami, Y., Philippsen, P., Tettelin, H. & Oliver, S. G. (1996). Life with 6000 genes. *Science* **274**, 563–7.

Gold, S. E., Garcia-Pedrajas, M. D. & Marinez-Espinoza, A. D. (2001). New and used approaches to the study of fungal pathogenicity. *Annual Review of Phytopathology* **39**, 337–65.

Gowda, M., Jantasuriyarat, C., Dean, R. A. & Wang, G. L. (2004). Robust-LongSAGE (RL-SAGE): a substantially improved LongSAGE method for gene discovery and transcriptome analysis. *Plant Physiology* **134**, 890–7.

Gustin, M. C., Albertyn, J., Alexander, M. & Davenport, K. (1998). MAP kinase pathways in the yeast *Saccharomyces cerevisiae*. *Microbiology and Molecular Biology Reviews* **62**, 1264–300.

Hall, J. P., Cherkasova, V., Elion, E., Gustin, M. C. & Winter, E. (1996). The osmoregulatory pathway represses mating pathway activity in *Saccharomyces cerevisiae*: isolation of a FUS3 mutant that is insensitive to the repression mechanism. *Molecular and Cellular Biology* **16**, 6715–23.

Hamer, J. E., Howard, R. J., Chumley, F. G. & Valent, B. (1988). A mechanism for surface attachment in spores of a plant pathogenic fungus. *Science* **239**, 288–90.

Hammond-Kosack, K. E. & Parker, J. E. (2003). Deciphering plant-pathogen communication: fresh perspectives for molecular resistance breeding. *Current Opinion in Biotechnology* **14**, 177–93.

Harrison, P. M. & Gerstein, M. (2002) Studying genomes through the aeons: protein families, pseudogenes and proteome evolution. *Journal of Molecular Biology* **318**, 1155–74.

Heath, M. C., Valent, B., Howard, R. J. & Chumley, F. G. (1990). Interactions of two strains of *Magnaporthe grisea* with rice, goosegrass and weeping lovegrass. *Canadian Journal of Botany* **68**, 1627–37.

Howard, R. J., Ferrari, M. A., Roach, D. H. & Money, N. P. (1991). Penetration of hard substrates by a fungus employing enormous turgor pressures. *Proceedings of the National Acadamy of Sciences of the USA* **88**, 11 281–4.

Irie, T., Matsumura, H., Terauchi, R. & Saitoh, H. (2003). Serial analysis of gene expression (SAGE) of *Magnaporthe grisea*: genes involved in appressorium formation. *Molecular Genetics and Genomics* **270**, 181–9.

Iyer, V. R., Horak, C. E., Scafe, C. S., Botstein, D., Snyder, M. & Brown, P. O. (2001). Genomic binding sites of the yeast cell-cycle transcription factors SBF and MBF. *Nature*, **409**, 533–8.

Kamakura, T., Xiao, J.-Z., Choi, W.-B., Kochi, T., Yamaguchi, S., Teraoka, T. & Yamaguchi, I. (1999). cDNA subtractive cloning of genes expressed during early stage of appressorium formation by *Magnaporthe grisea*. *Bioscience Biotechnology and Biochemistry* **63**, 1407–13.

Kershaw, M. J. & Talbot, N. J. (1998). Hydrophobins and repellents: Proteins with fundamental roles in fungal morphogenesis. *Fungal Genetics and Biology* **23**, 18–33.

Kershaw, M. J., Wakley, G. E. & Talbot N. J. (1998). Complementation of the *Mpg1* mutant phenotype in *Magnaporthe grisea* reveals functional relationships between fungal hydrophobins. *EMBO Journal* **17**, 3838–49.

Kershaw, M. J., Thornton, C. R., Wakley, G. E. & Talbot N. J. (2005). Four conserved intra-molecular disulphide linkages are required for secretion and cell wall localisation of a hydrophobin during fungal morphogenesis. *Molecular Microbiology* **56**, 117–25.

Kihara, J., Sato, A., Okajima, S. & Kumagai, T. (2001). Molecular cloning, sequence analysis and expression of a novel gene induced by near-UV light in *Bipolaris oryzae*. *Molecular Genetics and Genomics* **266**, 64–71.

Kim, S., Ahn, I. P. & Lee, Y. H. (2001). Analysis of genes expressed during rice-*Magnaporthe grisea* interactions. *Molecular Plant-Microbe Interactions* **14**, 1340–6.

Kim, S. T., Yu, S., Kim, S. G., Kim, H. J., Kang, S. Y., Hwang, D. H., Jang, Y. S. & Kang, K. Y. (2004). Proteome analysis of rice blast fungus (*Magnaporthe grisea*) proteome during appressorium formation. *Proteomics* **4**, 3579–87.

Kulkarni, R. D. & Dean, R. A. (2004). Identification of proteins that interact with two regulators of appressorium development, adenylate cyclase and cAMP-dependent protein kinase A, in the rice blast fungus *Magnaporthe grisea*. *Molecular Genetics and Genomics* **270**, 497–508.

Lau, G. W. & Hamer, J. E. (1998). *Acropetal*: a genetic locus required for conidiophore architecture and pathogenicity in the rice blast fungus. *Fungal Genetics and Biology* **24**, 228–39.

Lee, K. S., Irie, K., Gotoh, Y., Watanabe, Y., Araki, H., Nishida, E., Matsumoto, K. & Levin, D. E. (1993). A yeast mitogen-activated protein kinase homolog (Mpk1p) mediates signalling by protein kinase C. *Molecular and Cellular Biology* **13**, 3067–75.

Li, L., Xue, C. Y., Bruno, K., Nishimura, M. & Xu, J. R. (2004). Two PAK kinase genes, *CHM1* and *MST20*, have distinct functions in *Magnaporthe grisea*. *Molecular Plant-Microbe Interactions* **17**, 547–56.

Liu, S. & Dean, R. A. (1997). G protein α subunit genes control growth, development, and pathogenicity of *Magnaporthe grisea*. *Molecular Plant-Microbe Interactions* **10**, 1075–86.

Lu, J.-P., Liu, T.-B. & Lin, F.-C. (2005). Identification of mature appressorium-enriched transcripts in *Magnaporthe grisea*, the rice blast fungus, using suppression subtractive hybridisation. *FEMS Microbiology Letters* **245**, 131–7.

Lugones, L. G., Bosscher, J. S., Scholtmeyer, K., De Vries, O. M. H. & Wessels, J. G. H. (1996). An abundant hydrophobin (ABH1) forms hydrophobic rodlet layers in *Agaricus bisporus* fruiting bodies. *Microbiology* **142**, 1321–9.

McNally, M. T. & Free, S. J. (1988). Isolation and characterization of a *Neurospora* glucose repressible gene. *Current Genetics* **14**, 545–51.

Madden, K., Sheu, Y. J., Baetz, K., Andrews, B. & Snyder, M. (1997). SBF cell cycle regulator as a target of the yeast PKC-MAP kinase pathway. *Science* **275**, 1781–4.

Matsumura, H., Reich, S., Ito, A., Saitoh, H., Kamoun, S., Winter, P., Kahl, G., Reuter, M., Kruger, D. H. & Terauchi, R. (2003). Gene expression analysis of plant host-pathogen interactions by SuperSAGE. *Proceedings of the National Academy of Sciences of the USA* **100**, 15 718–23.

Matsumura, H., Ito, A., Saitoh, H., Winter, P., Kahl, G., Reuter, M., Kruger, D. H. & Terauchi, R. (2005). SuperSAGE. *Cellular Microbiology* **7**, 11–18.

Mendgen, K. & Hahn, M. (2002). Plant infection and the establishment of fungal biotrophy. *Trends in Plant Sciences* **7**, 352–6.

Mitchell, T. K. & Dean, R. A. (1995). The cAMP-dependent protein kinase catalytic subunit is required for appressorium formation and pathogenesis by the rice blast fungus *Magnaporthe grisea*. *Plant Cell* **7**, 1869–78.

Nasmyth, K. & Dirick, L. (1991). The role of Swi4 and Swi6 in the activity of G1 cyclins in yeast. *Cell* **66**, 995–1013.

Nevoigt, E. & Stahl, U. (1997). Osmoregulation and glycerol metabolism in the yeast *Saccharomyces cerevisiae*. *FEMS Microbiology Reviews* **21**, 231–41.

Ohno, S. (1970). *Evolution by Gene Duplication.* New York: Springer.

Oliver, R. P. & Ipcho, S. V. S. (2004). *Arabidopsis* pathology breathes new life into the necrotrophs-vs.-biotrophs classification of fungal pathogens. *Molecular Plant Pathology* **4**, 347–52.

O'Rourke, S. M. & Herskowitz, I. (1998). The Hog1 MAPK prevents cross talk between the HOG and pheromone response MAPK pathways in *Saccharomyces cerevisiae. Genes and Development* **12**, 2874–86.

O'Rourke, S. M., Herskowitz, I. & O'Shea, E. K., (2002). Yeast go the whole HOG for the hyperosmotic response. *Trends in Genetics* **18**, 405–12.

Ou, S. H. (1985). *Rice Diseases.* Kew, UK: Commonwealth Mycological Institute, Commonwealth Agricultural Bureau.

Park, S., Zarrinpar, A. & Lim, A. W. (2003). Rewiring MAP kinase pathways using alternative scaffold assembly mechanisms. *Science* **299**, 1061–4.

Rathour, R., Singh, B. M., Sharma, T. R. & Chauhan, R. S. (2004). Population structure of *Magnaporthe grisea* from North-western Himalayas and its implications for blast resistance breeding of rice. *Journal of Phytopathology* **152**, 304–12.

Rauyaree, P., Choi, W., Fang, E., Blackmon, B. & Dean, R. A. (2001). Genes expressed during early stages of rice infection with the rice blast fungus *Magnaporthe grisea. Molecular Plant Pathology* **2**, 347–54.

Reuber, T. L., Plotnikova, J. M., Dewdney, J., Rogers, E. E., Wood, W. & Ausubel, F. M. (1998). Correlation of defense gene induction defects with powdery mildew susceptibility in *Arabidopsis* enhanced disease susceptibility mutants. *Plant Journal* **16**, 473–85.

Rosewich, U. L. & Kistler, H. C. (2000). Role of horizontal gene transfer in the evolution of fungi. *Annual Review of Phytopathology* **38**, 325–63.

Rossman, A. Y., Howard, R. J. & Valent, B. (1990). *Pyricularia grisea,* the correct name for the rice blast fungus. *Mycologia* **82**, 509–12.

Schaad, N. W., Frederick, R. D., Shaw, J., Schneider, W. L., Hickson, R., Petrillo, M. D. & Luster, D. G. (2003). Advances in molecular-based diagnostics in meeting crop biosecurity and phytosanitary issues. *Annual Review of Phytopathology* **41**, 305–24.

Schenk, P. M., Kazan, K., Wilson, I., Anderson, J. P., Richmond, T., Somervilee, S. C. & Manners, J. M. (2000). Coordinated plant defence responses in *Arabidopsis* revealed by microarray analysis. *Proceedings of the National Academy of Sciences of the USA* **97**, 11 655–60.

Sesma, A. & Osbourn, A. (2004). The leaf rice blast pathogen undergoes developmental processes typical of root-infecting fungi. *Nature* **431**, 582–6.

Skinner, W., Keon, J. & Hargreaves, J. (2001). Gene information for fungal plant pathogens from expressed sequences. *Current Opinion in Microbiology* **4**, 381–6.

Soanes, D. M., Kershaw, M. J., Cooley, R. N. & Talbot, N. J. (2002a). Regulation of the MPG1 hydrophobin gene in the rice blast fungus *Magnaporthe grisea. Molecular Plant-Microbe Interactions* **12**, 1253–67.

Soanes, D. M., Skinner, W., Keon, J., Hargreaves, J. & Talbot, N. J. (2002b). Genomics of phytopathogenic fungi and the development of bioinformatic resources. *Molecular Plant-Microbe Interactions* **15**, 421–7.

Stringer, M. A., Dean, R. A., Sewell, T. C. & Timberlake, W. E. (1991). *Rodletless*, a new *Aspergillus* developmental mutant induced by directed gene inactivation. *Genes Development* **5**, 1161–71.

Sweigard, J. A., Chumley, F. G. & Valent, B. (1992). Disruption of a *Magnaporthe grisea* cutinase gene. *Molecular and General Genetics* **232**, 183–90.

Talbot, N. J. (1995). Having a blast: exploring the pathogenicity of *Magnaporthe grisea. Trends in Microbiology* **3**, 9–16.

Talbot, N. J. (1997). Fungal biology: growing into the air. *Current Biology* **7**, R78–R82.

Talbot, N. J. (2003). On the trail of a cereal killer: Exploring the pathology of *Magnaporthe grisea*. *Annual Review of Microbiology* **57**, 177–202.

Talbot, N. J., Ebbole, D. J. & Hamer, J. E. (1993). Identification and characterization of *MPG1*, a gene involved in pathogenicity from the rice blast fungus *Magnaporthe grisea*. *Plant Cell* **5**, 1575–90.

Talbot, N. J., Kershaw, M. J., Wakley, G. E., De Vries, O., Wessels, J. G. H. & Hamer, J. E. (1996). MPG1 encodes a fungal hydrophobin involved in surface interactions during infection-related development of *Magnaporthe grisea*. *Plant Cell* **6**, 985–99.

Thines, E., Weber, R. W. S. & Talbot, N. J. (2000). MAP Kinase and protein kinase A-dependent mobilization of triacylglycerol and glycogen during appressorium turgor generation by *Magnaporthe grisea*. *Plant Cell* **12**, 1703–18.

Thinlay, X., Finckh, M. R., Bordeos, A. C. & Ziegler, R. S. (2000). Effects and possible causes of an unprecedented rice blast epidemic on the traditional farming system of Bhutan. *Agriculture, Ecosystems and Environment* **78**, 237–48.

Thomma, B. P. H. J., Pennickx, I. A. M. A., Broekaert, W. F. & Cammue, B. P. A. (2001). The complexity of disease signaling in *Arabidopsis*. *Current Opinion in Immunology* **13**, 63–8.

Tunlid, A. & Talbot, N. J. (2002). Genomics of parasitic and symbiotic fungi. *Current Opinion in Microbiology* **5**, 513–19.

van den Brink, H. M., van Gorcom, R. F., van den Hondel, C. A. & Punt, P. J. (1998). Cytochrome P450 enzyme systems in fungi. *Fungal Genetics and Biology* **23**, 1–17.

Valent, B. & Chumley, F. G. (1991). Molecular genetic analysis of the rice blast fungus, *Magnaporthe grisea*. *Annual Review of Phytopathology* **29**, 443–67.

Velculescu, V. E., Zhang, L., Vogelstein, B. & Kinzler, K. W. (1995). Serial analysis of gene expression. *Science* **270**, 484–7.

Veneault-Fourrey, C. & Talbot, N. J. (2005). Moving toward a systems biology approach to the study of fungal pathogenesis in the rice blast fungus *Magnaporthe grisea*. *Advances in Applied Microbiology* **57**, 177–215.

Voegele, R. T. & Mendgen, K. (2003). Rust haustoria: nutrient uptake and beyond. *New Phytologist* **159**, 93–100.

Wessels, J. G. H. (1994). Developmental regulation of fungal cell wall formation. *Annual Review of Phytopathology* **32**, 413–37.

Wessels, J. G. H. (1997). Hydrophobins: proteins that change the nature of the fungal surface. *Advances in Microbial Physiology* **38**, 1–45.

Wood, V., Gwilliam, R., Rajandream, M. A., Lyne, M., Lyne, R., Stewart, A., Sgouros, J., Peat, N., Hayles, J., Baker, S., Basham, D., Bowman, S., Brooks, K., Brown, D., Brown, S., Chillingworth, T., Churcher, C., Collins, M., Connor, R., Cronin, A., Davis, P., Feltwell, T., Fraser, A., Gentles, S., Goble, A., Hamlin, N., Harris, D., Hidalgo, J., Hodgson, G., Holroyd, S., Hornsby, T., Howarth, S., Huckle, E. J., Hunt, S., Jagels, K., James, K., Jones, L., Jones, M., Leather, S., McDonald, S., McLean, J., Mooney, P., Moule, S., Mungall, K., Murphy, L., Niblett, D., Odell, C., Oliver, K., O'Neil, S., Pearson, D., Quail, M. A., Rabbinowitsch, E., Rutherford, K., Rutter, S., Saunders, D., Seeger, K., Sharp, S., Skelton, J., Simmonds, M., Squares, R., Squares, S., Stevens, K., Taylor, K., Taylor, R. G., Tivey, A., Walsh, S., Warren, T., Whitehead, S., Woodward, J., Volckaert, G., Aert, R., Robben, J., Grymonprez, B., Weltjens, I., Vanstreels, E., Rieger, M., Schafer, M., Muller-Auer, S., Gabel, C., Fuchs, M., Fritzc, C., Holzer, E., Moestl, D., Hilbert, H., Borzym, K., Langer, I., Beck, A., Lehrach, H., Reinhardt, R., Pohl, T. M., Eger, P., Zimmermann, W., Wedler, H., Wambutt, R., Purnelle, B., Goffeau, A., Cadieu, E., Dreano, S., Gloux, S., Lelaure, V., Mottier, S., Galibert, F., Aves, S. J., Xiang, Z., Hunt, C., Moore, K., Hurst, S. M., Lucas, M., Rochet, M., Gaillardin, C., Tallada, V. A., Garzon, A., Thode, G., Daga, R. R., Cruzado, L., Jimenez, J., Sanchez, M., del Rey, F., Benito, J., Dominguez, A., Revuelta, J. L., Moreno, S.,

Armstrong, J., Forsburg, S. L., Cerrutti, L., Lowe, T., McCombie, W. R., Paulsen, I., Potashkin, J., Shpakovski, G. V., Ussery, D., Barrell, B. G. & Nurse, P. (2002). The genome sequence of *Schizosaccharomyces pombe*. *Nature* **415**, 871–80.

Wösten, H. A. (2001). Hydrophobins: multipurpose proteins. *Annual Review of Microbiology* **55**, 625–46.

Wösten, H. A. B., de Vries, O. M. H. & Wessels, J. G. H. (1993). Interfacial self-assembly of a fungal hydrophobin into a hydrophobic rodlet layer. *Plant Cell* **5**, 1567–74.

Wösten, H. A. B., Schuren, F. H. J. & Wessels, J. G. H. (1994). Interfacial self-assembly of a hydrophobin into an amphipathic membrane mediates fungal attachment to hydrophobic surfaces. *EMBO Journal* **13**, 5 848–54.

Xu, J. (2000). MAP kinases in fungal pathogens. *Fungal Genetics and Biology* **31**, 137–52.

Xu, J. R. & Hamer, J. E. (1996). MAPK and Cyclic AMP signalling regulate infection structure formation and pathogenic growth in the rice blast fungus, *Magnaporthe grisea*. *Genes and Development* **10**, 2696–706.

Xu, J., Staiger, C. J. & Hamer, J. E. (1998). Inactivation of the mitogen-activated protein kinase Mps1 from the rice blast fungus prevents penetration of host cells but allows activation of plant defence responses. *Proceedings of the National Academy of Sciences of the USA* **95**, 12 713–18.

Xue, C., Park, G., Choi, W., Zheng, L., Dean, R. A. & Xu, J.-R. (2002). Two novel fungal virulence genes specifically expressed in appressoria of the rice blast fungus. *Plant Cell* **14**, 2107–19.

Zarzov, P., Mazzoni, C. & Mann, C. (1996). The SLT2(MPK1) MAP kinase is activated during periods of polarized cell growth in yeast. *EMBO Journal* **15**, 83–91.

Zeigler, R. S. (1998). Recombination in *Magnaporthe grisea*. *Annual Review of Phytopathology* **36**, 249–75.

Zhao, X., Kim, Y., Park, G. & Xu, J.-R. (2005). A mitogen-activated protein kinase cascade regulating infection-related morphogenesis in *Magnaporthe grisea*. *Plant Cell* **17**, 1317–29.

12

Exploring the interaction between nematode-trapping fungi and nematodes by using DNA microarrays

ANDERS TUNLID
Department of Microbial Ecology, Lund University

Introduction

Soils contain a diverse range of fungi that are parasites on nematodes. They include more than 200 species representing all major taxonomic groups of fungi including deuteromycetes, basidiomycetes, chytridiomycetes and zygomycetes. Nematophagous fungi are found in all regions of the world, from the tropics to Antarctica. They are present in all sorts of soil environments, including agricultural and forest soils (Barron, 1977).

Based on the infection mechanisms, three broad groups can be recognized among the nematophagous fungi: the nematode-trapping and the endoparasitic fungi that attack free-living nematodes by using specialized structures, and the egg- and cyst-parasitic fungi that infect these stages with their hyphal tips (Barron, 1977). The nematode-trapping fungi are the best-known group, probably owing to their remarkable morphological adaptations and their dramatic infection of nematodes. With few exceptions, including the mushroom *Hohenbuehelia* (asexual state *Nematoctonus*) (Barron & Dierkes, 1977), the majority of the identified species of nematode-trapping fungi belong to a monophyletic clade among the apothecial ascomycetes (Liou & Tzean, 1997; Ahrén *et al.*, 1998; Hagedorn & Scholler, 1999).

Nematode-trapping fungi can grow as saprophytes in soils. They enter the parasitic stage by developing specific morphological structures called traps. The traps develop from hyphal branches; they can either be formed spontaneously or be induced in response to signals from the environment, including peptides and other compounds secreted by the host nematode (Dijksterhuis *et al.*, 1994). There is large variation in the morphology of trapping structures, even between closely related

Fungi in the Environment, ed. G. M. Gadd, S. C. Watkinson & P. S. Dyer. Published by Cambridge University Press.

species (Fig. 12.1). In some species, the trap consists of an erect branch that is covered with an adhesive material. In other species, such as in the well-studied *Arthrobotrys oligospora,* the trap is a complex three-dimensional adhesive net. A third type of trap is the adhesive knob. The knob is a morphologically distinct cell, often produced on the apex of a slender

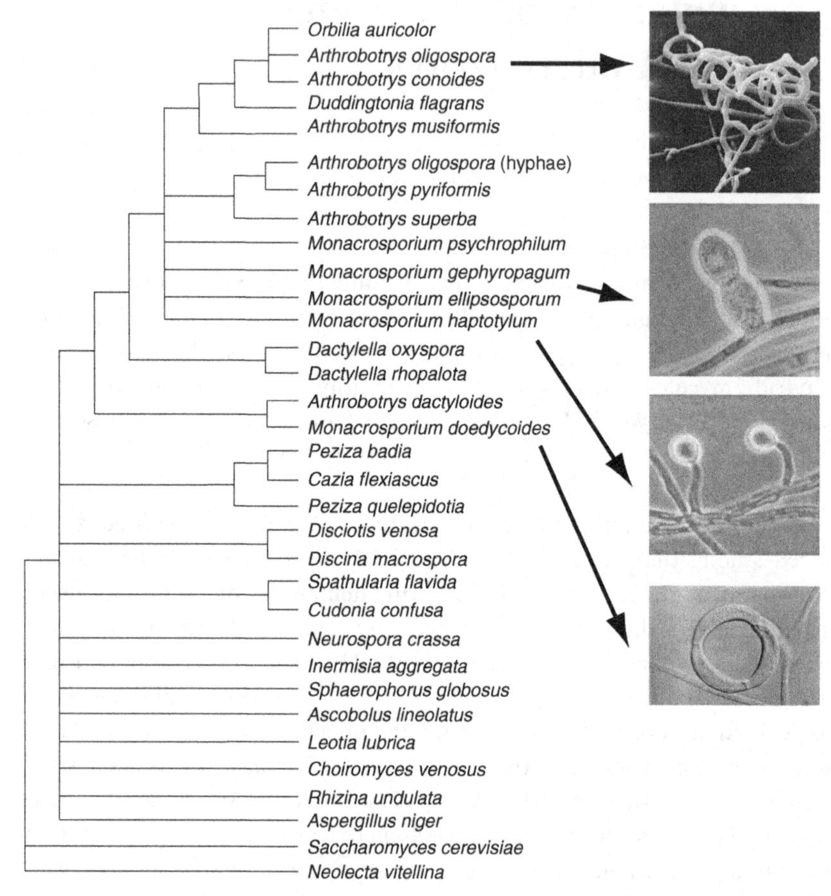

Fig. 12.1. A cladistic tree showing the relationship between various nematode-trapping fungi and other ascomycetes, based on 18*S* rDNA sequences (redrawn from Ahrén *et al.*, 1998). All branches shown have bootstrap support values above 50 (bootstrap values for important nodes are given in Ahrén *et al.*, 1998). Note that the nematode-trapping fungi form a monophyletic clade among an unresolved cluster of apothecial ascomycetes. The phylogenetic pattern within the clade of nematode-trapping fungi is concordant with the morphology of the traps. The pictures are reproduced from Nordbring-Hertz *et al.* (1995), courtesy of Birgit Nordbring-Hertz and IWF, Göttingen.

hyphal stalk. A layer of adhesive polymers covers the knob; this layer is not present on the support stalks. Finally, there are some species of nematode-trapping fungi that capture nematodes in mechanical traps called constricting rings (Barron, 1977). Following the development of the trap, the nematode-trapping fungi infect their hosts through a sequence of events: attachment of the trap cells to the surface of the nematode, penetration of the cuticle, digestion, and assimilation of the nutrients from the killed nematode.

Although there are numerous detailed ultrastructural studies following the infection of nematodes by fungi (Dijksterhuis *et al.*, 1994), the molecular background to these interactions are not well known. So far most studies on molecular mechanisms have used *A. oligospora* as a model system. For example, the function of a serine protease (a subtilisin designated PII) and a carbohydrate-binding protein (a lectin designated AOL) have been examined in detail. PII was isolated from culture filtrates of *A. oligospora*. The protein was sequenced and the corresponding gene (*PII*) was cloned (Åhman *et al.*, 1996). The importance of the activity of PII for the pathogenicity of *A. oligospora* was studied by constructing various PII mutants by using a transformation protocol for *A. oligospora* (Tunlid *et al.*, 1999; Åhman *et al.*, 2002). ΔPII deletion mutants produced fewer infection structures (traps) and had a slightly lower virulence than the wild type. Mutants containing additional copies of the genomic *PII* gene and over-expressing PII developed an increased number of infection structures and captured and killed nematodes more rapidly compared with the wild type. The toxic activity of PII was verified by demonstrating that a heterologously produced PII (in *Aspergillus niger*) had a nematicidal activity when added to free-living nematodes (Åhman *et al.*, 2002). AOL belongs to a family of low-molecular-mass, saline-soluble lectins that appear to be unique to filamentous fungi (Rosén *et al.*, 1996). Although deletion of the *AOL* gene did not affect the phenotype of *A. oligospora* (Balogh *et al.*, 2003), expression studies suggest that one of the functions of AOL is to store the nitrogen obtained from infected nematodes. Large amounts of AOL are synthesized during the digestion of the nematode; the lectin is later transported from the infected nematode to other parts of the mycelium (Rosén *et al.*, 1997).

Methods developed within functional genomics, including DNA microarrays, open up new possibilities for obtaining information on the global patterns of genes being expressed during the development of traps and infection of nematodes by nematode-trapping fungi. So far this technique has mainly been used for analysing gene expression in organisms with fully sequenced genomes, including the fungi *Saccharomyces cerevisiae*, *Schizosaccharomyces pombe* and *Candida albicans* (Brown & Botstein, 1999; Mata *et al.*, 2002; Bensen *et al.*, 2004). However, even in the absence of complete genome sequences, information from large sets of expressed sequence tags (ESTs) is well suited for construction of cDNA arrays (Johansson *et al.*, 2004). ESTs are single-pass, partial sequence reads from either the 5′- or the 3′-end of a cDNA clone and thus represent a survey of the transcribed portion of the genome. Furthermore, the nematode–fungus systems are excellent for following gene expression during fungal infection, as they provide the possibility of using the nematode *Caenorhabditis elegans* as a host. Owing to the fact that the genome of *C. elegans* has been fully sequenced, the transcripts expressed by the fungus and the nematode can easily be separated, even if the EST sequences are generated from cDNA libraries containing both fungal and worm transcripts. In addition to identifying fungal genes, analyses of the fungus – *C. elegans* system will give information on the defence systems that are activated as a response to fungal infections. The tractability of using *C. elegans* as a pathogenesis model has recently been demonstrated in studies of bacterial and fungal virulence (Mylonakis *et al.*, 2002; Couillault *et al.*, 2004; Sifri *et al.*, 2005).

An EST database for Monacrosporium haptotylum

We have generated an EST database for the nematode-trapping fungus *Monacrosporium haptotylum* (syn. *Dactylaria candida*) (Ahrén *et al.*, 2005). This fungus infects nematodes by using an adhesive knob (Fig. 12.1). The advantage of using *M. haptotylum* is that during growth in liquid cultures with heavy aeration the connections between the traps (knobs) and mycelium can be broken easily and the knobs can be separated from the mycelium by filtration (Friman, 1993). The isolated knobs retain their function as infection structures, i.e. they can 'capture' and infect nematodes. We constructed four directional cDNA libraries from mycelium, knobs, and knobs infecting *C. elegans* for 4 and 24 h, respectively. The knobs were isolated from liquid-grown mycelium of *M. haptotylum* and then incubated with *C. elegans* on water agar plates. After 4 h of infection, a number of the knobs had adhered and had started to penetrate the cuticle of *C. elegans*. Approximately 30%–35% of the nematodes were non-motile and considered to have been killed by the fungus. After 24 h of infection, approximately 80%–85% of the nematodes were immobile, and fungal hyphae were growing inside the infected nematodes. In total, 8463 ESTs were sequenced from the four cDNA libraries. The sequences were assembled into 3121 contigs that putatively represent unique genes and/or transcripts. Between 5% and 37% of the assembled sequences displayed a high degree of similarity to sequences in the GenBank nr protein database. A large fraction (38%–60%) of the assembled sequences (orphans) showed no homology to protein sequences in the GenBank nr protein database; the proportion of orphans was highest in the knob (58%) and 24 h infection (60%) libraries. Based on sequence homology, the ESTs were assigned functional roles and EC numbers (where applicable). Functional roles were classified into categories according to catalogues used for the *S. cerevisiae* genome, provided by the Munich Information Centre for Protein Sequences (MIPS) (Mewes *et al.*, 2002). All EST information was stored in an MySQL database and processed by using the PHOREST tool (Ahrén *et al.*, 2004).

Based on the information, a cDNA microarray was constructed. In total, 3518 clones were amplified by PCR and spotted on the array: 2822 of fungal, 540 of *C. elegans* and 156 of unknown origin.

Comparison of gene expression in trap cells and vegetative hyphae

The first experiment in which the cDNA array was used aimed at analysing the global patterns of genes expressed in traps (knobs) and mycelium of *M. haptotylum* (Ahrén *et al.*, 2005). RNA was extracted from three biological replicates of knobs and mycelium growing in liquid cultures. Following amplification, the RNA samples were labelled with fluorescent dyes (Cy3 and Cy5) and hybridized on the arrays.

Despite the fact that the knobs and mycelium were grown in the same medium, there were substantial differences in the patterns of genes expressed in the two cell types. Following a statistical analysis (mixed-model ANOVA) (Wolfinger *et al.*, 2001) and using a significance level $p < 0.05$, we identified that 23.3% (657 of 2822) of the putative genes were differentially expressed in knobs versus mycelium (Fig. 12.2). There was a large difference in the functional distribution of genes being differentially regulated. A significant proportion of the genes that had putative roles in 'transcription', 'cellular transport and transport mechanisms', 'cellular communication or signal transduction', 'cell rescue, defence, cell death and ageing' and 'cell growth, cell division and DNA synthesis' were expressed at lower levels in the knobs than in the mycelium.

The trap in *M. haptotylum* is a spherical cell, which develops at the tip of an apically growing hyphal branch (Fig. 12.1). This change in morphology represents a shift in the polarity of the cells. Several of the genes that were differentially expressed in knobs and mycelium displayed significant sequence similarities to genes known to be involved in modifying cell polarity in fungi, including *S. cerevisiae*. A key component of this system is the actin cytoskeleton (Pruyne & Bretscher, 2000). During apical growth, the actin cytoskeleton polarizes growth by assembling into cortical patches and actin cables at the tip. During isotropic growth, the proteins of the cap are more diffusely distributed and actin cables form a meshwork. Among the regulated genes in knobs and mycelium of *M. haptotylum* were homologues to the actin-binding proteins profilin and cofilin. Profilin is a small actin-binding protein, which was originally

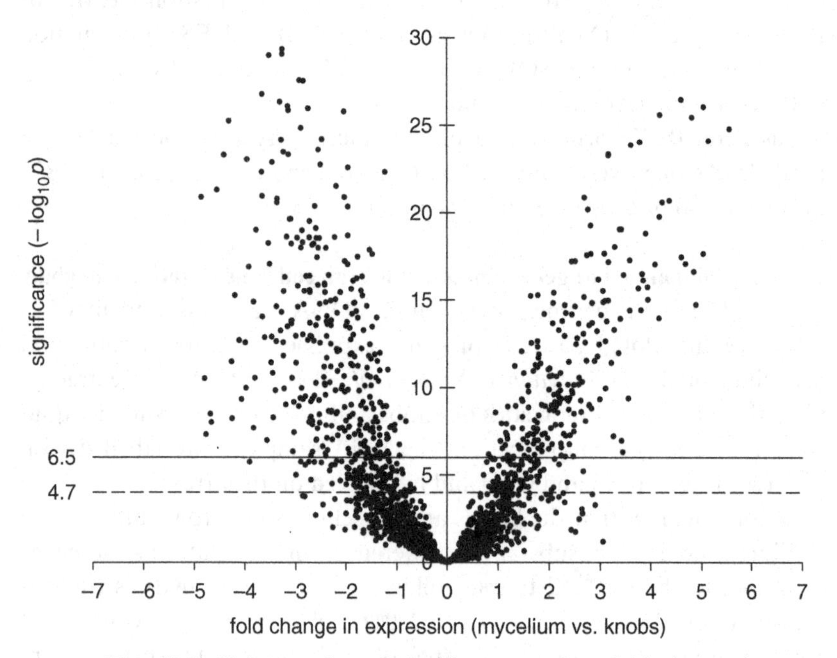

Fig. 12.2. Volcano plot of significance against fold change in gene expression, in mycelium (left) versus knobs (right). Each point represents a single gene analysed by the mixed-model ANOVA (Wolfinger *et al.*, 2001). The negative $\log_{10}$ of the p-value is plotted against the difference between least square means of $\log_2$-normalized expression values. Highly significant values are located towards the top, and small expression differences are located at the centre of each plot. The horizontal dashed line represents the test-wise threshold of $p = 0.05$ (corresponding to a Bonferroni corrected value of $-\log_{10} = 4.7$), and the solid line represents a threshold of $p = 0.001$ ($-\log_{10} = 6.5$). Using a cut-off value of $p = 0.05$, the total number of differentially expressed genes in knobs versus mycelium were 657 (259 up- and 398 down-regulated). The total number of fungal gene reporters spotted on the arrays was 2822. The figure is reproduced from Ahrén *et al.* (2005) with permission.

identified as an actin monomer sequestering protein that can inhibit the growth of actin filaments and prevent polymerization. Studies have also indicated that profilin can have a role in promoting actin polymerization (Ayscough, 1998). Cofilin binds to actin monomers and its activity is important for rapid turnover of actin filaments in *S. cerevisiae* (Lappalainen & Drubin, 1997). Regulated in knobs and mycelium of *M. haptotylum* were also several homologues to members of the rho and ras family of small GTPases that are known to play an important role in regulating the actin part of the cytoskeleton. These molecules act as molecular switches through their ability to hydrolyse GTP. In the GTP-bound form, these signalling proteins are active and exert a positive signal on proteins associated with polarized growth (Pruyne & Bretscher, 2000).

There are several similarities in the structure and function of knobs of nematode-trapping fungi and that of appressoria formed by plant pathogenic fungi. Like a knob, an appressorium is a specialized infection structure, which develops as a spherical cell at the tip of a hypha (germ tube). Both structures contain an adhesive layer on the outside, which binds to the surface of the host. Furthermore, both appressoria and knobs form a hypha that penetrates the host by using a combination of physical force and extracellular enzymic activities (Tucker & Talbot, 2001). Comparison of data from the transcriptional profiling of knobs with similar data of appressoria in *Magnaporthe grisea* and *Blumeria graminis* shows that there are also many similarities in the patterns of gene regulation in the infection structures of nematode-trapping and plant parasitic fungi (Table 12.1). Genes involved in stress and defence responses are one of the largest classes of genes that are differentially expressed during appressorium formation in *M. grisea* (Rauyaree *et al.*, 2004). Several such genes, including cyclophilins, peptidyl-prolyl *cis–trans* isomerases, metallothionein and thioredoxins, were differentially expressed in the knobs of *Monacrosporium haptotylum.* A number of genes involved in protein synthesis (such as homologues for ribosomal proteins and translation elongation factor), protein destination and degradation (such as homologues for ubiquitin, ubiquitin-conjugating enzyme and proteosome components) were differentially expressed in both knobs and appressoria. This result suggests that development of the infection structures in both nematode-trapping and plant pathogenic fungi is associated with an extensive synthesis and turnover of proteins (McCafferty & Talbot, 1998).

In *Magnaporthe grisea,* the physical force needed for penetration of the plant cuticle is produced by generating a high turgor pressure. The turgor pressure results from a rapid accumulation of glycerol, and there are

Table 12.1. *Examples of genes being differentially regulated in knobs or appressoria of* Monacrosporium haptotylum, Magnaporthe grisea *and* Blumeria graminis, *as compared with mycelium, conidia, or germinating conidia, respectively*

Gene	Knob *M. haptotylum*[a]	Appressoria *M. grisea*[b]	Appressoria *B. graminis*[c]
gEgh16	up	up	up
Chitinase	down	down	up
Serine protease (subtilisin)	up	up	up
Proteasome subunit	down	up	—
Cyclophilin	up/down	—	down
Peptidyl-prolyl *cis–trans* isomerase	up	up	up
Metallothionein	down	up	up
Thioredoxin	down	—	up
Translation elongation factor (TEF1)	down	down	—
14-3-3-like protein (BMH2)	down	up	down
Triose phosphate isomerase	down	—	up
Glyceraldehyde 3-phosphate dehydrogenase	down	down	down
Enolase	down	—	down
Fructose bisphosphate aldolase	down	—	down
GTPase Rho1	down	—	down
Profilin	down	—	down

Notes:
[a] Fold values of expression levels in knobs (trap cells) compared with mycelium.
[b] Fold values of expression levels in appressorium as compared with differentiating conidia (Takano *et al.*, 2004).
[c] Fold values calculated from serial analysis of gene expression (SAGE) comparing appressoria and germinated conidia (Thomas *et al.*, 2002).
Source: Reprinted from Ahrén *et al.* (2005).

different lines of evidence suggesting that the production of glycerol is achieved by the mobilization of energy reserves such as glycogen and neutral lipids (Thines *et al.*, 2000). Notably, one of the most up-regulated genes in the knobs compared with the mycelium was a glycogen phosphorylase (*gph1*) gene homologue. This enzyme catalyses and regulates the degradation of glycogen to glucose-1-phosphate. This product is further metabolized in the glycolytic pathway and glycerol can be synthesized from several of the intermediates in this pathway.

Gene expression during infection of nematodes

More recently, we have used the constructed *Monacrosporium haptotylum* – *C. elegans* cDNA microarray to analyse the regulation of both fungal and worm genes during infection (unpublished data: C. Fekete, M. Tholander, D. Ahrén, B. Rajashekar, E. Friman, K. Eriksen, T. Johansson and A. Tunlid). Four different time points (0, 4, 16 and 24 h of interaction) were analysed, representing different stages of the infection, including the adhesion of knobs to the nematode surface, penetration of the cuticle, digestion and assimilation of nutrients of the nematode tissues. At each time point, we identified sets of genes that were significantly up- or down-regulated in the fungus and nematodes compared with non-infected samples, i.e. when the fungus or nematodes were grown in axenic cultures. Clearly, a large fraction of the fungal genes (39.8%) were regulated during the infection (Fig. 12.3). Notably, during the early stages of the infection, concomitant with penetration of the nematode, a majority of the genes annotated to the functional categories of 'transcription', 'metabolism', 'energy', 'cellular transport and transport mechanisms', 'cell rescue, defence, cell death and ageing' and 'cell growth, cell division and DNA synthesis' were significantly down-regulated. However, later on during the stages of digestion and assimilation of the killed nematodes a majority of the fungal genes in all functional categories were significantly up-regulated.

Conclusions and perspectives

Since their introduction in the mid-1990s, DNA microarrays have become major tools in functional genomics for exploring global patterns of gene expression in an organism (Brown & Botstein, 1999). The method was developed for transcriptional analyses in model organisms with complete genome sequences. By generating a database containing a limited set of EST sequences, we have here demonstrated that the DNA microarray technology can be applied to examine gene expression in the nematode-trapping fungus *M. haptotylum*. This is evidently a 'non-model' fungus where the background information on genetics is limited.

DNA microarray experiments typically generate a large amount of data. How can such complex set of data be meaningfully interpreted and validated in a poorly characterized fungus such as *M. haptotylum*? Clearly, the interpretation will to a large extent depend on the knowledge of the many genes that have been characterized in model fungi such as *Saccharomyces cerevisiae*, *Schizosaccharomyces pombe* and *Neurospora crassa* by classical methods of genetics, molecular biology and biochemistry. Fortunately, the rate of evolution of many genes with respect to both sequence and function has been so slow that characterization in these organisms can suffice for other species such as the nematode-trapping fungi. This suggests that

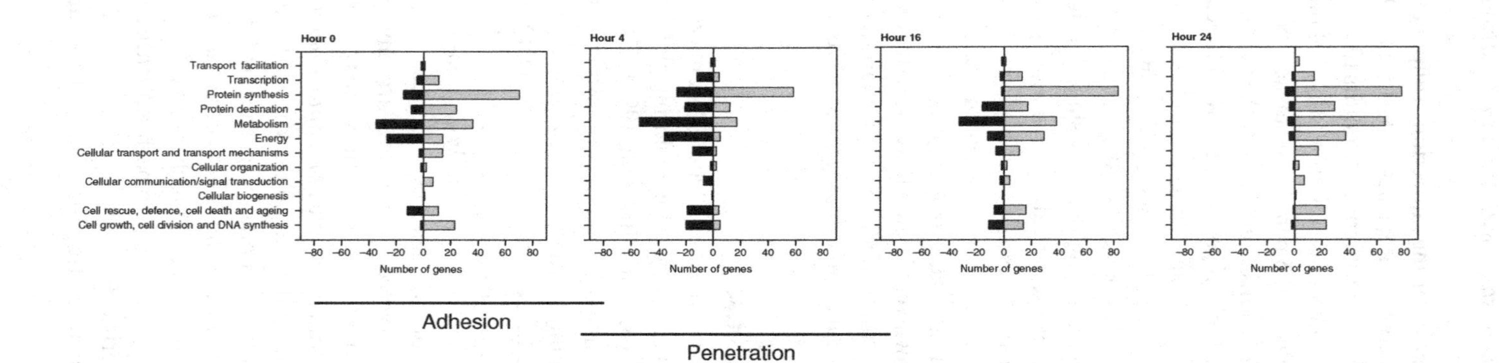

Fig. 12.3. The number of down-(left) and up-(right) regulated genes in *M. haptotylum* during the infection of the nematode *C. elegans*. Comparisons were made between infected and non-infected samples (fungus grown in axenic culture). Data are derived from cDNA microarray analysis: a significance level, $p < 0.05$. The genes are classified in the functional categories of MIPS (Mewes *et al.*, 2002). The annotation was based on the sequence similarity to information in the GenBank nr protein database. In total 1124 fungal genes (out of 2822) were found to be significantly regulated at one or more time points.

clusters of co-regulated genes identified in DNA array experiments on fungus–nematode interactions will, in many cases, contain at least some genes that encode proteins with orthologues that have been functionally characterized in other fungi. In many cases, this information can serve as a starting point for generating new hypotheses for the development of traps and parasitism of nematode-trapping fungi. One example of such a comparative approach is the identification of genes putatively involved in the change in morphology and polarity accompanying the development of knobs in *M. haptotylum*. However, in many cases the function of regulated genes identified by the microarray experiments of *M. haptotylum* cannot be inferred by sequence similarities to well-characterized genes. The functions of these genes have to be analysed by genetic and molecular methods. A transformation system like the one developed for *A. oligospora* that can be used for generating knockout and conditional mutants provides an excellent tool for more detailed functional analyses (Tunlid *et al.*, 1999). Furthermore, data from DNA microarray experiments needs to be verified by using independent methods. Regulated genes should be re-sequenced to confirm their identity, and expression levels should be verified by using RT–PCR or Northern blots (Ahrén *et al.*, 2005).

Owing to the fact that DNA microarray resources are presently set up for a number of different pathogenic fungi, it will be possible to compare gene expression in various fungus–host interactions in the near future. There are a number of common steps that these fungi need to accomplish for successful infections: development of infection structures, adhesion, penetration of the host surface and suppression of the host defence system. Comparative analyses of gene expression data will become an important tool for elucidating the molecular background to these steps, to identify the common as well as the unique strategies used by pathogenic fungi. For such analysis, it is essential that researchers have access to one another's raw transcriptome data. The work on establishing international standards for depositing microarray data in public databases is therefore very welcome (Ball *et al.*, 2004) (http://www.mged.org/).

Species of nematode-trapping fungus have successfully been used in biological control of plant and animal parasitic nematodes (Larsen, 2000; Jansson & Lopez-Llorca, 2004). The application of the tools of functional genomics, including EST sequencing and DNA microarray analyses, can provide important information for enhancing the biocontrol potential of nematode-trapping fungi. DNA microarrays can be used for identifying differences in gene expression and genome composition between closely related strains of fungi that differ in virulence, including host preferences (Le Quéré *et al.*, 2004). cDNA array experiments will also generate a number of 'candidate' genes that could be modified by genetic engineering to produce strains with an enhanced capacity to capture and kill nematodes. So far this approach has only been used for over-expressing the subtilisin PII in *A. oligospora* (Åhman *et al.*, 2002).

Despite the fact that the biology of nematode-trapping fungi has been studied in the laboratory for almost half a century, very little is known about their growth and parasitic activities in soils. We have recently demonstrated that RNA can be extracted, amplified and analysed by cDNA microarrays of an ectomycorrhizal fungus growing in soil microcosms (Wright *et al.*, 2005). A similar approach could probably be used to follow the activity of nematode-trapping fungi *in situ*.

Acknowledgements

The research on nematode-trapping fungi in my laboratory has been supported by grants from the Swedish Natural Science Research Council, the Swedish Research Council for Environment, Agricultural Sciences and Spatial Planning, and the Knut and Alice Wallenberg Foundation through the SWEGENE consortium.

References

Åhman, J., Ek, B., Rask, L. & Tunlid, A. (1996). Sequence analysis and regulation of a gene encoding a cuticle-degrading serine protease from the nematophagous fungus *Arthrobotrys oligospora*. *Microbiology* **142**, 1605–16.

Åhman, J., Johansson, T., Olsson, M., Punt, P. J., Van den Hondel, C. A. & Tunlid, A. (2002). Improving the pathogenicity of a nematode-trapping fungus by genetic engineering of a subtilisin with nematotoxic activity. *Applied and Environmental Microbiology* **68**, 3408–15.

Ahrén, D., Ursing, B. M. & Tunlid, A. (1998). Phylogeny of nematode-trapping fungi based on 18 S rDNA sequences. *FEMS Microbiology Letters* **158**, 179–84.

Ahrén, D., Troein, C., Johansson, T. & Tunlid, A. (2004). PHOREST: a web-based tool for comparative analyses of expressed sequence tag data. *Molecular Ecology Notes* **4**, 311–14.

Ahrén, D., Tholander, M., Fekete, C., Rajashekar, B., Friman, E., Johansson, T. & Tunlid, A. (2005). Comparison of gene expression in trap cells and vegetative hyphae of the nematophagous fungus *Monacrosporium haptotylum*. *Microbiology* **151**, 789–803.

Ayscough, K. R. (1998). In vivo functions of actin-binding proteins. *Current Opinion in Cell Biology* **10**, 102–11.

Ball, C. A., Brazma, A., Causton, H. & 15 other authors. (2004). Submission of microarray data to public repositories. *PLoS Biology* **2**, E317.

Balogh, J., Tunlid, A. & Rosén, S. (2003). Deletion of a lectin gene does not affect the phenotype of the nematode-trapping fungus *Arthrobotrys oligospora*. *Fungal Genetics and Biology* **39**, 128–35.

Barron, G. L. (1977). *The Nematode-Destroying Fungi*. Guelph, Canada: Lancester Press.

Barron, G. L. & Dierkes, Y. (1977). Nematophagous fungi: *Hohenbuehelia* the perfect state of *Nematoctonus*. *Canadian Journal of Botany* **55**, 3054–62.

Bensen, E. S., Martin, S. J., Li, M., Berman, J. & Davis, D. A. (2004). Transcriptional profiling in *Candida albicans* reveals new adaptive responses to extracellular pH and functions for Rim101p. *Molecular Microbiology* **54**, 1335–51.

Brown, P. O. & Botstein, D. (1999). Exploring the new world of the genome with DNA microarrays. *Nature Genetics* **21**, 33–7.

Couillault, C., Pujol, N., Reboul, J., Sabatier, L., Gnichou, J. F., Kohara, Y. & Ewbank, J. J. (2004). TLR-independent control of innate immunity in *Caenorhabditis elegans* by the TIR domain adaptor protein TIR-1, an ortholog of human SARM. *Nature Immunology* **5**, 488–94.

Dijksterhuis, J., Veenhuis, M., Harder, W. & Nordbring-Hertz, B. (1994). Nematophagous fungi: physiological aspects and structure-function relationships. *Advances in Microbial Physiology* **36**, 111–43.

Friman, E. (1993). Isolation of trap cells from the nematode-trapping fungus *Dactylaria candida*. *Experimental Mycology* **17**, 368–70.

Hagedorn, G. & Scholler, M. (1999). A reevaluation of predatory orbiliaceous fungi. I: Phylogenetic analysis using rDNA sequence data. *Sydowia* **51**, 27–48.

Jansson, H.-B. & Lopez-Llorca, L. V. (2004). Control of nematodes by fungi. In *Fungal Biotechnology in Agriculture, Food and Environmental Applications*, ed. D. K. Arora, pp. 205–15. New York: Marcel Dekker.

Johansson, T., Le Quéré, A., Ahrén, D., Söderström, B., Erlandsson, R., Lundeberg, J., Uhlén, M. & Tunlid, A. (2004). Transcriptional responses of *Paxillus involutus* and *Betula pendula* during formation of ectomycorrhizal root tissue. *Molecular Plant-Microbe Interactions* **17**, 202–15.

Lappalainen, P. & Drubin, D. G. (1997). Cofilin promotes rapid actin filament turnover in vivo. *Nature* **388**, 78–82.

Larsen, M. (2000). Prospects for controlling animal parasitic nematodes by predacious micro fungi. *Parasitology* **120**, S121–31.

Le Quéré, A., Schützendübel, A., Rajashekar, B., Canbäck, B., Hedh, I., Erland, S., Johansson, T. & Tunlid, A. (2004). Divergence in gene expression related to variation in host specificity of an ectomycorrhizal fungus. *Molecular Ecology* **13**, 3809–19.

Liou, G. Y. & Tzean, S. S. (1997). Phylogeny of the genus *Arthrobotrys* and allied nematode-trapping-fungi based on rDNA sequences. *Mycologia* **89**, 876–84.

Mata, J., Lyne, R., Burns, G. & Bahler, J. (2002). The transcriptional program of meiosis and sporulation in fission yeast. *Nature Genetics* **32**, 143–7.

McCafferty, H. R. & Talbot, N. J. (1998). Identification of three ubiquitin genes of the rice blast fungus *Magnaporthe grisea*, one of which is highly expressed during initial stages of plant colonisation. *Current Genetics* **33**, 352–61.

Mewes, H. W., Frishman, D., Guldener, U. & 27 other authors. (2002). MIPS: a database for genomes and protein sequences. *Nucleic Acids Research* **30**, 31–4.

Mylonakis, E., Ausubel, F. M., Perfect, J. R., Heitman, J. & Calderwood, S. B. (2002). Killing of *Caenorhabditis elegans* by *Cryptococcus neoformans* as a model of yeast pathogenesis. *Proceedings of the National Academy of Sciences of the USA* **99**, 15 675–80.

Nordbring-Hertz, B., Jansson, H.-B., Persson, Y., Frimans, C. & Dackman, C. (1995). *Nematophagous Fungi*. Film C1851. Göttingen: Institut für den Wissenschaftlichen Film.

Pruyne, D. & Bretscher, A. (2000). Polarization of cell growth in yeast. *Journal of Cell Science* **113**, 571–85.

Rauyaree, P., Choi, W., Fang, E., Blackmon, B. & Dean, R. A. (2004). Genes expressed during early stages of rice infection with the rice blast fungus *Magnaporthe grisea. Molecular Plant Pathology* **2**, 347–54.

Rosén, S., Kata, M., Persson, Y., Lipniunas, P. H., Wikström, M., van den Hondel, C. A. M. J. J., van den Brink, J. M., Rask, L., Hedén, L.-O. & Tunlid, A. (1996). Molecular characterization of a saline soluble lectin from a parasitic fungus. Extensive sequence similarity between fungal lectins. *European Journal of Biochemistry* **238**, 822–9.

Rosén, S., Sjollema, K., Veenhuis, M. & Tunlid, A. (1997). A cytoplasmic lectin produced by the fungus *Arthrobotrys oligospora* functions as a storage protein during saprophytic and parasitic growth. *Microbiology* **143**, 2593–604.

Sifri, C. D., Begun, J. & Ausubel, F. M. (2005). The worm has turned – microbial virulence modeled in *Caenorhabditis elegans. Trends in Microbiology* **13**, 119–27.

Takano, Y., Choi, W., Mitchell, T. K., Okuno, T. & Dean, R. A. (2004). Large scale parallel analysis of gene expression during infection-related morphogenesis of *Magnaporthe grisea. Molecular Plant Pathology* **4**, 337–46.

Thines, E., Weber, R. W. & Talbot, N. J. (2000). MAP kinase and protein kinase A-dependent mobilization of triacylglycerol and glycogen during appressorium turgor generation by *Magnaporthe grisea. Plant Cell* **12**, 1703–18.

Thomas, S. W., Glaring, M. A., Rasmussen, S. W., Kinane, J. T. & Oliver, R. P. (2002). Transcript profiling in the barley mildew pathogen *Blumeria graminis* by serial analysis of gene expression (SAGE). *Molecular Plant-Microbe Interactions* **15**, 847–56.

Tucker, S. L. & Talbot, N. J. (2001). Surface attachment and pre-penetration stage development by plant pathogenic fungi. *Annual Review of Phytopathology* **39**, 385–417.

Tunlid, A., Åhman, J. & Oliver, R. P. (1999). Transformation of the nematode-trapping fungus *Arthrobotrys oligospora. FEMS Microbiology Letters* **173**, 111–16.

Wolfinger, R. D., Gibson, G., Wolfinger, E. D., Bennett, L., Hamedeh, H., Buchel, P., Afshari, C. & Paules, R. S. (2001). Assessing gene significance from cDNA microarray expression data via mixed models. *Journal of Computational Biology* **8**, 625–37.

Wright, D. P., Johansson, T., Le Quéré, A., Söderström, B. & Tunlid, A. (2005). Spatial patterns of gene expression in the extramatrical mycelium and mycorrhizal root tips formed by the ectomycorrhizal fungus *Paxillus involutus* in association with birch (*Betula pendula* Roth.) seedlings in soil microcosms. *New Phytologist* **167**, 579–96.

13

Role of α(1-3) glucan in *Aspergillus fumigatus* and other human fungal pathogens

ANNE BEAUVAIS
Aspergillus *Unit, Institut Pasteur, Paris*
DAVID S. PERLIN
Public Health Research Institute, Newark, New Jersey
JEAN PAUL LATGÉ
Aspergillus *Unit, Institut Pasteur, Paris*

Introduction

The fungal cell wall has been considered for a long time as an inert organelle but recent studies, mainly based on the analysis of the yeast cell wall, suggest that it is indeed a dynamic structure where constitutive polymers are continuously chemically modified and rearranged during morphogenesis. The cell wall plays an essential role in sensing adverse or favourable environments. In particular, it provides the fungus with adaptative responses to variable osmotic pressures and other stress factors including host defence reactions. The cell wall is continuously in contact with the host and acts also as a sieve and a reservoir for molecules such as enzymes, antigens, and elicitors or toxins that play an active role during infection (Mouyna & Latgé, 2001).

The major component of the cell wall is polysaccharide. It accounts for over 90% of the cell wall mass and and consists of three basic components: glucans, mannan and chitin (Fig. 13.1). The fibrillar skeleton of the cell wall is considered to be the alkali-insoluble fraction, whereas the material in which the fibrils are embedded is alkali-soluble (Fontaine *et al.*, 2000). The central core of the cell wall is branched β1,3, β1,6 glucans that are linked to chitin via a β(1-4) linkage (Fontaine *et al.*, 2000). This core is present in most fungi, and at least in all ascomycetes and basidiomycetes. The alkali-soluble amorphous cement varies with the fungal species; its composition has been analysed in few fungal species. Moreover, the alkali solubility of the extracted fraction is not associated with a specific polysaccharide composition. In *Candida albicans* and *Saccharomyces cerevisiae*

Fungi in the Environment, ed. G. M. Gadd, S. C. Watkinson & P. S. Dyer. Published by Cambridge University Press.

it is composed of complex mannans and branched β1,3, β1,6 glucans (Manners *et al.*, 1973). In *Aspergillus* species and *Schizosaccharomyces pombe* it is composed of α1,3 glucan and galactomannan (Grün *et al.*, 2004; Beauvais *et al.*, 2005). An α1,3 glucan component has also been identified in other ascomyetes, such as *Blastomyces dermatitidis*, *Histoplasma capsulatum* and *Paracoccidioides brasiliensis*, and basidiomycetes such as *Cryptococcus neoformans* (Hogan & Klein, 1994; Klimpel & Goldman, 1988; Rappleye, *et al.*, 2004; Reese & Doering, 2003; San-Blas *et al.*, 1976). In many human pathogens, α1,3 glucan is present in high concentration; early studies have shown that it plays a role in fungal virulence (San-Blas *et al.*, 1976; Klimpel & Goldman, 1988).

It is therefore of importance to gain a better understanding of the structure and function of α1,3 glucans and to characterize the proteins and genes involved in the biosynthesis of this constitutive cell wall polysaccharide in human fungal pathogens. *Saccharomyces cerevisiae* has been a primary model for the study of the cell wall β1,3 glucan and chitin biosynthesis, but it cannot be used to analyse α1,3 glucan synthesis because of the absence of this polymer from this fungus. In contrast, α1,3 glucan is found as a major polymer in *Schizosaccharomyces pombe*. This yeast was the first model used to explore the molecular details of α1,3 glucan synthesis including the elucidation of the genes involve in the synthesis of this polysaccharide (Hochstenbach *et al.*, 1998). Deletion or mutation of these genes, and their regulation, has now been extensively investigated in other

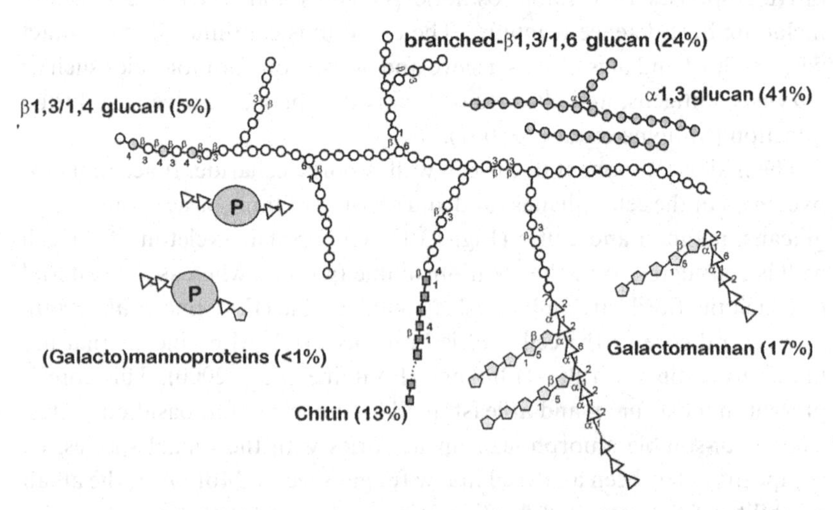

Fig. 13.1. Typical polysaccharide components of the cell wall of *Aspergillus fumigatus*.

fungi, such as *Aspergillus* spp., *H. capsulatum* or *C. neoformans*, that are human pathogens.

Structural organization of the α1,3 glucan in the cell wall

The cell wall is made of highly insoluble material; chemical analysis of cell-wall components first requires their solubilization. Sodium hydroxide treatment is needed to initially separate the alkali-insoluble β1,3 glucan and chitin and the alkali-soluble α1,3 glucan and galactomannan (Fontaine *et al.*, 2000) (Fig. 13.2). The structure of α1,3 glucan has been studied only in *S. pombe* and *A. fumigatus* (Grün *et al.*, 2004). It can be purified from the galactomannan following a periodic oxidation and a mild acid hydrolysis (Smith & Montgomery, 1956) (Fig. 13.2). Structure features of the fungal α1,3 glucan can then be analysed by classical carbohydrate chemistry methods.

α1,3 Glucan is one of the major polysaccharides present in the cell wall of *A. fumigatus* and the yeast pathogenic forms of *P. brasiliensis*, *H. capsulatum* and *B. dermatitidis*. It can account for up to 50% of the total cell-wall carbohydrates (Grün *et al.*, 2004; Beauvais *et al.*, 2005). In all these fungi, α1,3 glucan is alkali-soluble; immunolabelling studies using the monoclonal antibodies MOPC 104E (Sigma) or 401925 (Calbiochem), which recognize α1,3 glucan, show that α1,3 glucan can be found at the surface of both yeasts and *A. fumigatus* conidia (San-Blas *et al.*,

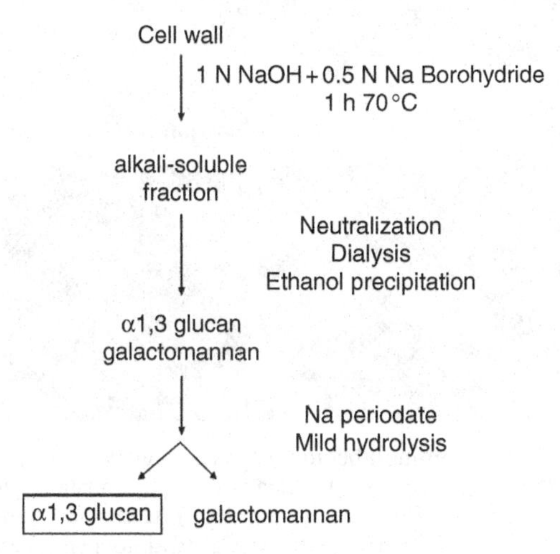

Fig. 13.2. Protocol for extraction of α1,3 glucan from *Aspergillus fumigatus* (W. Morelle *et al.*, unpublished).

1976; Klimpel & Goldman, 1988; Hogan & Klein, 1994) (Fig. 13.3). In germinating conidia of *A. fumigatus*, α1,3 glucan is mostly hidden from the surface of the cell wall by galactomannan and/or galactomanno-proteins (Klimpel & Goldman, 1988). α1,3 Glucan is also present in *C. neoformans*. Although the cell-wall composition of this pathogen has not been analysed in detail, α1,3 glucan also appears to be localized at the surface of the acapsular mutant cap59 of *C. neoformans* (Reese & Doering, 2003).

Most data have been obtained to date on the fission yeast *S. pombe*. In this fungus, 28% of α1,3 glucans are found in the cell wall, in which they are present as either fibrillar alkali-insoluble or amorphous alkali-soluble material (Sugawara *et al.*, 2003). They consist mainly of 1,3-linked α-glucose with *c.* 7% of 1,4-glycosidic linkages. These 1,4-linked glucose residues are located at the reducing end and in the centre of the polysaccharide, where they interconnect two chains of *c.* 120 1,3-linked glucose residues (Grün *et al.*, 2004). α1,3 Glucan is essential for the rigidity and shape of the *S. pombe* cells (Hochstenbach *et al.*, 1998). Immunolabelling studies show that this polysaccharide is present on the surface and interior of the cell wall but most labelling occurred along the yeast plasma

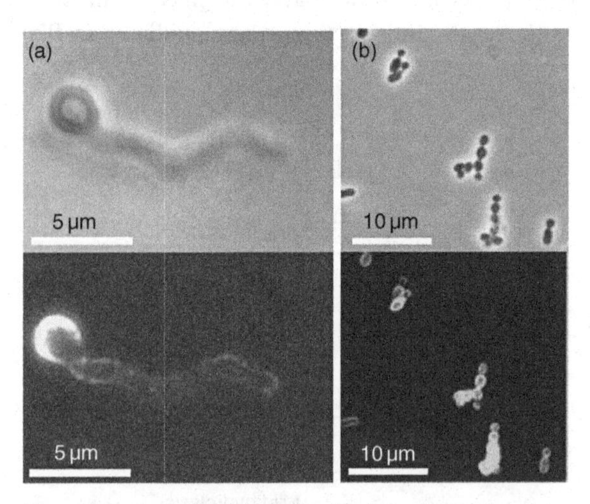

Fig. 13.3. Immunolabelling of α1,3 glucan in *Aspergillus fumigatus* (a) (M. Bernard & A. Beauvais, unpublished) and *Histoplasma capsulatum* (b) (Rappleye *et al.*, 2004), using the monoclonal antibody MOPC 104E (Sigma) and 401925 (Calbiochem) respectively. Upper panel: phase contrast micrograph; lower panel: fluorescence micrograph.

membrane (Sugawara *et al.*, 2003). However, during the regeneration of *S. pombe* protoplasts, the first fibrillar network observed was not α1,3 glucan but β1,3 glucan (Osumi *et al.*, 1995).

In *A. fumigatus*, preliminary studies suggest that α1,3 glucan is a branched structure (W. Morelle & J. P. Latgé, personnal communication), and that α1,3 glucan is the first polysaccharide deposited during cell-wall regeneration of protoplast (M. Costachel & J. P. Latgé, personnal communication). A change in the α1,3 glucan content can compensate for the reduction of another cell-wall component. For example, a 50% reduction of cell-wall chitin content in a double chitin synthase chsE-chsG mutant can be compensated for by an increase in the content of the amorphous α1,3 glucan (Mellado *et al.*, 2003). Interestingly, this report was the first to demonstrate that a compensatory effect in cell-wall biosynthesis is associated with an amorphous component. Previous studies have shown that cell-wall mutations addressing amorphous compounds (mannan in the *MNN9* deletion mutant of *Saccharomyces cerevisiae*) or fibrillar material (β1,3 glucan in the *FKS1* deletion mutant of *S. cerevisiae*) were always compensated for by an increase in chitin (Osmond *et al.*, 1999; Garcia-Rodriguez *et al.*, 2000). Our recent data on the compensatory activity of α1,3 glucan suggest that it plays a more important role in the cell wall than was originally thought.

α(1-3) Glucan synthesis

Putative α1,3 glucan synthases

One to five *AGS* genes (for alpha 1,3 glucan synthase) encoding putative α(1-3) glucan synthases have been found in various fungal species that are known to contain α1,3 glucan in their cell wall. In *C. neoformans* and *H. capsulatum*, only one *AGS* gene was found (Reese & Doering, 2003; Rappleye *et al.*, 2004), whereas three and five *AGS* genes are present in *A. fumigatus* and *Schizosaccharomyces pombe*, respectively (Katayama *et al.*, 1999; Beauvais *et al.*, 2005). *AGS* genes are the largest genes (*c.* 8 kb) involved in cell wall synthesis. They encoded predicted proteins of *c.* 2400 amino acids (aa) with estimated molecular masses of *c.* 272 kDa and pI values of 6.4–6.8. The amino acid sequences of the different species are highly similar between species, with 39%–48% identity (Table 13.1). The highest homology between all *AGS* genes is found in the region shown in Fig. 13.4 (Hochstenbach *et al.*, 1998; Beauvais *et al.*, 2005). A hydrophobicity profile of the predicted Ags proteins (Fig. 13.5) shows five structural domains (Hochstenbach *et al.*,

Table 13.1. *Percentage identity between the amino acid sequence of* A. fumigatus *(Af) Ags1p, Ags2p, Ags3p,* S. pombe *(Sp) Ags1p,* H. capsulatum *(Hc) Ags1p and* C. neoformans *(Cn) Ags1p*

	Af Ags1p	Af Ags2p	Af Ags3p	Sp Ags1p	Hc Ags1p	Cn Ags1p
Af Ags1p	100	72	57	46	67	46
Af Ags2p		100	56	48	69	46
Af Ags3p			100	46	57	46
Sp Ags1p				100	48	47
Hc Ags1p					100	45
Cn Ags1p						100

1998; Katayama *et al.*, 1999; Beauvais *et al.*, 2005). The first is the NH_2-terminal 30 aa residues, which are highly hydrophobic and may act as a signal peptide. The second region contains *c.* 1000 aa residues and has three putative transmembrane domains. It also shows 23%–32% identity with glycosylhydrolases, including bacterial amylase. Domains two and three are connected by a transmembrane segment of *c.* 20 aa. The third region comprises the next *c.* 1000 aa residues with 25%–30% identity to both bacterial glycogen synthases and plant starch synthases. The last domain is the COOH-terminal 400 aa residues that predict 12 membrane-spanning domains. The consensus UDP-glucose binding sequence (Lys/Arg-X-Gly-Gly, with X representing any residue) was found three times in Ags proteins: one consensus sequence is present in the glycoside hydrolase homologous domain and two in the glycogen synthase homologous region. However, no point mutations have been engineered to investigate whether any of these triads are involved in catalysis. It has been proposed that the intracellular synthase domain may produce α1,3 glucan homopolymers at the plasma membrane. The polysaccharide would be transported across the membrane by the multipass C-terminal domain. The Agsp extracellular glycosylhydrolase domain could cross-link the α glucan chain to another one or to other cell-wall carbohydrates (Hochstenbach *et al.*, 1998) (Fig. 13.6). In the absence of new experimental evidence, this model remains hypothetical.

Only *AGS1* of *S. pombe* is an essential gene. Deletion or over-expression of this gene is lethal for the cells (Hochstenbach *et al.*, 1998; Katayama *et al.*, 1999). At a semi-permissive temperature, the thermosensitive mutant ags1.1 contains only 7% α1,3 glucan in its cell wall (Hochstenbach *et al.*, 1998). The cells are rounded and pear-shaped and the cell wall becomes looser and thicker relative to those of the parental strain (Fig. 13.7a). Ags1p has been localized in the growing tips of the cells and inside the primary septum (Katayama *et al.*, 1999; Konomi *et al.*, 2003) (Fig. 13.7b). RNA interference or deletion of *AGS1* (ags1-i and ags1 mutants, respectively) of *H. capsulatum* or *C. neoformans* leads to a loss of α1,3 glucan

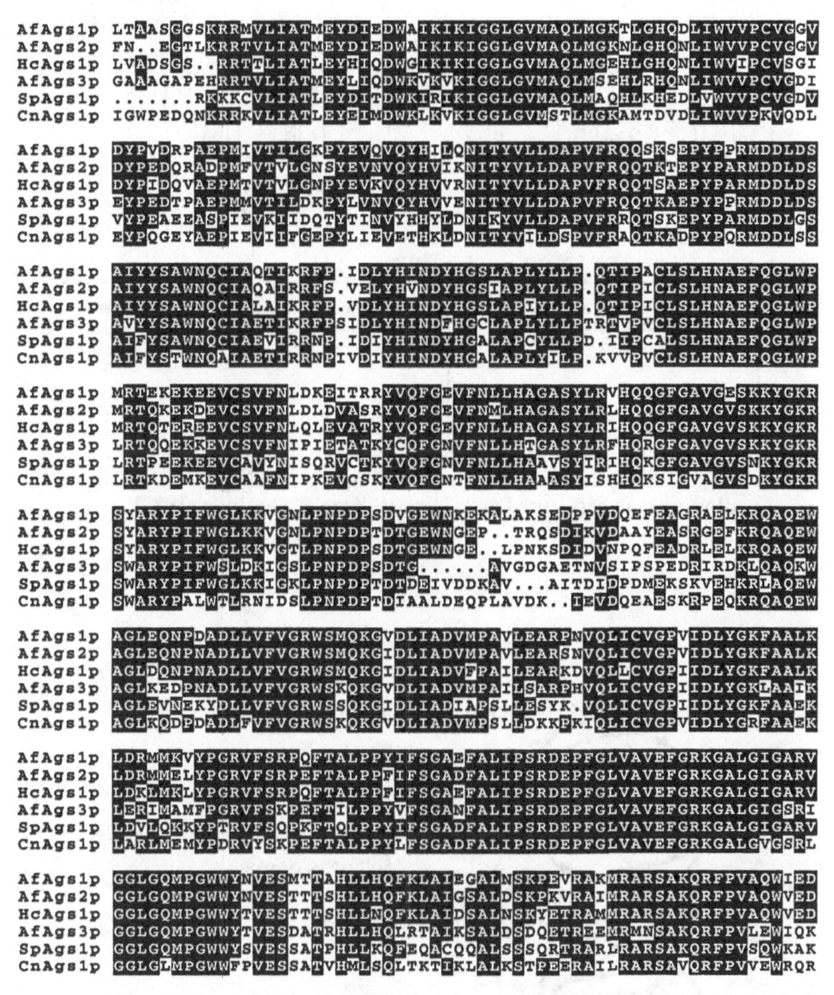

Fig. 13.4. Box–shade representation of the highest amino acid (aa) sequence similarities found in a synthase domain fragment of the Ags proteins of *Aspergillus fumigatus* (Af), *Cryptococcus neoformans* (Cn), *Histoplasma capsulatum* (Hc) and *Schizosaccharomyces pombe* (Sp). AfAgs1p, aa 1155–1632; AfAgs2p, aa 1152–11625; AfAgs3p, aa 1156–1629; CnAgs1p, aa 1156–1632; HcAgs1p, aa 1152–1625; SpAgs1p, aa 1141–1607.

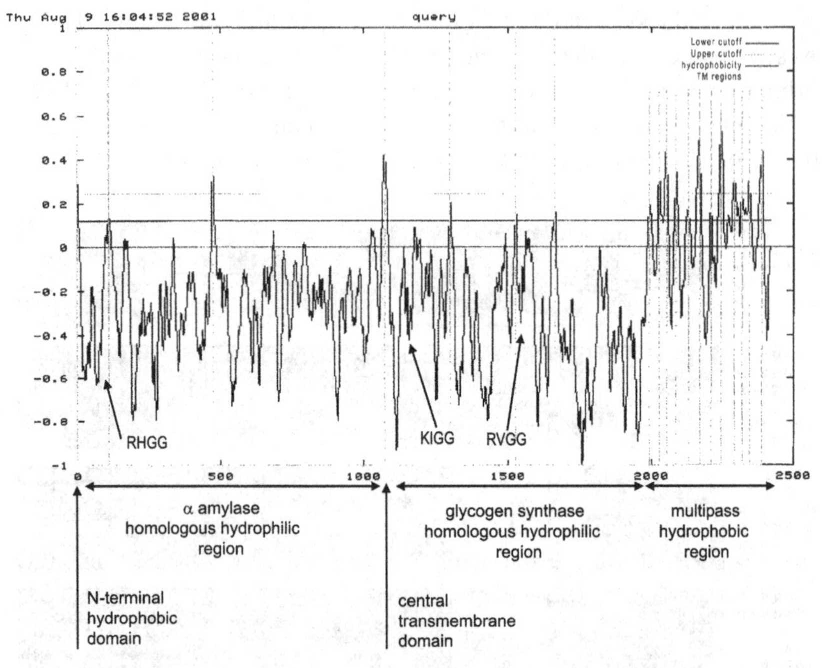

Fig. 13.5. Hydrophobicity profile of Ags1p (see Fig. 13.4). Hydropathy plotting was performed according to Kyte & Doolittle (1982).

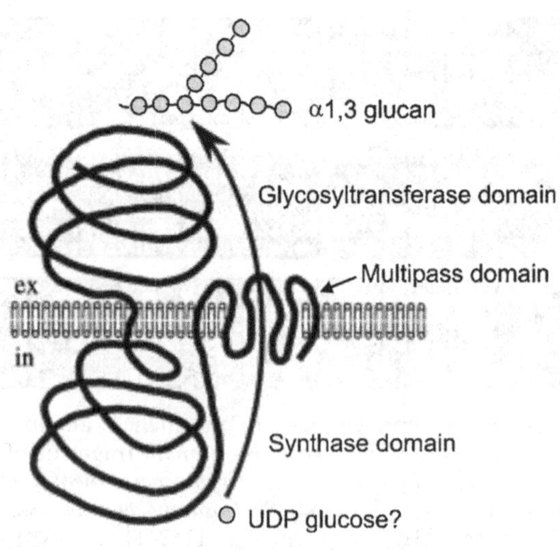

Fig. 13.6. Speculative model for Ags1p function (Hochstenbach *et al.*, 1998).

in the cell wall of both species. *H. capsulatum* ags1 became smooth; *C. neoformans* ags1-i cells had a severely depressed growth at 37 °C and no capsule formation (Reese & Doering, 2003; Rappleye *et al.*, 2004).

In *A. fumigatus*, three homologous *AGS* genes have been identified (Mouyna & Latgé, 2001; Beauvais *et al.*, 2005). All genes are expressed during vegetative growth; none is essential. In spite of the high homology between the three genes, a 50% reduction in α1,3 glucan content is only seen in the cell wall of the ags1 deletion mutant, whereas no reduction in α1,3 glucan is observed in ags2 and ags3 mutants. Accordingly, Ags1p seems to play the major role in the synthesis of cell-wall α1,3 glucan. This is confirmed by the cellular localization of the encoded Ags1 protein, which is present at the cell-wall level of the germinating conidia, on the apical region, and at the septum (Fig. 13.8a). Quantitative RT–PCR analysis of the expression of all *AGS* genes in *A. fumigatus* ags mutants and the parental strain suggests, however, that all Agsp are interconnected (Maubon *et al.*, 2006). For example, the deletion of *AGS3* results in a 2.5-fold increase in the

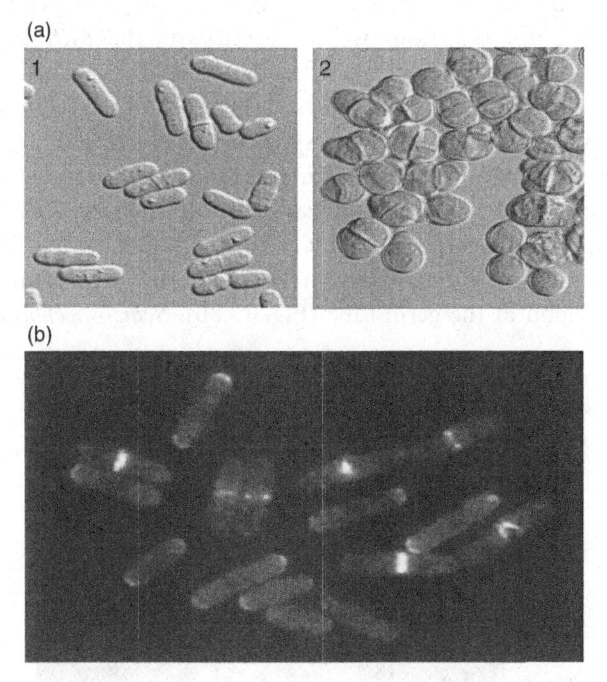

Fig. 13.7. (a) Cell morphology of wild type (1) and ags1 mutant (2) of *Schizosaccharomyces pombe* (Hochstenbach *et al.*, 1998). (b) Immunolabelling of *S. pombe* yeasts with an anti-Ags1p antibody, showing that Ags1p is located at the septum and apex of the cell (Katayama *et al.*, 1999).

expression of *AGS1*, suggesting that *AGS1* can compensate for the lack of *AGS3* (Fig. 13.9). Deletion of *AGS1* also results in an increase in *AGS3* expression, suggesting that *AGS3* is also involved in the synthesis of the cell-wall α1,3 glucan (Fig. 13.9). In comparison with the *AGS1* over-expression in the ags3 mutant, the over-expression of *AGS3* is limited in the ags1 mutant and cannot functionally compensate for the lack of *AGS1*. As seen in *S. pombe*, the ags1 mutant of *A. fumigatus* shows an altered hyphal morphology compared with the parental strain (Fig. 13.10a). The parental strain produced long hyphae with unique apices at the ends, whereas ags1 produces hyphae with excessive branching and dichotomous apices. The ags1 mutant had a 3.5-fold reduction in conidiation, owing to the formation of altered phialides on the *Aspergillus* head (Fig. 13.10b).

The role of *A. fumigatus AGS2* or *S. pombe AGS2–5* remains unknown, because deletion of these genes does not affect the viability of the cells (Katayama *et al.*, 1999; Beauvais *et al.*, 2005). The function of the proteins may not be the synthesis of α1,3 glucan but its regulation, as suggested for *A. fumigatus AGS2*, which is expressed during mycelium growth (Beauvais *et al.*, 2005). Similar behaviour was observed with the *FKS* genes in *Saccharomyces cerevisiae* (Lesage *et al.*, 2004). *FKS1* and *FKS2* are involved in β1,3 glucan synthase whereas, despite homologous sequences, *FKS3* is not associated with β1,3 glucan synthesis. In *A. fumigatus*, the ags2 mutant shows the same altered hyphal morphology and reduced conidiation, but no cell-wall defect. The localization of Ags2p in the wild type is different from that of Ags1p; this result confirms a different function for Ags1p. Ags2p is located intracellularly in the conidia, in the germ tube, and at the apical region of the germ tube (Fig. 13.8b). *Schizosaccharomyces*

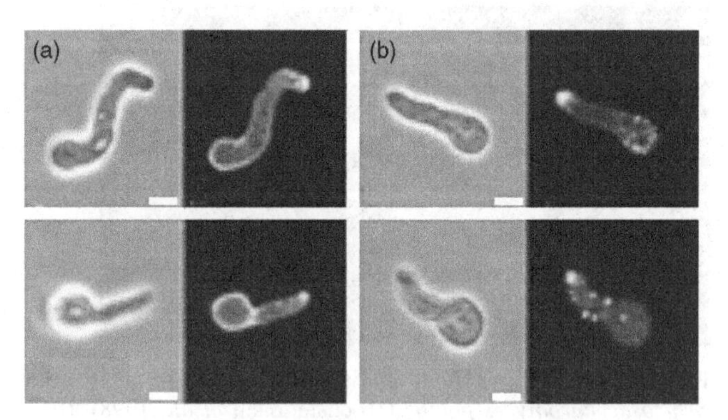

Fig. 13.8. Immunolocalization of Ags1p (a) and Ags2p (b) in germ tube of *Aspergillus fumigatus* with anti-Ags1p and anti-Ags2p antiserum (Beauvais *et al.*, 2005).

pombe AGS2–5 have not been localized, although Katayama *et al.* (1999) suggest that these homologues may become important in certain growth and stress conditions.

Regulation of α1,3 glucan synthesis

The regulation of α1,3 glucan synthesis has been studied in *S. pombe* and has been shown to be under the regulation of a Rho-GTPase, which is related to but different from the Rhop involved in β1,3 glucan synthesis. Over-expression of *RHO2* is lethal and caused an increase in the level of cell-wall α1,3 glucan (Calonge *et al.*, 2000). Rho2p appears to be a positive regulator of α1,3 glucan synthesis. Over-expression of *AGS1* also causes an increase in the α1,3 glucan fraction. In

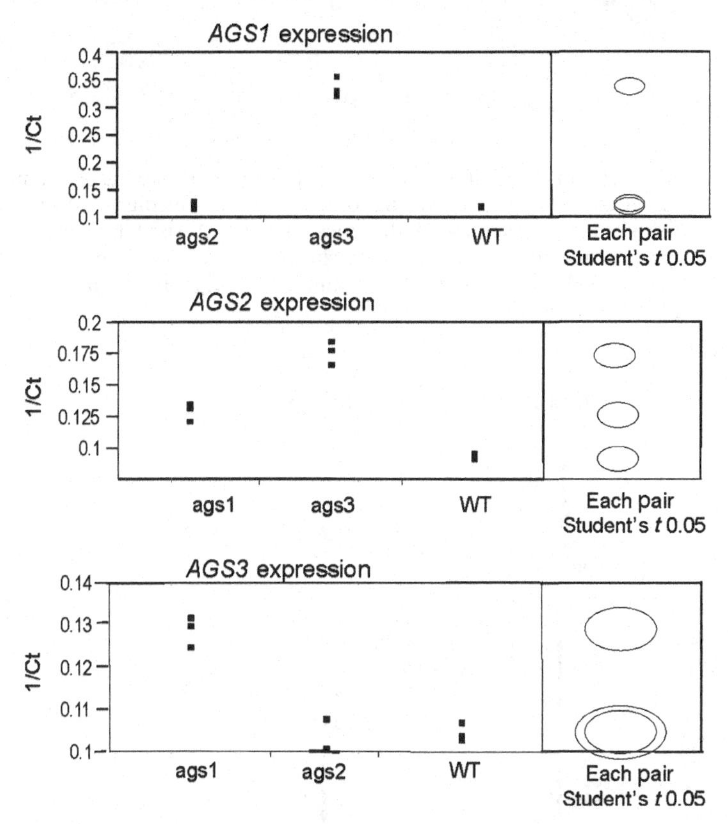

Fig. 13.9. Respective levels of expression of *AGS1*, *AGS2* and *AGS3* in wild type (WT) and ags1, ags2 and ags3 mutants in *Aspergillus fumigatus.* 1/Ct (Ct, threshold cycle) values are shown. Separate circles in the right panel indicate significant differences in the levels of expression. Dots represent three separate experiments.

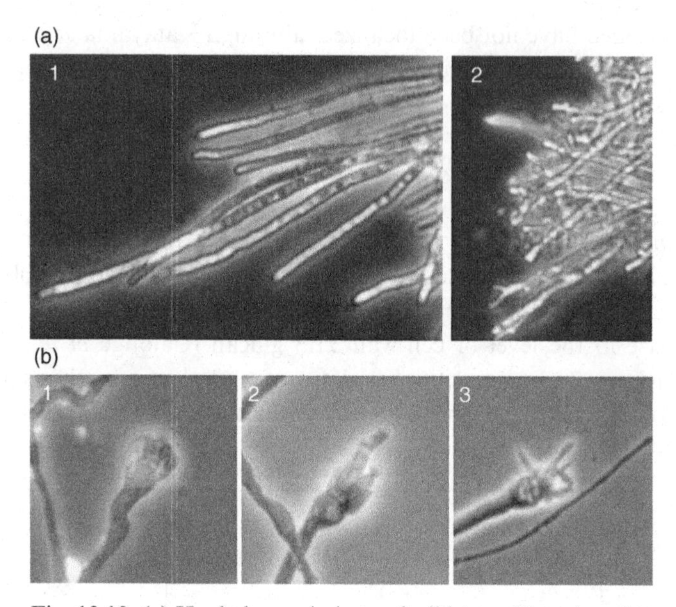

Fig. 13.10. (a) Hyphal morphology of wild-type (1) and ags2 mutant (2) *Aspergillus fumigatus* in Sabouraud liquid medium. Note the dichotomous branching of the apex of ags2. A similar phenotype is observed in the ags1 mutant (not shown). (b) Conidiation of the wild type (1) and ags2 (2) and ags1 (3) mutants of *A. fumigatus* in malt solid medium (Beauvais *et al.*, 2005).

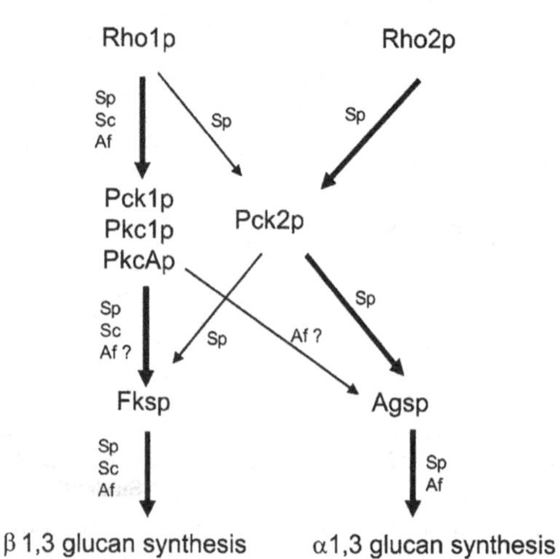

Fig. 13.11. Putative regulation of β1,3 and α1,3 glucan synthases in fungi. Sp, *Schizosaccharomyces pombe*; Af, *Aspergillus fumigatus*; Sc, *Saccharomyces cerevisiae*. Wide arrows indicate the main pathway.

contrast, deletion of *RHO2* induces a reduction in the level of α1,3 glucan. The deleted mutant is not lethal and can be compensated for by *RHO1*. Rho2p is localized with Ags1p to the site of growth. Rho2p regulates Ags1p through the protein kinase Pck2: the increase of α1,3 glucan caused by overexpression of *RHO2* does not occur in mutant cells with a pck2 deletion. These data indicate that the Pck2p kinase is required for Rho2p to regulate the biosynthesis of cell-wall α glucan (Fig. 13.11). Pck2p also co-localizes with Ags1p and Rho2p at the growing tips of the cell.

Loss of Ags1p results in randomization of F-actin, whereas overproduction of Ags1p results in asymmetrical cell shape and accumulation of both actin and cell-wall material. In addition, Ags1p localization is dependent on the integrity of the F-actin cytoskeleton (Katayama *et al.*, 1999). If Ags1p and actin are interconnected, the nature of their interaction has not been elucidated yet.

The regulation of α1,3 glucan synthesis in *S. pombe* appears similar to the regulation of β1,3 glucan synthesis (Arellano *et al.*, 1999a, b; Delley & Hall, 1999). In both cases, Rho GTPases, Rho1p and Rho2p, act as positive regulators of β1,3 and α1,3 glucan synthesis. Both GTPases use the protein kinases Pck1p or Pck2p to coordinately regulate the biosynthesis of the two main cell-wall polymers. Similar regulation exists in *Saccharomyces cerevisiae* for the synthesis of β1,3 glucan (Delley & Hall, 1999). Rho1p positively regulates the putative β1,3 glucan synthases Fks1p and Fks2p through Pkc1p; it is the only Pckp homologous kinase found in *S. cerevisiae*. The F-actin cytoskeleton also plays a crucial role in determining the cellular localization of Fksp, which helps establish their function.

In *A. fumigatus*, five Rho proteins and one PkcAp (homologous to *Schizosaccharomyces pombe* Pkcp) exist; and the regulation of Fksp by one Rho protein has been determined (Beauvais *et al.*, 2001). The role of the Rho GTPases or kinases in the α1,3 glucan synthesis has not been investigated. Based on the *S. pombe* and *Saccharomyces cerevisiae* models, it would have been expected that two protein kinases regulate glucan synthesis: one for β1,3 and one for α1,3. The regulatory role of PkcAp in glucan synthesis in *A. fumigatus* is under investigation.

Role of α1,3 glucan in host defence reactions

General host defence reactions

The α1,3 glucan-rich human pathogens *A. fumigatus*, *H. capsulatum*, *P. brasiliensis*, *C. neoformans* and *B. dermatitidis* infect their host through inhalation. The alveolar macrophage is the first line of defence against yeasts and conidia that are inhaled. After engulfment, the fungal elements are normally killed inside the phagolysosome. The cell wall plays an essential role in protecting the fungal cell against the reactive oxidants and other molecules produced by the host in response to the infection. Because α1,3 glucan is a main component of the outer layer of the cell wall, its role in fungus–host interactions has been investigated in several fungal models. The α1,3 glucan interacts with host cells in several ways. It can bind to immunoreactive molecules, as in *C. neoformans* or *A. fumigatus*, and/or it can form a protective layer at the host – fungal cell interface as in *B. dermatitidis*, *H. capsulatum* and *P. brasiliensis*.

C. neoformans is one of the few fungal pathogens that possess clearly proven virulence factors, two of which involve capsule and melanin formation. The polysaccharide capsule produced by *C. neoformans* is the hallmark of cryptococcosis and is probably the major virulence factor of this fungus. Acapsular mutants are avirulent (Chang & Kwon-Chung, 1994). Capsular material impedes the host immune response by altering defence processes including phagocytosis, proinflammatory cytokine production and leucocyte migration (Buchanan & Murphy, 1998). α1,3 Glucan mediates the interaction between capsule components and the cell-wall surface (Reese & Doering, 2003). Electron microscopy studies have shown that ags1-i mutant cells have no capsule and do not react with anticapsular antibody, confirming the lack of capsule (Fig. 13.12). The role of α1,3 glucan in capsule cell-wall association is confirmed by the use of *H. capsulatum* in capsule binding assays. This organism contains surface-accessible α1,3 glucan in its cell wall. Cryptococcal capsule

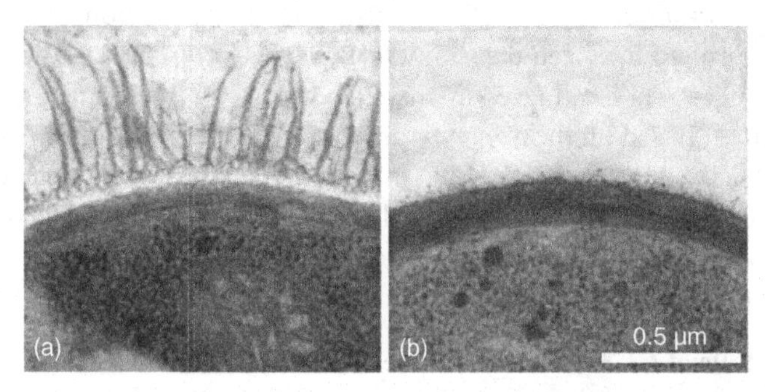

Fig. 13.12. Thin-section electron micrographs showing portions of the edge of a wild-type cell (a) and an ags1-i mutant cell (b) of *Cryptococcus neoformans*. Note the absence of capsule on the ags1-i cell (Reese & Doering, 2003).

polysaccharides bound extensively to wild-type *H. capsulatum* but not to the mutant lacking α1,3 glucan (Reese & Doering, 2003). In addition to the loss of the capsule, the *C. neoformans* ags1-i mutant grows poorly at 37 °C in culture and is therefore likely to be avirulent owing to low fitness.

Melanin is a second well-characterized virulence factor in many human and plant fungal pathogens (Nosanchuk & Casadevall, 2003). In *C. neoformans*, *Exophiala dermatitidis* and *A. fumigatus*, melanin-deficient mutants display reduced virulence compared with the wild type (Jahn *et al.*, 2002; Nosanchuk & Casadevall, 2003). Melanin appears to play a role in protection from oxidative killing and phagocytosis (Jahn *et al.*, 2002; Nosanchuk & Casadevall, 2003). In *A. fumigatus* (Beauvais *et al.*, 2005) a 50% reduction of the total α1,3 glucan of the cell wall (ags1 mutant) was not associated with a decrease in melanin production or in fungal virulence. Conversely, deletion of the *AGS3* gene leads to a more aggressive strain, compared with its parental wild type, that appears to be related to melanin formation (Maubon *et al.*, 2006) (Fig. 13.13). However, this ags3 mutant does not show any cell-wall modification in terms of the

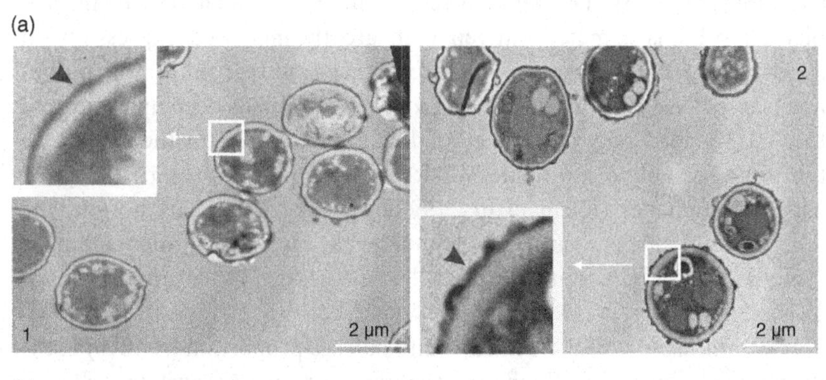

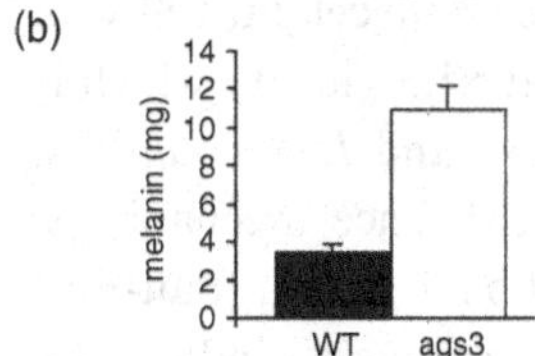

Fig. 13.13. (a) Ultrastructure of conidia of the wild type (1) and an ags3 mutant (2) of *Aspergillus fumigatus*. Note that the electron-dense layer of melanin (arrowheads) was thinner in the wild type than in the mutant. (b) Amount of melanin (expressed in milligrams of lyophilized material) in *A. fumigatus* ags3 mutant and wild type (Maubon *et al.*, 2006).

composition and quantification of each polysaccharide. Rather, the apparently enhanced aggressiveness of the ags3 mutant is due to its specific conidial phenotype. A two–three-fold increase in melanin production is observed in the ags3 mutant compared with the wild type; electron microscopic observations confirm that the dense outer layer of melanin at the surface of the cell wall is twice as large in the ags3 mutant than in the wild type (Fig. 13.13a). Accordingly, the melanin mass per conidium is two–three-fold higher in the ags3 mutant than in the parental strain (Fig. 13.13b). In agreement with the increased melanin content, the ags3 mutant shows heightened resistance to H_2O_2. An unique structure of α1,3 glucan in ags3 may anchor more melanin, which would confer a better resistance of the mutant to reactive oxidants. In addition to the melanin phenotype, the ags3 mutant conidia germinated faster; it is known in *A. fumigatus* that virulence is directly correlated with growth rate (Latgé, 2003).

Another virulence factor, WI-1 (Klein *et al.*, 1994), is a 120 kDa cell-wall adhesin that has been isolated from the surface of strains of *B. dermatitidis* (Klein *et al.*, 1994). The molecule contains 34 copies of a 25 aa tandem repeat, which are highly homologous to invasin, an adhesin of *Yersinia* (Klein *et al.*, 1993). The tandem repeat mediates attachment to macrophages. WI-1 is a major antigen with greatly increased expression on surfaces, which can promote enhanced phagocyte recognition in spontaneous hypovirulent mutants lacking α1,3 glucan. It has been hypothesized that α1,3 glucan may mask the WI-1 adhesin/antigen on the surface of the yeast, or act to modulate WI-1 mediated interactions of the yeast with macrophages (Klein *et al.*, 1994). In support of this notion, the α1,3 glucan-deficient mutants stimulate a respiratory burst that is 20-fold greater than seen with the wild-type yeasts (Morrison & Stevens, 1991).

In *P. brasiliensis*, α1,3 glucan is found in large proportion (*c.* 40%) in the pathogenic yeast cell wall whereas the non-pathogenic mycelium cell wall contains only β1,3 glucan. It has been proposed that α1,3 glucan, which is present at the surface of the cell wall of *P. brasiliensis* and *H. capsulatum*, plays a role as a protective layer against the host defence mechanisms, relying on the inability of phagocytic cells to digest α1,3 glucan (San-Blas *et al.*, 1977). Moreover, α1,3 glucan synthesis in *P. brasiliensis* is induced by fetal calf serum and is lost when *P. brasiliensis* strains are cultivated for long periods *in vitro* (San-Blas & Vernet, 1977). The α1,3 glucan-less mutants obtained by chemical mutagenesis are avirulent, like the ags1 mutant or the spontaneous α1,3 glucan-less mutants of *H. capsulatum* (Klimpel &

Goldman, 1988; Rappleye *et al.*, 2004). *In vitro* infections of macrophage-like cells with *H. capsulatum* wild type and ags1-i mutant show that the wild type causes substantial destruction of the macrophage monolayer, whereas the mutant is defective in killing the macrophages. This loss of virulence in macrophages correlated with a reduced ability of ags1-i mutant to colonize lungs after *in vivo* infection (Rappleye *et al.*, 2004).

Conclusions

There is a growing recognition of the importance of α1,3 glucan in the life of human fungal pathogens, and much new information has emerged in recent years. Yet in spite of this progress, a great deal remains to be understood about the biochemistry of α1,3 glucan and its interaction with host immune defences. No studies exist on the chemical and/or molecular relationship between α1,3 glucan and capsule, melanin and WI-1 protein. Although the genes involved in α1,3 glucan synthesis are known, the substrate and enzymatic partial reactions leading to the synthesis of α1,3 glucan are unknown. Accordingly, inhibitors of α1,3 glucan synthase cannot be suitably investigated. In *A. fumigatus*, the *AGS* genes belong to a gene family; expression studies show an interaction between all members of the family, yet the specific physiological role of each one remains to be elucidated. A similar puzzle exists for several gene families involved in cell-wall synthesis (chitin synthase and glucosyltransferase families) (Mouyna & Latgé, 2001). This point also raises the question of the regulation of the synthesis of this polysaccharide in *A. fumigatus*: only one Pkcp is known.

It is not known whether α1,3 glucan helps subvert host defences by playing a passive role as an anchor of surface molecules implicated in virulence or as a protective layer against host enzymes or antifungal molecules, as has been described for some *Aspergillus* species. In *A. flavus*, an amphotericin B-resistant mutant shows an increase in α1,3 glucan (43% of the mutant cell wall, and 39% in the parental strain) which adsorbs amphotericin B (Seo *et al.*, 1999). In *A. niger*, expression of *AGSA*, one of five putative α1,3 glucan-synthase-encoding genes, is 20-fold reduced in response to cell-wall stress induced by agents such as calcofluor SDS or caspofungin (Damveld *et al.*, 2005). Alternatively, α1,3 glucan can be specifically synthesized in response to the host immune system, as in *P. brasiliensis*. There is a great deal to learn about the complex interplay of α1,3 glucan and the immune system. Presently, no studies have investigated the putative role of α1,3 glucan on the production of cytokines such as TNFα and their role in the host defence reaction.

References

Arellano, M., Coll, P. M. & Perez, P. (1999a). RHO GTPases in the control of cell morphology, cell polarity, and actin localization in fission yeast. *Microscopy Research Technique* **47**, 51–60.

Arellano, M., Valdivieso, M. H., Calonge, T. M., Coll, P. M., Duran, A. & Perez, P. (1999b). *Schizosaccharomyces pombe* protein kinase C homologues, pck1p and pck2p, are targets of rho1p and rho2p and differentially regulate cell integrity. *Journal of Cell Science* **112**, 3569–78.

Beauvais, A., Bruneau, J. M., Mol, P. C., Buitrago, M. J., Legrand, R. & Latgé, J. P. (2001). Glucan synthase complex of *Aspergillus fumigatus*. *Journal of Bacteriology* **183**, 2273–9.

Beauvais, A., Maubon, D. & Park, S. (2005). Two α(1–3) glucan synthases with different functions in *Aspergillus fumigatus*. *Applied and Environmental Microbiology* **71**, 1531–8.

Buchanan, K. L. & Murphy, J. W. (1998). What makes *Cryptococcus neoformans* a pathogen? *Emerging Infectious Diseases* **4**, 71–83.

Calonge, T. M., Nakano, K., Arellano, M., Arai, R., Katayama, S., Toda, T., Mabuchi, I. & Perez, P. (2000). *Schizosaccharomyces pombe* Rho2p GTPase regulates cell wall α-glucan biosynthesis through the protein kinase Pck2p. *Molecular Biology of the Cell* **11**, 4393–401.

Chang, Y. C. & Kwon-Chung, K. J. (1994). Complementation of a capsule-deficient mutation of *Cryptococcus neoformans* restores its virulence. *Molecular and Cell Biology* **14**, 4912–19.

Damveld, R. A., vanKuyk, P. A., Arentshorst, M., Klis, F. M., Van den Hondel, C. A. & Ram, A. F. (2005). Expression of agsA, one of five 1,3-alpha-D-glucan synthase-encoding genes in *Aspergillus niger*, is induced in response to cell wall stress. *Fungal Genetics and Biology* **42**, 165–77.

Delley, A. P. & Hall, M. N. (1999). Cell wall stress depolarizes cell growth via hyperactivation of RHO1. *Journal of Cell Biology* **147**, 163–74.

Fontaine, T., Simenel, C., Dubreucq, G., Adam, O., Delepierre, M., Lemoine, J., Vorgias, C. E., Diaquin, M. & Latge, J. P. (2000). Molecular organization of the alkali-insoluble fraction of *Aspergillus fumigatus* cell wall. *Journal of Biological Chemistry* **275**, 27 594–607.

Garcia-Rodriguez, L. J., Trilla, J. A., Castro, C., Valdivieso, M. H., Duran, A. & Roncero, C. (2000). Characterization of the chitin biosynthesis process as a compensatory mechanism in the *fks1* mutant of *Saccharomyces cerevisiae*. *FEBS Letters* **478**, 84–8.

Grün, C. H., Hochstenbach, F., Humbel, B. M., Verkleij, A. J., Sietsma, J. H., Klis, F. M., Kamerling, J. P. & Vliegenthart, J. F. G. (2004). The structure of cell wall α-glucan from fission yeast. *Glycobiology* **15**, 245–57.

Hochstenbach, F., Klis, F. M., Van Den Ende, H. *et al.* (1998). Identification of a putative alpha-glucan synthase essential for cell wall construction and morphogenesis in fission yeast. *Proceedings of the National Academy of Sciences of the USA* **95**, 9161–6.

Hogan, L. H. & Klein, B. S. (1994). Altered expression of surface alpha-1,3-glucan in genetically related strains of *Blastomyces dermatitidis* that differ in virulence. *Infection and Immunity* **62**, 3543–6.

Jahn, B., Langfelder, K., Schneider, U., Schindlel, C. & Brakhage, A. A. (2002). PKSP-dependent reduction of phagolysosome fusion and intracellular kill of *Aspergillus fumigatus* conidia by human monocyte-derived macrophages. *Cellular Microbiology* **4**, 793–803.

Katayama, S., Hirata, D., Arellano, M., Pérez, P. & Toda, T. (1999). Fission yeast α-glucan synthase *mok1* requires the actin cytoskeleton to localize the sites of growth and plays an essential role in cell morphogenesis downstream of protein kinase C funtion. *Journal of Cell Biology* **144**, 1173–86.

Klein, B. S., Hogan, L. H. & Jones, J. M. (1993). Immunologic recognition of a 25-amino acid repeat arrayed in tandem on a major antigen of *Blastomyces dermatitidis*. *Journal of Clinical Investigation* **92**, 330–7.

Klein, B. S., Chaturvedi, S., Hogan, L. H., Jones, M. & Newman, S. L. (1994). Altered expression of surface protein WI-1 in genetically related strains of *Blastomyces dermatitidis* that differ in virulence regulates recognition of yeasts by human macrophages. *Infection and Immunity* **62**, 3536–42.

Klimpel, K. R. & Goldman, W. E. (1988). Cell walls from avirulent variants of *Histoplasma capsulatum* lack alpha-(1,3)-glucan. *Infection and Immunity* **56**, 2997–3000.

Konomi, M., Fujimoto, K., Toda, T. & Osumi, M. (2003). Characterization and behaviour of alpha-glucan synthase in *Schizosaccharomyces pombe* as revealed by electron microscopy. *Yeast* **20**, 427–38.

Kyte, J. & Doolittle, R. F. (1982). A simple method for displaying the hydrophathic character of a protein. *Journal of Molecular Biology* **157**, 105–32.

Latgé, J. P. (2003). *Aspergillus fumigatus*, a saprophytic pathogenic fungus. *Mycologist* **17**, 56–61.

Lesage, G., Sdicu, A. M., Menard, P. *et al.* (2004). Analysis of beta-1,3-glucan assembly in *Saccharomyces cerevisiae* using a synthetic interaction network and altered sensitivity to caspofungin. *Genetics* **167**, 35–49.

Manners, D. J., Masson, A. J. & Patterson, J. C. (1973). The structure of a ß-(1-3)-D-glucan from yeast cell walls. *Biochemical Journal* **135**, 19–30.

Maubon, D., Park, S., Tanguy, M. *et al.* (2006). Ags3p, a third putative α(1-3)glucan synthase in *Aspergillus fumigatus*, is involved in virulence. *Fungal Genetics and Biology* **43**, 366–75.

Mellado, E., Dubreucq, G., Mol, P., Sarfati, J., Paris, S., Diaquin, M., Holden, D. W., Rodriguez-Tudela, J. L. & Latgé, J. P. (2003). Cell wall biogenesis in a double chitin synthase mutant (chsG-/chsE-) of *Aspergillus fumigatus*. *Fungal Genetics and Biology* **38**, 98–109.

Morrison, C. J. & Stevens, D. A. (1991). Mechanisms of fungal pathogenicity: correlation of virulence *in vivo*, susceptibility to killing by polymorphonuclear neutrophils *in vitro*, and neutrophil superoxide anion induction among *Blastomyces dermatitidis* isolates. *Infection and Immunity* **59**, 2744–9.

Mouyna, I. & Latgé, J. P. (2001). Cell wall of *Aspergillus fumigatus*: Structure, biosynthesis and role in host-fungus interactions. In *Fungal Pathogenesis: Principles and Clinical Applications*, ed. R. Cilhar & R. Calderone, pp. 515–37. New York: Marcel Dekker.

Nosanchuk, J. D. & Casadevall, A. (2003). The contribution of melanin to microbial pathogenesis. *Cellular Microbiology* **5**, 203–23.

Osmond, B. C., Specht, C. A. & Robbins, P. W. (1999). Chitin synthase III: synthetic lethal mutants and "stress related" chitin synthesis that bypasses the *CSD3/CHS6* localization pathway. *Proceedings of the National Academy of Sciences of the USA* **96**, 11 206–10.

Osumi, M., Yamada, N., Yaguchi, H., Kobori, H., Nagatani, T. & Sato, M. (1995). Ultrahigh-resolution low-voltage SEM reveals ultrastructure of the glucan network formation from fission yeast protoplast. *Journal of Electron Microscopy (Tokyo)* **44**, 198–206.

Rappleye, C. A., Engle, J. T. & Goldman, W. E. (2004). RNA interference in *Histoplasma capsulatum* demonstrates a role for alpha-(1,3)-glucan in virulence. *Molecular Microbiology* **53**, 153–65.

Reese, A. J. & Doering, T. L. (2003). Cell wall alpha-1,3-glucan is required to anchor the *Cryptococcus neoformans* capsule. *Molecular Microbiology* **50**, 1401–9.

San-Blas, F., San-Blas, G. & Cova, L. J. (1976). A morphological mutant of *Paracoccidioides brasiliensis* strain IVIC Pb9. Isolation and wall characterization. *Journal of Genetics Microbiology* **93**, 209–18.

San-Blas, G. & Vernet, V. (1977). Induction of the synthesis of cell wall α-1,3-glucan in the yeastlike form of *Paracoccidioides brasiliensis* strain IVIC Pb9 by fetal calf serum. *Infection and Immunity* **15**, 897–902.

San-Blas, G., San-Blas, F. & Serrano, L. E. (1977). Host-parasite relationships in the yeastlike form of *Paracoccidioides brasiliensis* strains IVIC Pb9. *Infection and Immunity* **15**, 343–6.

Seo, K., Akiyoshi, H. & Ohnishi, Y. (1999). Alteration of cell wall composition leads to amphotericin B resistance in *Aspergillus flavus*. *Microbiology and Immunology* **43**, 1017–25.

Smith, F. & Montgomery, R. (1956). End group analysis of polysaccharides. In *Methods of Biochemical Analysis*, ed. D. Glick, pp. 153–212. New York: Interscience Publishers.

Sugawara, T., Sato, M., Takagi, T., Kamasaki, T., Ohno, N. & Osumi, M. (2003). *In situ* localization of cell wall alpha-1,3-glucan in the fission yeast *Schizosaccharomyces pombe*. *Journal of Electron Microscopy (Tokyo)* **52**, 237–42.

14
Plagues upon houses and cars: the unnatural history of *Meruliporia incrassata*, *Serpula lacrymans* and *Sphaerobolus stellatus*

NICHOLAS P. MONEY
Department of Botany, Miami University, Oxford, Ohio

Dry rot in the twenty-first century

Santa Barbara in California is such a stunningly beautiful place that it seems reasonable that this apparent utopia is troubled by one or two things. In the context of fungi in the indoor environment, residents have encountered a pestilence that is transforming homes into rubble and sawdust. Its name is *Meruliporia incrassata* and it causes dry rot (Figs. 14.1, 14.2). I'll begin with the thoughts of journalist Matt Kettmann, who wrote a story for Santa Barbara's newspaper, *The Independent*, titled, 'Invasion of the House-Eating Fungus' (Kettmann, 2002). Matt interviewed the Kastner family, who had moved into a US $1.2 million ranch-style house. The nastiness began when, 'a mysterious foam slowly started peeking out of the laundry room walls'. Dry rot was diagnosed and treated by an extermination company. Feeling confident that the fungus was gone, the family left for a vacation. When they returned, Matt explained, nearly a quarter of the house 'was eaten in a mere week'. Christina Kastner remarked, 'It came back with a vengeance – like it got mad.' The problem was solved once the exterminators found a 'taproot with the diameter of a grapefruit in the rear of the house'. This umbilical cord was cut, a ditch was dug around the foundations and filled with concrete, and the damaged home was refurbished. The Kastners' insurance policy did not cover any of the US$100 000 repair bill.

Many species of fungus cause the type of wood decay encountered in homes, but most of them prefer to work outdoors in woodlands. Very few

This contribution is adapted, with permission, from a chapter published in Money, N. P. (2004). *Carpet Monsters and Killer Spores: A Natural History of Toxic Mold.* New York: Oxford University Press.

Fungi in the Environment, ed. G. M. Gadd, S. C. Watkinson & P. S. Dyer. Published by Cambridge University Press.

Fig. 14.1. Basidiomes of *Meruliporia incrassata* on door frame in Californian home damaged by dry rot. Photograph courtesy of Luis De La Cruz (De La Cruz Wood Preservation Services, Van Nuys, CA).

specialize in indoor work. Much of the horrific damage reported in California, as I have mentioned, is caused by *Meruliporia incrassata*. The basidiomycete *Serpula lacrymans* is responsible for the same kind of destruction in Britain, elsewhere in Europe, and in other parts of the world, including Australia and Japan. *Serpula* means serpent or worm, evoking the taproots of the fungus that slither indoors. The specific name, *lacrymans*, is the Latin term for weeping and refers to the appearance of globules of fluid on the surface of the fungus. Mordecai Cooke (1871) wrote that the fungus 'is often dripping with moisture, as if weeping in regret for the havoc it has made'. Until a few years ago, the fungus was called *Merulius lacrymans*. (How do taxonomists ever settle upon names for their children or pets?) *Merulius* referred to the yellow colour of the mycelium, which was seen as comparable to the beak of a male blackbird

Fig. 14.2. Extensive dry rot of timber frame of home in California caused by *Meruliporia incrassata*. Damage occurred during construction. Photograph courtesy of Luis De La Cruz (De La Cruz Wood Preservation Services, Van Nuys, CA).

(*Turdus merula*). The Latin names offer a vivid picture of the glistening, yellowish colonies of these fungi and their serpentine taproots. To complete this exercise in etymology, I must deconstruct *Meruliporia incrassata*, which was first described in 1849 (Verrall, 1968). The *Meruli* part has been covered; *poria* refers to the porous nature of the fruiting bodies, and *incrassata* refers to their thickness.

Strands and rhizomorphs

The warmest invitation for dry rot is a house in which some wooden structure makes direct contact with the soil. A post sunk into the ground without the protection of concrete is a choice entry point. Dry rot also intrudes when soil accumulates under the bottom edge of a stucco exterior, or contacts the exterior particle board underneath vinyl siding. Many cases of dry rot could be avoided if builders ensured that a few inches of bare concrete showed all the way around a home exterior. This is a good pointer for inspecting any home before purchase. Even in regions without dry rot, piling soil against a house is an inexcusable mistake, because ants and termites can use the same pathways into homes as fungi. Other defects are less easily spotted. Cracks in concrete slabs and any holes for pipes or electrical work also serve as gateways for the fungus. Terry Amburgey, Professor of Forest Products at Mississippi State University, says that the dry rot fungus 'will infiltrate a foundation, wood, or concrete, and pretty soon the entire house goes' (www.aquarestoration.com). Comparing dry rot to 'a horde of army ants in search of food', mycologist David Arora (1996) adds that the fungus will 'overrun everything in [its] way: bricks, stones, tiles, plaster, drainpipes, wires, leather boots, cement floors, books, tea kettles, even corpses'.

Indoor moulds like *Stachybotrys chartarum* get their water from the dampness of their surroundings, accounting for their appearance in water-damaged homes. Dry rot operates in a completely different fashion. Its common name refers to the ability of the fungus to destroy wood in dry buildings, but ironically *Meruliporia* and *Serpula* are very sensitive to dehydration. They thrive in the wet interior of a beam, but only emerge on surfaces if they remain damp and shaded from sunlight (which partly explains some of their fondness for basements). The key to understanding the extraordinary destructive abilities of these organisms lies in the taproot structure mentioned above. *Meruliporia*, *Serpula* and other wood-decay fungi form two types of root-like structure: strands (or chords) and rhizomorphs (Moore, 1998). Strands and rhizomorphs develop as assemblages of hyphae; the rhizomorph is the more complex of the two organs and grows only at its tip. *Serpula* forms strands but does not seem to produce rhizomorphs (Fig. 14.3); *Meruliporia* generates both kinds of structure. Young strands show little internal differentiation, although a thick 'parent' hypha occupies the core of the strand and is surrounded by thinner hyphae growing along it in both directions. To further consolidate the strand, the parent produces tendrils that wind around the other hyphae. Many fungi form these cables, even in culture plates. Experiments suggest that strands begin to develop in response to nutrient exhaustion. According to this idea, strands are exploratory structures that the fungus can send out from the colony in search of new food sources. As soon as the scouting strand hits some nutrients, hyphae fan out to form new mycelia and begin extracting nutrients. It seems that as long as food is limited the hyphae will cooperate with one another to grow across barren terrain, but as soon as they reach an 'oasis' the strand disassembles.

Rhizomorphs can be much larger than strands, have a waterproofed surface, and enclose a central pipe for the transmission of water and

Fig. 14.3. Branched strand of *Serpula lacrymans* attached to tile and connected to wood food source. Diameter of primary axis of strand is approximately 1 mm. Photograph courtesy of Jim Worrall (USDA Forest Service).

dissolved nutrients. These are the structures that invade Californian homes (Fig. 14.4). The rhizomorph can grow for many metres, piping water from its base (in wet soil) to its tip (buried in dry wood or a slab of masonry). In some situations, air can pass along the rhizomorph so that the fungus can avoid suffocation while exploring, for example, the anoxic interior of a painted beam. Rhizomorphs bear some similarities to mushrooms: both are complex, multi-celled structures and develop through the interweaving and adhesion of scores of hyphae. Interestingly, a cut rhizomorph exudes a distinctive mushroomy smell. Or, I suppose, one might say that a cut mushroom has a distinctive rhizomorphy smell. The rest of the fungus has a similarly strong odour, which has led some home inspectors in Scandinavia to use dogs to detect dry rot before too much damage has occurred. A recent study in Finland evaluated the skills of Labrador retrievers (Kauhanen *et al.*, 2002). The investigators concealed small pieces of pine colonized by *Serpula* and other wood-rotting fungi, plus cultures of indoor moulds and bacteria. Clean pieces of wood were hidden as controls. The dogs were very effective at finding the contaminated materials and cultures, but failed to discriminate between the various microbes (which is probable cause for celebration among human mycologists). Though difficult to detect during the early stages of installation, advanced cases of dry rot are obvious. Nothing else produces white mats of mycelium

Fig. 14.4. Rhizomorphs of *Meruliporia incrassata* entering basement of home. Photograph courtesy of Luis De La Cruz (De La Cruz Wood Preservation Services, Van Nuys, CA).

clinging to the surfaces of beams, strands and rhizomorphs dangling between different areas of decay, and revolting fruiting bodies.

Once in a basement, hyphae of dry rot fungi forge through the microscopic cells of hardwood beams, digesting their cellulose walls and transforming springy lumber into parched brown cubes that crumble when touched. Some contractors refer to dry rot as brown cubical rot. Brown rot fungi in forests leave the same signature in decaying logs. The brown colour is due to the progressive concentration of the dark-pigmented lignin which the fungus leaves behind as it digests the cellulose. By contrast, white rot fungi extract the lignin and other pigmented components of the wood, leaving white cellulose behind. Unlike indoor moulds that rely upon spores to move from one location to another, dry rot fungi can colonize an entire home without forming a single spore. If it can conduct enough water indoors through its rhizomorph, or if the home is saturated by plumbing leaks, the fungus will emerge from the interior of a wooden beam and encase the surface in a white cobweb of mycelium. Once this platform is established, the fungus can use hyphal strands to bridge gaps between beams. The strands can also span concrete or plastics that offer nothing edible to the fungus, allowing the rot to jump from meal to meal, continually nurtured by water piped from the backyard by the rhizomorph. This mycological nightmare can keep getting worse. A single, well-established mycelium can also colonize other locations by developing additional indoor rhizomorphs to transfer water, and thereby set up areas of decay all over a house. *Meruliporia* and *Serpula* crawl upward from the basement, issuing between floorboards and cracks in walls, spreading like pancake mix over the home. One Californian homeowner said that her infestation looked like Camembert cheese, which is an appropriate description because the Camembert rind is a living mat of hyphae, just like a dry-rot colony (Kettmann, 2002). In a dry home, every area of decay is dependent upon the original umbilical connection to the backyard provided by an invading rhizomorph. If this cable can be found and cut, the entire mega-fungus will wither. In a wet home, the dry rot fungus will flourish along with indoor moulds, and the original connection to the outdoors may be severed without consequence. In these situations, the fungus can also spread via spores, which brings me to dry-rot fruiting bodies, or basidiomes.

Basidiomes

In a basement suffering from advanced decay, basidiomes as big as surfboards burgeon as crusts on the surface of beams and exude an unmistakably fungal smell. These excrescences are yellow to orange-brown in colour, with a white margin, and form masses of rusty spores (Fig. 14.5). Spores are formed in quartets on basidia, from which they catapult themselves into the air. In this manner, sizeable fruiting bodies can fill the atmosphere of a rotting basement with mind-boggling numbers of spores. A single healthy basidiome the size of this book can shed billions of spores. In basements where there is little air circulation, the spores collect beneath the crusts, covering furniture, boxes, and sundry clutter with a reddish shroud. Fruiting bodies of *Meruliporia* and *Serpula* can extend all the way along beams, creating colossal platforms for spore production. A Belgian company called Alert-Pest measured a single fruiting body that spread over an area of 6 square metres (or 65 square feet) in a property blanketed with a total of 78 square metres (840 square feet) of spore-producing rot!

As they mature, dry rot basidiomes develop ridges, then become pitted with tubes as the ridges extend and fuse with one another, and finally sport a honeycombed or toothed appearance. This process of maturation elaborates an ever greater fertile surface for spore production. Older fruiting bodies can also extend short shelves into the air, like those of bracket fungi

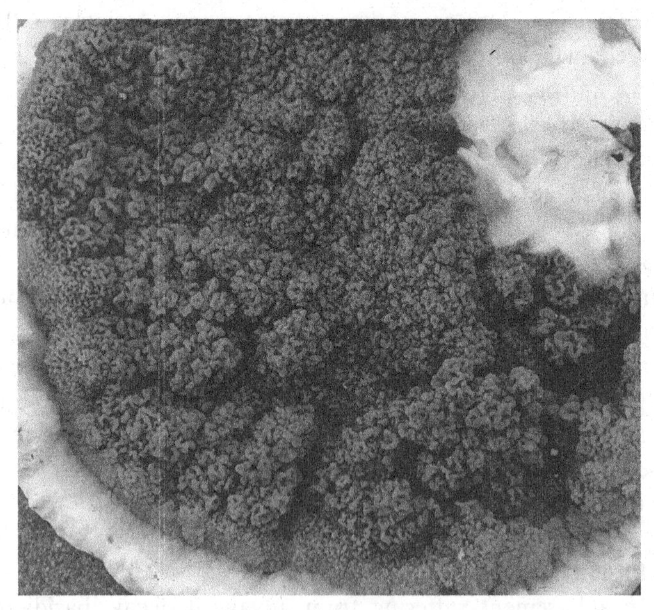

Fig. 14.5. Basidiome of *Serpula lacrymans*. White periphery is approximately 2.5 cm in width. Photograph courtesy of Jim Worrall (USDA Forest Service).

on diseased trees and decaying logs. Dry rot basidiomes and edible mushrooms have similar anatomy and identical function, but nobody would serve dry rot at a dinner party. Nobody, that is, with the possible exception of eccentric mycologist Captain Charles McIlvaine (1840–1909). McIlvaine did not sample dry rot crusts, but he did taste related fungi with resupinate fruiting bodies, and wrote, 'I have tasted, raw, every species I have found. They are all more or less woody in flavour, and I believe them to be edible' (McIlvaine, 1900). He concluded that these should be eaten only in an emergency, though he failed to specify the nature of the emergency that would justify breaking one's teeth for little, if any, nutritional reward. McIlvaine also tried a number of fleshier bracket-forming fungi. Here are three of his reviews: 'undoubtedly tough, but cut fine and stewed slowly for half an hour it is quite as tender as the muscle of an oyster and has a pleasant flavor'; 'pleasantly crisp when stewed', and 'edible if chopped fine and very well cooked'. Among all of the bracket-forming fungi, he reserved scorn for *Polyporus heteroclitus*, saying, 'As it ages it becomes offensive.' David Arora (1996) describes the dry rot fungus as 'utterly and indisputably inedible'.

Dry rot in a changing environment

Meruliporia caused a great deal of destruction in the southeastern United States in the early part of the twentieth century (Verrall, 1968), but has only become a serious problem on the West Coast in the past decade. The reason that dry rot is wrecking homes in Santa Barbara and elsewhere in California may be quite straightforward. To begin transforming productive agricultural land into a sprawling housing development in this part of America, fragrant orchards of fruit and nut trees are obliterated with the aid of bulldozers. Trees are often snapped aside, rather than pulled out whole, leaving a stump and intact roots in the soil, and the dying tissues are swiftly colonized by fungi. Buried wood will keep the fungi occupied for a while, but once they have rendered all the cellulose into sugar they begin to search for more food by sending out rhizomorphs. The rhizomorphs extend through the soil, around drainage pipes, and into basements. Sometimes they hit the jackpot: the wood frame of a house. When this happens, mycelia sprout from the ends of the rhizomorphs and penetrate their favourite food source. If the basement is dry as a bone, the fungus imports water as necessary, keeping the wood-chewing mycelia soggy. Imported topsoil and wood mulch brought in for landscaping serve as additional sources of dry rot inoculum. Homeowners can worsen the situation by keeping rhizomorphs soaked with lawn sprinklers, and turning on the air conditioner to cool the fungus to its optimum growth temperature of around 23 °C (73 °F). It would be difficult to design a more perfect situation for *Meruliporia*.

Changing climatic conditions may aid *Meruliporia* in California and elsewhere, particularly in areas that experience an increase in rainfall, because anything that elevates soil moisture is good for wood-rotting fungi. Dry rot tends to be more common in wet years, because a high water table will maintain the *outdoor* water supply for the *indoor* fungus. But dry rot is not a new phenomenon in other parts of the world.

Dry rot history

In the Old Testament book of Leviticus, in which Moses is instructed on practical issues ranging from kosher methods of animal slaughter and cooking, to the various punishments for having sex with animals or one's mother-in-law, there is a lengthy passage that some interpret as the description of an action plan for treating problems with fungi in homes (Heller *et al.*, 2003). The Bible identifies the nuisance as a 'fretting leprosy of the house', which betrays itself when the homeowner concludes that 'It seemeth to me that there is as it were a plague in the house' (Leviticus 14: 33–53). 'Fretting' refers to a gradual or insidious destruction, and 'leprosy' is applied loosely here to imply a disease. Once the homeowner admits his or her fears to a priest, an ordained course of action is initiated. First, the priest must carry out an inspection. If he discovers that 'the plague be in the walls of the house with hollow strakes [streaks], greenish or reddish', the home must be closed up for a week. By restricting airflow and elevating indoor humidity this measure would promote fungal growth and sharpen the mushroom odour, which would help the priest appraise the problem. If, after the trial period, the 'plague be spread in the walls of the house', the suggested remedy includes scraping the interior of the home and dumping 'the dust that they scrape off without the city into an unclean place'. As a last resort, the priest must order the destruction of the house and have the timber and mortar carted off to the (increasingly) unclean place beyond the city walls. This edict would have severely limited mould and dry rot claims by uninsured homeowners in the ancient world. Contemporary building consultants may see some logic in all this, but I will never be convinced that killing a bird 'in an earthen vessel over running water' (the details of the method – shaking or drowning? – are not specified in the Bible) is an effective approach to certifying that a home is purified. Some mycologists have suggested that Leviticus refers to indoor moulds, but the colour and behaviour of the plague is undeniably more suggestive of dry rot or other wood-decay fungi.

Dry rot has ravaged the British more than any other people, or perhaps it would be more accurate to say that the British have been most vocal about its offences. The fungal culprit for dry rot in Europe has a much longer criminal record than the Californian menace. *Serpula* has lived with the British for so long, and is so common, that home inspectors expect to find it in older properties. Like *Meruliporia, Serpula* prefers cooler temperatures than many indoor moulds, perhaps explaining the prevalence of dry rot, and absence of the warmth-loving mould *Stachybotrys chartarum*, in Britain. The outdoor biology of *Serpula* is a mystery, because the fungus has never been found growing on tree stumps anywhere in the British Isles. It does show up, however, in forests in India and in central Europe. Beginning in 1929, an Indian mycologist called Bagchee spent more than 20 years looking for the fungus in the western Himalayas (Bagchee, 1954). He found it on tree stumps and fallen logs, and also in buildings. Fruiting bodies developed on the shaded underside of logs in contact with soil, echoing their appearance in basements. The same species was also reported in the 1970s, emerging from stone walls, and was relocated in the Himalayas by a multinational expedition in 1992 (Singh *et al.*, 1993). Might dry rot have moved from India to Britain? Singh and colleagues suggested that the fungus could have been brought home on the possessions of Victorians returning from India, and then spread to other countries. British imports of Himalayan timber between 1850 and 1920 may have provided an additional shot of dry rot, particularly since the logs were floated down the Ganges to Calcutta and exported without proper drying and treatment. This Indian origin might seem compelling, but it is not tenable because the British were cursed by dry rot a long time before the Raj.

Beyond India, fruiting bodies of *Serpula lacrymans* have been found on the roots of damaged Norway spruce trees in the Czech Republic, suggesting that the fungus sometimes survives as a wound parasite in nature (Bech-Andersen, 1999). But many of the sightings of the fungus in central Europe since the nineteenth century are cases of mistaken identity. The 'true' dry rot fungus is easily confused with a related beast called *Serpula himantioides*, and is as rare in European forests as it is in India (White *et al.*, 1997). It is interesting that both *Stachybotrys* and the dry rot fungus torment homeowners but seem to be weak competitors against closely related fungi in the wild (explaining their rarity). Home construction has been highly beneficial to the biological success of these fungi. The mystery of dry rot's immigration to Britain is unlikely to be solved now. Centuries ago, infected timber must have been used to build someone a refuge, and the fungus embraced the opportunity to become a plague upon our houses.

Dry rot of buildings has been tolerated as a fact of life in Britain for centuries but, unlike the American preoccupation with indoor moulds, *Serpula* is rarely mentioned by the media. However, the fungus was a celebrity in the seventeenth century when it ravaged the Royal Navy. As Secretary of Admiralty Affairs in the 1680s, Samuel Pepys surveyed new ships under construction at the Royal Dockyard at Chatham and discovered that 'many of them ... lye in danger of sinking at their very Moorings.' The planks were 'in many places perished to powder', and, 'Their Holds not clean'd nor air'd, but ... suffer'd to heat and moulder, till I have with my own Hands gather'd Toadstools ... as big as my Fists' (Pepys, 1690). *Serpula* was not the only fungus attacking the ships at dock. The common woodland species *Laetiporus sulphureus*, or sulphur shelf, ate at the oak planks and formed staircases of bright yellow brackets inside the rotting holds. Wood-boring beetles followed the fungi, and their burrows further weakened the timber. Similar secondary damage of wood by death watch beetles, woodworm, and other insects is also common in homes (Ridout, 2000). After assessing 'how deeply the Ships were infected with that evil', Pepys concluded that a horrifying proportion of the fleet required complete refitting. This led to concern, not only about the cost of the exercise but also about the availability of timber.

Severe timber shortages in Britain began after the Reformation, when the dissolution of the monasteries led to wholesale destruction of their vast wooded estates (Albion, 1926). Various timber preservation acts were passed, but supplies of serviceable oak trees for shipbuilding continued to diminish. This led to an ever-increasing reliance on timber imported from Eastern Europe, which probably caused the dry rot epidemic. Domestic hardwood tended to be better seasoned, drier and more rot-resistant than imported logs, which were often soaked for weeks by floating in rivers or were rotten on arrival after confinement in the holds of transport ships. The link between the use of unseasoned timber and dry rot was recognized by Pepys, but the need for warships far outstripped the availability of quality lumber. Interestingly, contemporary use of unseasoned 'green', or already infected wood for home construction has been implicated in the recent increase in wood rot in California caused by the fungus *Gloeophyllum trabeum*. Little attention has been paid to this fungus in comparison with *Meruliporia*, even though it may be responsible for substantial damage. In the eighteenth century, unseasoned wood was used for repairs to the hull of the *Royal George*, which sank with several hundred crew members at Portsmouth in 1782. A subsequent court martial revealed that the wood had been so rotten that it failed to hold a nail. The earlier loss of Henry VII's flagship, the *Mary Rose*, off Portsmouth in 1545 was probably due to the same cause.

In the seventeenth century, warships were expected to last for 25–30 years, but dry rot began to severely abbreviate the useful life of a new vessel: by the eighteenth century, the average life of a ship was cut to 12 years, and dwindled to 'no duration' after the Battle of Trafalgar in 1805. In practical terms, ships became disposable items, which pleased neither crews nor politicians. Nelson's *Victory* and other famous ships from the time of the Napoleonic Wars (1799–1815) were heavily colonized by wood-rotting fungi. The 110-gun battleship *Queen Charlotte* deteriorated so swiftly during construction that the navy was forced to rebuild her in 1810 *before* she could set sail! In his lively description of the impact of dry rot, Ramsbottom (1941) cited a cost of £287 837 for the repair of the *Queen Charlotte* before she could be launched, and noted that after refitting 'her name was changed to the *Excellent* – a whimsical choice'. Moore (2001) estimated that at today's prices the repairs would cost US$2.5 billion. The financial impact of the damage was alarming, but the government had no choice but to pay. After all, 'The royal navy of England hath ever been its greatest defence and ornament; it is its ancient and natural strength; the floating bulwark of the island' (Blackstone, 1765).

Dry rot remediation

Michael Faraday became involved in the search for a solution to the dry rot problem and delivered a lecture on the subject to the Royal Institution in 1833 (Faraday, 1836). He was impressed with a remedy for dry rot patented by Dublin-born inventor John Howard Kyan in 1832. The method became known as wood-kyanizing. Kyan thought, that 'the evil might be stopped; that the commencement even might be prevented by the application of corrosive sublimate'. Previously, corrosive sublimate, or mercuric chloride, had been used to preserve brain tissue and other delicate body parts for scientific study, and also to purge bookworm from library books. Faraday was concerned that the preventive effect against dry rot would be short-lived. Once 'the timber of the vessels which were exposed to the bilgewater, and other water, where vessels were not coppered', he wondered, 'was the sublimate not likely to be removed, and its effects be destroyed?' The Admiralty ordered a trial in the 'fungus pit' at Woolwich. This was 'a pit dug in the [naval] yard, and enclosed by wood on all sides, having a double wooden cover; it was damp of itself, and into this were put various kinds of wood, of which they wished to make a trial' (Faraday, 1836). Timber was submerged in a tank filled with the corrosive sublimate, dried, transferred to the pit, and left for 5 years. The experiments showed that the treatment was effective against dry rot, and Faraday concluded that, 'The process here employed completely, and with certainty, prevents the possibility of the destructive effects of the active principle which Nature employs to cause decomposition and decay . . . Corrosive sublimate neutralizes this primary element of fermentation . . . rendering [wood] as indestructible as charred timber'.

Faraday recognized, however, that Kyan's solution had some limitations: 'was there not a fatal injury that might arise from the production of a noxious atmosphere?' This had been a concern for users of kyanized library books too. Answering his own question, Faraday concluded that once it combined with the treated wood, corrosive sublimate would not pose a serious problem. But his confidence was qualified: 'it would be found useful in a far higher degree, in the construction of cottages and outhouses, than palaces; for it is of far more importance to those whose means are small, that they should have that duration given to their timber which would extend the application of their means, and give permanency to their comforts'. (To which he should have added, 'even if they forget who they are and their hair falls out'.) The method was not embraced to any great degree by the Navy or the public.

A tremendously successful and less noxious preventive for wood decay was introduced in 1838, when the use of creosote as a wood preservative was patented by John Bethell. Creosote is a blend of thousands of compounds derived from coal tar and is effective at stalling the growth of all wood-decay fungi. It seemed like the perfect remedy for galleon-rot, especially when the alternative wood treatments were so obviously incompatible with human health. Some of the constituents of creosote are very nasty chemicals, including trimethylbenzene, naphthalene and pyrene, but once the wood is impregnated and coated with the stuff the only danger comes from skin contact. It is interesting to consider that many of the same cancer-causing chemicals found in creosote are supplied by cigarette smoke. Besides its immediate application in shipbuilding, creosote proved an impressive defence against rot in fence posts, railway sleepers (railroad ties) and telephone poles. The basidiomycete *Lentinus lepideus* shows high tolerance to creosote and attacks wood products that are not impregnated with a sufficient quantity of water-repelling creosote at the time of treatment. The common name of this mushroom is the train wrecker. The fungus is probably implanted in the softwood timber before treatment, and forms its fruiting bodies when the moisture content of the wood increases through contact with soil.

Soil and groundwater contamination around creosote treatment plants is problematic, however, and after a century and a half of use, the European Union introduced a wholesale ban in 2003. Following an agreement with the wood-preservative industry in 2003, the US Environmental Protection Agency (EPA) banned the use of chromated copper arsenate after 2003, but did not extend the ban to creosote. American creosote advocates argue that a ban would lead to a new epidemic of wood decay. The debate about creosote continues in the United States, where the EPA has yet to make a final ruling on the issue. Extermination companies would be happy to see the elimination of creosote, and other businesses also view the potential opportunity for house-destroying fungi as a blessing. Commenting on trends in home construction, an article in *U.S. News and World Report* (Benjamin *et al.*, 2000) gave the following plug for the steel industry: 'Want a house that's immune to dry rot, moisture, and termites, and resistant to hurricanes and earthquakes? Hire a steel house framer.'

Today's steel armadas are resistant to dry rot, but fungi continue to attack wood-hulled pleasure boats. Freeway galleons, otherwise known as RVs, are another target. If a wood-decay fungus can become established in a wet location within the vehicle (flooring beneath a plumbing leak is a likely starting place) it can use its rhizomorphs to colonize the rest of the structure. But stationary homes remain the staple food for dry rot fungi. Insurance companies have used the term dry rot in a very general sense to describe the damage caused by any of the wood-decay fungi that can appear in homes, and a blanket exclusion for dry rot is written into most homeowners policies. Nevertheless, some victims of *Meruliporia* have challenged this rule in court by arguing that the damage caused by this fungus represents a new and unforeseen disaster deserving special attention. The California home of Joseph and Jodeanna Glaviano was severely damaged by *Meruliporia*, but the couple were blissfully unaware of this until their hardwood floor collapsed, revealing fungal destruction of the sub-flooring. Arguing the distinction between dry rot, and dry rot caused by *Meruliporia*, attorneys representing the Glavianos were successful in prying some compensation from Allstate Insurance in California (Joseph Glaviano *et al.* v. Allstate Insurance Co., No. 00-56754, 9th Cir.; 2002 U.S. App. LEXIS 9324). The appeals court ruled that the losses due to 'collapse caused by hidden decay' were covered.

Health concerns

Although they can collapse homes, *Meruliporia* and *Serpula* do not raise the health concerns associated with indoor moulds: their spores do not carry toxins, so they cannot be blamed for lung bleeding or brain damage. Dry rot spores are certainly allergenic, however, and their tremendous numbers in contaminated homes are of concern for anyone exposed to them. Besides asthma, dry rot spores are known to have caused hypersensitive pneumonitis, which is a type of immune response involving gamma globulins. One published case history described a 32-year-old schoolteacher who developed severe breathing difficulties and joint pains and began losing weight (O'Brien *et al.*, 1978). His condition improved during a brief spell in hospital, but worsened again when he returned home. Blood tests showed that he had formed high levels of antibodies (IgE and IgG) to *Serpula* spores, which was not surprising because his home suffered from extensive dry rot. The relationship between dry rot and allergic disease leaves open the possibility of future personal injury lawsuits as add-ons to claims for property damage.

Dry rot in fiction

As a metaphor, dry rot is used frequently as a reflection of human misery, self-imposed or otherwise. In *Great Expectations*, rot served to garnish Miss Havisham's monument to the misfortune of her betrayal at the altar:

> *... dry rot and wet rot and all the silent rots that rot in neglected roof and cellar ... addressed themselves faintly to my sense of smell ...* Charles Dickens, *Great Expectations* (1860–1861)

This queen of self-pity presided over the biodegradation of her uncelebrated wedding banquet, when an afternoon's work with rubber gloves and trash cans would have allowed the woman to reclaim her life. In Poe's *The Fall of the House of Usher*, dry rot is promoted to a major character. The narrator of the story describes Usher's mansion in the following manner:

> *The discoloration of ages had been great. Minute fungi overspread the whole exterior, hanging in a fine tangled web-work from the eaves ... there was much that reminded me of the specious totality of old wood-work which has rotted for long years in some neglected vault, with no disturbance of breath of the external air. Beyond this indication of extensive decay, however, the fabric gave little token of instability.*
>
> Edgar Allen Poe, *The Fall of the House of Usher* (1839)

The specious nature of rotten wood is familiar to home inspectors. *Meruliporia* and *Serpula* can turn the interior of a plank into dust before anything more than bubbling of paintwork is visible. But once the cellulose is extracted by the fungus, a screwdriver can be pushed through a wood beam without any resistance.

Wet rot and the artillery fungus

Scientific interest centres on the dry rotters, but wet rot fungi that corrode indoor wood that is pre-soaked for the fungus by natural flooding, plumbing disasters, or persistent leaks in the building exterior (or 'envelope' in technical parlance) are more common (Fig. 14.6). The cellar rot fungus, *Coniophora puteana*, causes a lot of mess in damp basements, and ink cap mushrooms (*Coprinus* and *Coprinopsis*) are common on sodden door and window frames. Other fungi are content to blemish the outsides of homes and, more importantly, cars. The enemy of a beautifully enamelled finish is called *Sphaerobolus stellatus*, better known as the artillery fungus. This is a coprophilous fungus, adapted for growth on the partly digested cellulose fibres defecated by herbivores, or on sodden wood shavings in gardening mulch. Like other kinds of dung fungi, it faces a severe challenge once it has exhausted its food source: How can I escape the dung heap? This has led to the evolution of a series of intriguing mechanical devices that shoot spores onto grass blades surrounding the dung. *Sphaerobolus* uses a miniature springboard to eject a single 1 mm diameter black ball (gleba) filled with spores. There is nothing else like this in the living world.

Fig. 14.6. Wood decay in recently constructed home by unidentified wet rot fungus. Photograph courtesy of Luis De La Cruz (De La Cruz Wood Preservation Services, Van Nuys, CA).

The operation of the fungal springboard was first described by Pier Antonio Micheli in his *Nova Plantarum Genera*, published in 1729 (Fig. 14.7). The device develops as a sphere on the surface of the dung, and cracks open to reveal an inner cup that glistens with fluid and cradles the gleba. Held under increasing tension, the inner cup finally everts, propelling the gleba up to 5.5 metres from the fruiting body and setting the record for fungal propulsion. An old, slightly flaccid tennis ball serves as a good model: form a dish by pressing one half into the other, then watch it flip outward to restore the sphere. The rapid motion of the inner

Fig. 14.7. The earliest image of the artillery fungus, *Sphaerobolus stellatus* (referred to as *Carpobolus*), by Pier Antonio Micheli, in *Nova Plantarum Genera*, published in 1729. Courtesy of the Lloyd Library, Cincinnati, Ohio.

cup of the *Sphaerobolus* fruiting body is audible as a popping sound, as is the impact of the gleba on the lid of a culture dish. For details of the mechanism, I refer you to a lovely article written by Terence Ingold (1972). To achieve an arching trajectory for its spore mass, *Sphaerobolus* aims for the sun and hits anything in its path, including automobiles. This is a problem, because the glebal mass is very sticky, and as it dries it becomes glued to the paintwork. Vigorous rubbing will dislodge the black blob, but a circular depression is left in the paint, deep enough that it cannot be buffed out by polishing with a cloth. Cars parked next to thousands of the trampolines in a flower bed become conspicuously dalmatianed, potentially ruining the day for the car owner, the owner's insurance company, the mulch supplier, the landscaping company, the owner of the parking lot, etc. (while brightening the day for the local repair shop). One study at Pennsylvania State University estimated that US$1 million in claims for automobile damage were filed in a single year in Pennsylvania (www.aerotechlabs.com). Ohio is another hot spot for *Sphaerobolus*, but it grows in most parts of the United States during periods of warm, wet weather. House exteriors are another common target for the fungus, although it's difficult to imagine one so spotless that it would seem marred by the artillery fungus. Bird's nest fungi do precisely the same damage, although their spore masses are propelled by raindrops splashing into their cup-shaped fruiting bodies. These are coprophilous fungi too, and the reach of their discharge mechanism comes close to the performance of the artillery fungus.

By hitting a car, the artillery fungus is condemned to travel thousands of miles with no prospect of germination. The gleba remains fastened to the vehicle indefinitely, and the spores inside can survive for years with little opportunity to settle on wet mulch or animal dung where they can raise their own families of miniature springboards. Two centuries ago, the artillery fungus travelled across oceans on the sopping wood of decaying warships. It was figured by the botanical artist James Sowerby during his investigation of the disintegrating *Queen Charlotte* in 1812 (Ramsbottom, 1941). Shooting from ship to ship was far more effective than lodging on painted metal, because the fungus had some prospect of continuing to grow and reproduce at sea and, one day, of making it back to land when the timber was recycled. This may have enabled *Sphaerobolus* to spread far beyond its prehistoric distribution.

Russian scientists studying samples taken from Mir since the 1980s have documented more than 100 different species of fungi growing in the spacecraft (Karash, 2000). Cosmonauts and astronauts have also noticed that fungal colonies have been growing on the portholes of the International Space Station, prompting extreme efforts to sterilize incoming cargo, and aggressive housekeeping in space. The fungi on the Space Station are the same species that grow in homes on Earth, although the subjects of this chapter have not been found up there yet. After millions of years of unwitnessed toil, the biological careers of fungi happen to have intersected with ours. They followed us indoors from the woods, joined us as cabin-mates across oceans and into orbit, and when human history comes to a close, a deluge of their spores will help erase the record of our presence on this planet.

References

Albion, R. G. (1926). *Forests and Sea Power*. Cambridge, MA: Harvard University Press.

Arora, D. (1996). *Mushrooms Demystified: A Comprehensive Guide to the Fleshy Fungi*, 2nd edn. Berkeley, CA: Ten Speed Press.

Bagchee, K. (1954). *Merulius lacrymans* (Wulf.) Fr. in India. *Sydowia* **8**, 80–5.

Bech-Andersen, J. (1999). Comparison between *Serpula lacrymans* found in the Indian Himalayas and Mount Shasta, California. In *From Ethnomycology to Fungal Biotechnology: Exploiting Fungi from Natural Resources for Novel Products*, ed. J. Singh & K. R. Aneja, pp. 279–86. New York: Kluwer Academic/Plenum Publishers.

Benjamin, M., Curry, A., Fischman, J. *et al.* (2000). Flip-of-the-coin-jobs. *U.S. News and World Report* (November 6).

Blackstone, W. (1765). *Commentaries on the Laws of England.* Oxford: Clarendon Press.

Cooke, M. C. (1871). *A Plain and Easy Account of the British Fungi.* London: Robert Hardwicke.

Faraday, M. (1836). *On the Prevention of Dry Rot in Timber*. London: John Weale.

Heller, R. M., Heller, T. W. & Sasson, J. M. (2003). Mold: “tsara’at,” Leviticus, and the history of a confusion. *Perspectives in Biology and Medicine* **46**, 588–91.

Ingold, C. T. (1972). *Sphaerobolus*: The story of a fungus. *Transactions of the British Mycological Society* **58**, 179–95.

Karash, Y. (2000). Space fungus: A menace to orbital habitats. www.space.com (July 27).

Kauhanen, E., Harri, M., Nevalainen, A. & Nevalainen, T. (2002). Validity of detection of microbial growth in buildings by trained dogs. *Environment International* **28**, 153–7.

Kettmann, M. (2002). Invasion of the house-eating fungus. Mysterious mushroom munches through south coast homes. *The Independent* (Santa Barbara), July 18.

McIlvaine, C. (1900). *One Thousand American Fungi.* Indianapolis: Bowen-Merill.

Moore, D. (1998). *Fungal Morphogenesis.* Cambridge: Cambridge University Press.

Moore, D. (2001). *Slayers, Saviors, Servants and Sex. An Exposé of Kingdom Fungi.* New York: Springer.

O’Brien, I. M., Bull, J., Creamer, B., Sepulveda, R., Harries, M., Burge, P. S. & Pepys, J. (1978). Asthma and extrinsic allergic alveolitis due to *Merulius lacrymans. Clinical Allergy* **8**, 535–42.

Pepys, S. (1690). *Memoires Relating to the State of the Royal Navy of England, for Ten Years, Determin’d December 1688*. London: Ben Griffin.

Ramsbottom, J. (1941). Dry rot in ships. *Essex Naturalist* **25**, 231–67.

Ridout, B. (2000). *Timber Decay in Buildings. The Conservation Approach to Treatment.* London: E. & F. N. Spon.

Singh, J., Bech-Andersen, J., Elborne, S. A., Singh, S., Walker, B. & Goldie, F. (1993). The search for wild dry rot fungus (*Serpula lacrymans*) in the Himalayas. *Mycologist* **7**, 124–30.

Verrall, A. F. (1968). *Poria Incrassata Rot: Prevention and Control in Buildings.* USDA Forest Service Technical Bulletin No. 1385. Washington D.C.: U.S. Government Printing Office.

White, N. A., Low, G. A., Singh, J., Staines, H. & Palfreyman, J. W. (1997). Isolation and environmental study of 'wild' *Serpula lacrymans and Serpula himantioides* from the Himalayan forests. *Mycological Research* **101**, 580–4.

V

Environmental population genetics of fungi

15

Fungal species: thoughts on their recognition, maintenance and selection

JOHN W. TAYLOR, ELIZABETH TURNER, ANNE PRINGLE

Department of Plant and Microbial Biology, University of California, Berkeley

JEREMY DETTMAN

Department of Botany, University of Toronto

HANNA JOHANNESSON

Department of Evolutionary Biology, Uppsala University

When it comes to fungal species and speciation, it is hard to find anything to say that has not already been said in several excellent recent reviews. The most comprehensive source of information is Burnett's recent book (Burnett, 2003), which expands upon the themes from his British Mycological Society Presidential Address (Burnett, 1983). In addition to reviewing mycological species concepts and speciation, he describes enough about basic mycology and the methodology of evolutionary studies to make chapters on defining fungal individuals and populations, or on the processes of evolution in fungi, useful for mycologists interested in evolution and for evolutionary biologists interested in fungi. Burnett's review of the early literature in fungal speciation is particularly helpful in the present age, when it seems as if literature that is not online is forgotten. A second source of information is Brasier (1997), who explored three of what he considered to be the four main elements contributing to fungal speciation: original interbreeding populations, natural selection on populations and reproductive isolation between populations. He left a discussion of mating systems to others. Brasier's discussion of natural selection is particularly good, and his figure comparing the narrow range of growth rates of dikaryotic hyphae taken from *Schizophyllum commune* fruiting bodies to the much broader range of growth rates for dikaryons synthesized from their haploid progeny is as clear a demonstration of the effects of selection as one could want. In the same era, Natvig & May (1996) considered biological species in terms of the physical scales and life histories

Fungi in the Environment, ed. G. M. Gadd, S. C. Watkinson & P. S. Dyer. Published by Cambridge University Press.

appropriate to fungi. Their discussion of the great variety of fungal mating systems (including pseudohomothallism) and the ability of fungi to make both clonal and recombined offspring is particularly useful when trying to fit fungal data into the Procrustean bed of evolutionary thought emerging from studies of other eukaryotes. Petersen & Hughes (1999) then focused on Agaricales and the biological species concept with an emphasis on practical aspects of fungal species recognition. When comparing species recognized by phylogenetic and biological approaches, they noted that many genetically and geographically distinct species are not reproductively isolated, a point that we will explore below. An additional comprehensive source of information is Worrall's book (Worrall, 1999), and especially the chapter on defining fungal species contributed by Harrington & Rizzo (1999). These authors also discuss the problem of assigning a taxonomic rank to populations that are genetically isolated but kept from mating only by geography. Harrington & Rizzo (1999) decided to recognize a taxon as a species only if it exhibits a unique, diagnosable, phenotypic character; this is another point that we will consider below. Finally, although it is based almost entirely on examples from the animal and plant kingdoms, we would be remiss to omit mention of Coyne & Orr's recent and thorough volume on speciation (Coyne & Orr, 2004). In light of the many recent contributions to knowledge of fungal species and speciation, commentary here will emphasize research that has been published subsequently to these works.

A quarter of a century ago, Burnett rendered a perhaps overly dismal picture of research on fungal speciation (Burnett, 1983), famously stating, 'Mycology and mycologists, on the whole, have contributed very little to the mainstream of ideas concerning the modes of origins of species.' Fourteen years later, Brasier echoed Burnett's sentiments before sounding a more hopeful note (Brasier, 1997), that '. . . fungi provide superb material with which to study evolution in action . . .' Just three years ago, Burnett again synthesized results from many studies (Burnett, 2003), but noted that most of them involved fungi having, '. . . populations whose structure is largely determined by agricultural activities'. Among the fungi that will be featured here, species of *Coccidioides, Histoplasma, Aspergillus* and *Neurospora* appear not to be so influenced. Species of the first three genera are important pathogens of animals and humans, but the evidence to date indicates that agriculture is more likely to influence fungal biogeography (and therefore speciation) than human disease (which is not to say that human activities have had no effect on the evolution of medically important fungi, as will be noted below).

Burnett lamented (Burnett, 1983) that, 'A real constraint is imposed by the incomplete understanding of the taxonomy, ecology and biogeography of the fungi compared with that available to those concerned with ... plants [and] animals ...' The situation is changing and, for some fungi, the phylogeny and biogeography are relatively well understood, although our understanding of fungal ecology is still lagging and has become a factor that limits our understanding of adaptation and speciation. In this chapter, we will begin by discussing recent work on species recognition, and then proceed to a discussion of speciation processes and processes that maintain species by reducing or ablating gene flow among populations. Finally we will consider the interplay of these processes and natural selection.

Species recognition

Species recognition is a human endeavour that is essential to subsequent study of the evolutionary processes leading to speciation. Placing individuals in genetically isolated species is a necessary prerequisite for studies of reproductive mode, hybridization, gene flow and selection. Traditionally, all fungal species were recognized by morphology. Where morphological species were broad, and where the need to distinguish species was great, mycologists employed other phenotypes for species recognition; for example, substrate utilization and growth rate on different media at different temperatures are important to systematics of yeasts and penicillia (Pitt, 1979; Kurtzman & Fell, 1998). Now that we know that multiple fungal species may share easily observed phenotypes and still be genetically isolated, other methods of species recognition are needed. An obvious choice would be species recognition by mating compatibility tests, or Biological Species Recognition (BSR). Unfortunately, given that only *c.* 11% of fungal species have been cultivated and that *c.* 20% do not reproduce sexually in cultivation (Ainsworth *et al.*, 2001), BSR by mating tests is not broadly applicable. In its place, measurement of genetic isolation by molecular Phylogenetic Species Recognition (PSR), using the concordance of gene genealogies, has provided a popular solution to the problem of fungal species recognition (Avise & Wollenberg, 1997; Taylor *et al.*, 2000).

To carefully implement PSR, it is necessary not only to find several genomic regions with sufficient polymorphism to build well-supported gene genealogies, but also to assemble a collection of individuals from throughout the range of the fungus. When the sequence data have been obtained, the resulting gene genealogies can be compared to discover the transition from concordance among gene genealogies (due to lineage specific loss of ancestral variation following genetic isolation) to conflict among gene genealogies (due to recombination within populations) and thereby recognize species. Species recognition by concordance of gene genealogies has had a dramatic impact on fungal taxonomy, particularly for socially important fungi. Its first application in mycology was with *Coccidioides immitis*, where comparison of five genes sequenced from 17 individuals showed two phylogenetic species in what had been considered to be one morphological species (Koufopanou *et al.*, 1997). Similar studies of medically and agriculturally important fungi have proven the wide applicability of PSR.

Recently, web-based Multilocus Sequence Typing (MLST) schemes (Maiden *et al.*, 1998; Taylor & Fisher, 2003) have been established to bring together PSR research conducted throughout the globe for a single fungal species. MLST works because a gene sequence is 'portable' in the sense that sequences determined in different laboratories can be combined, unlike RAPD or fingerprinting approaches. In addition, sequencing a region of DNA avoids ascertainment bias, that is, the problem that polymorphic loci discovered for one set of individuals may prove to be fixed (and, therefore, uninformative) in individuals used in subsequent studies. With DNA sequence, of course nucleotide positions polymorphic in the first population may be fixed in the second, but other nucleotide positions may be polymorphic in the second population, and so on. It seems very likely that each socially important fungus will have its own MLST scheme in the near future, which will enable truly global studies of fungal species and greatly enhance our knowledge of fungal biodiversity. A note of caution might be sounded concerning the generality of knowledge about species recognition, population structure, speciation, and species maintenance gained from socially important fungi, because many of them appear to rely heavily on clonal reproduction. For this reason it seems advisable also to experiment with fungi that are not human pathogens and that are not overly clonal, for example, outbreeding *Neurospora* species.

Intraspecific variation

Variation in DNA sequences can also be used to search for genetic variation within a species, but more variable markers are often needed. In this regard microsatellite loci have proven valuable. Fungal microsatellites were discovered in yeast (Field *et al.*, 1996; Field & Wills, 1996), but their first thorough application to elucidating fungal populations again was with *Coccidioides*. Here, with 9 microsatellite loci and nearly 170 individuals from throughout the New World range of the fungus, the two species proposed by MLST were confirmed, and populations were discovered in each species; at least two populations were found in *C. immitis* and at least three in the newly described species *C. posadasii* (Fisher *et al.*, 2002a). Microsatellites also are portable and can be used for Multilocus Microsatellite Typing (MLMT) schemes (see Chapter 16, this volume). Microsatellite loci can be used to discover populations by phylogenetic analysis, or by Bayesian assignment methods (Pritchard & Feldman, 1996; Falush *et al.*, 2003). In either case, once a MLMT scheme is in place, new individuals can be assigned to populations by using assignment methods (Rannala & Mountain, 1997), as has been demonstrated for isolates of *Coccidioides* species that were obtained from patients who lived far from the endemic areas (but whose travel histories showed that they had visited the endemic areas) (Fisher *et al.*, 2002b). Both MLST and MLMT schemes gain power as more researchers use them and they should prove very popular with both plant pathologists and medical mycologists. Prior to their broad application in *Coccidioides* species, the utility of microsatellies was evaluated by comparing them to sequence variation for *c.* 20 individuals from *C. posadasii* and *C. immitis* (Fisher *et al.*, 2000). An important result of that study was the finding that single microsatellites can be misleading owing to hypervariability and attendant homoplasy. The same conclusion was reached in a comparison of ten microsatellites in *Neurospora* with sequence variation for nearly 150 individuals (Dettman & Taylor, 2004). Therefore, microsatellies are best employed to extend multilocus sequence studies from species into populations; if the aim is to accomplish species recognition and population characterization simultaneously, many microsatellites should be used (Fisher and colleagues (Fisher *et al.*, 2004) used 20 in their study of *Penicillium marneffei*).

Challenging morphological species recognition with phylogenetic species recognition

A recent study that highlighted the potential for conflict between phenotypic and genotypic methods of species recognition used the fungus *Histoplasma capsulatum* (the mitosporic name for *Ajellomyces capsulatus*) (Kasuga *et al.*, 2003). This species was known to be phenotypically complex and was divided into three varieties based on host, geographic range and disease symptoms (Kwon-Chung & Bennett, 1992). The first variety, *Histoplasma capsulatum* var. *duboisii*, was confined to Africa and, in humans, caused bone and skin infections in addition to primary lung infections; *H. c.* var. *farciminosum* was found in Eurasia and caused skin lesions in horses and donkeys; *H. c.* var. *capsulatum* was found in the New World and caused pulmonary infections in humans. When the sequences for four loci were obtained from more than 130 individuals assigned to the three varieties, they were found to form at least seven genetically isolated clades. These seven as yet undescribed species showed a strong correlation with geography but not with host or symptoms (Fig. 15.1): North America 1 and North America 2, Latin America A (which includes a Eurasian subclade) and Latin America B, Africa, Australia, and Indonesia (Kasuga *et al.*, 2003). The African species was shown to harbour members formerly assigned by symptoms to varieties *H. c.* var. *duboisii* and *H. c.* var. *capsulatum*; therefore, *H. c.* var. *dubiosii* is not a monophyletic taxon. Individuals assigned by phenotype and host to *H. c.* var. *farciminosum* were found in three different clades; in the Eurasian clade, all individuals had the same genotype. *H. c.* var. *farciminosum* is a polyphyletic collection of clonally propagating lineages that have independently made the jump to horses or donkeys as hosts. Finally, individuals formerly assigned to *H. c.* var. *capsulatum* are found in all clades; again the historical variety *H. c.* var. *capsulatum* is not a monophyletic taxon. Although the phenotypes used to describe the old varieties are not useful for identifying the new clades, it is likely that other phenotypes may be found that are diagnostic of the new groups, as will be noted below for *Aspergillus flavus*.

Species divergence and geologic time

Sequence data can be used to estimate the dates of divergences among species and populations and to compare these dates to recent geologic, archaeological or historical events to make hypotheses about the evolutionary history of fungi. Estimates of geologic time have been applied to three studies of *Coccidioides* species. The first example concerned the divergence of the two *Coccidioides* species, which was estimated to have occurred between 10 and 12 million years ago (MYA) (Koufopanou *et al.*, 1997). A second study estimated divergence times between *Coccidioides* and the closely related genus *Uncinocarpus*, and among phylogenetic species recognized in both genera, and showed that the divergence between *Coccidioides* and *Uncinocarpus* was at least an order of magnitude older than divergences among phylogenetic species in either genus (Koufopanou *et al.*, 2001) (Fig. 15.2). Morphological species typically appear to harbour two or more cryptic species with divergences of the order of 3–10 MYA (Burt *et al.*, 2000) while, in these cases, the nearest morphologically distinct species seem to have diverged on the order of 30–100 MYA. If divergences recognizable by genetic

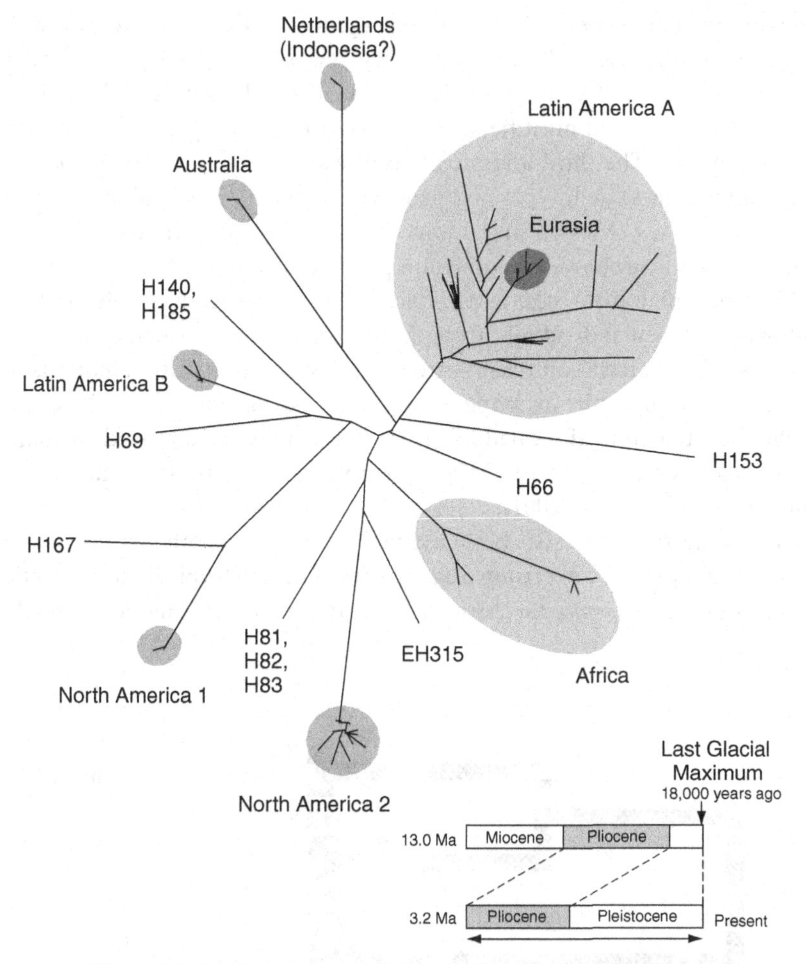

Fig. 15.1. Phylogenetic analysis of *Histoplasma capsulatum*, showing clades equivalent to cryptic species. The size of the clade is proportional to its genetic variation; the tropical clades, Latin America A and Africa, are the most diverse. This unrooted tree was made by neighbour-joining and shows a star phylogeny. The average genetic distance among distinct lineages (NAm 1, NAm 2, LAm A, LAm B, Australia + Indonesia, Africa, H81, and H153) was 2.8%. Half of this value corresponds to the time point of the radiation, which is 3.2 and 13.0 million years ago when DNA substitution rates of 4.3×10^{-9} and 1.1×10^{-9}, respectively, are used. Figure and legend from Kasuga *et al.* (2003) with the permission of the authors and publisher.

isolation occur every 3–10 MY, but divergences recognizable by morphology are found every 30–100 MY, very few genetically isolated clades persist to the point where they are morphologically distinct. The logical conclusion is that it must be far easier to form a new species than it is to maintain one. The third and most recent example concerns *C. posadasii.* A comparison of genotypes and genotypic diversity in populations from Arizona, Texas, Mexico and Latin America indicated that those from North American showed a significant positive correlation between genetic and geographic distances, as is typical of geographically stable species. However, when individuals from Latin America were included in the analysis, the correlation disappeared. This observation, and comparison of genotypic diversity in populations from North and South America, suggested that fungal individuals from Texas may have moved to Latin American as recently as 0.13 MYA, possibly in association with migrating humans (Fisher *et al.*, 2001).

Geologic time has also been useful in making hypotheses about the evolutionary history of *Histoplasma* species. The two tropical phylogenetic species of *Histoplasma*, the African clade and the Latin America clade A,

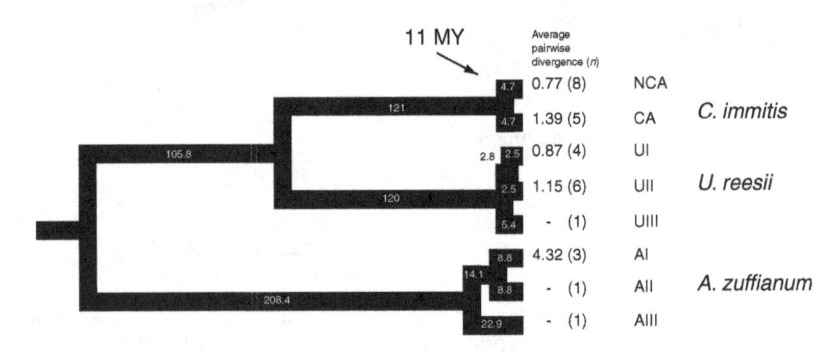

Fig. 15.2. Comparison of genetic distances among cryptic species and morphological species for *Coccidioides immitis, Auxarthron zuffianum* and *Uncinocarpus reesei.* If every cryptic species persisted, the tree ought to have far more branches and species. Estimates of the divergence among cryptic species (NCA and CA) in *Coccidioides immitis* is no greater than 11 MYA. Divergence between *Coccidioides* and *Uncinocarpus* is at least 20-fold greater than among their cryptic species. This maximum-parsimony tree is based on three genes with a 'molecular clock', enforced. Uncorrected average pairwise divergences within cryptic species are also shown for comparison. Both branch lengths and divergences within taxa are nucleotide substitions per 1000 sites; *n* is the number of isolates. Figure and legend adapted from Koufopanou *et al.* (2001) with the permission of the authors and publisher.

have a much greater diversity of genotypes than do the more temperate clades. One hypothesis that could account for the different genotypic diversities involves the climate cycles of ice age glaciations. By this hypothesis, the temperate clades would endure population bottlenecks associated with population expansions from equatorial refugia into newly available temperate habitat and contractions back to the refugia when temperate habitat chilled; tropical clades would escape these bottlenecks. By using the rate of nucleotide substitution determined for Ajellomycetaceae by Kasuga *et al.* (2002), the radiation of the seven clades was estimated to have occurred between 3.2 and 13 MYA, well before the most recent series of ice ages (Kasuga *et al.*, 2003). Therefore it seems likely that individuals in extant temperate clades were forced to migrate in north–south corridors during the ice ages, and plausible that temperate clades endured genetic bottlenecks that tropical species escaped, thereby accounting for differences in their diversity.

Curiously, the Eurasian *Histoplasma* clade emerges from the large, lowland tropical, Latin American A clade and is more distantly related to the geographically closer African clade (Fig. 15.1). If the Eurasian clade originated from a single Latin American A individual, then it must have existed for at least 1.7 million years (Kasuga *et al.*, 2003). If, however, the clade originated by the transportation to Europe of several genetically different Latin American individuals, the date of this event could have been much more recent. Given that the largest assemblage of clonal individuals from horses and other equids is found in the Eurasian clade, it is not completely ludicrous to speculate that the fungus was brought from Latin America to Eurasia by animals owned by the first European explorers of the New World, some 500 years ago. On the other hand, if a variety of genotypes were brought across the Atlantic Ocean on horses or donkeys, there is no evidence that they persisted on these animals.

A posteriori *correlation of phenotype and genotype*

Traditional morphological concepts of species often are proven wrong when phylogenetic data become available. However, once phylogenetic species are found, it becomes possible to find new phenotypes that correlate with the genetically isolated groups. A good example is provided by *Aspergillus flavus*, in which PSR with five DNA regions associated with nuclear genes found at least four species-level clades in what had been thought to be one morphological species (Geiser *et al.*, 1998b). As with *Histoplasma*, there proved to be a correlation with phylogenetic clades and geographic range in that the deepest divergence in the combined gene genealogies was consistent with a division between individuals collected in the Northern or Southern Hemispheres (Tran-Dinh *et al.*, 1999). Once the phylogenetic groups were identified, subsequent study of two phenotypes (sclerotium size and aflatoxin production) found unique combinations of these phenotypes for each clade (Geiser *et al.*, 2000). The ability to find taxonomically useful phenotypes once phylogenetic species are identified can allow species to be both evolutionarily sound and practical in the sense of Harrington & Rizzo (1999). In fact, a very recent example from Harrington's laboratory emphasizes this point (Engelbrecht & Harrington, 2005). The morphological species *Ceratocystis fimbriata* was shown to harbour several genetically and reproductively isolated groups, i.e. several phylogenetic and biological species. With this knowledge in hand, for each group phenotypic differences were found, i.e. the preferred host (sweet potato, sycamore or cacao) and the size and shape of meiosporic and mitosporic reproductive structures and spores. New species were described for the groups associated with sycamore (*C. platani*) and cacao (*C. cacaofunesta*). Interestingly, *C. cacaofunesta* encompasses two genetically and reproductively isolated groups that show some morphological differences but are kept together because they both infest the same agricultural host and the morphological differences were deemed trivial. When more is known about the natural hosts and additional phenotypes of *C. cacaofunesta*, it seems likely that another species will be recognized.

Correlation of phylogenetic species with geographic range

We have discussed examples of phylogenetic species that are strongly correlated with geography in the medically important species of *Coccidioides* and *Histoplasma*, and in the agriculturally important fungus *A. flavus*. There are many other such examples from species of the genera *Fusarium* (O'Donnell *et al.*, 1998), *Botrytis* (Giraud *et al.*, 1999), *Sclerotinia* (Carbone & Kohn, 2001a, b; Kohn & Carbone, 2001; Phillips *et al.*, 2002) and *Magnaporthe* (Couch & Kohn, 2002), as well as many examples from Basidiomycota (Vilgalys & Sun, 1994; Hibbett *et al.*, 1995; Petersen & Hughes, 1999; Johannesson & Stenlid, 2003). However, a strong geographic correlation, i.e. endemism, is not always found in fungal phylogenetic species. In *Aspergillus fumigatus*, PSR with five regions flanking nuclear microsatellites found two global species; one had many more isolates than the other, but neither showed any hint of endemism (Pringle *et al.*, 2005). This PSR study echoed a previous study that used DNA fingerprint data and a more limited sampling of isolates (Debeaupuis *et al.*, 1997). In terms of its global geographic range,

A. fumigatus clearly stands in contrast to the other species examined to date. How this fungal species maintains a global geography is not yet understood.

Reproductive mode certainly plays a role in maintaining species. A strictly clonal species would have to experience periodic global sweeps of a single genotype to avoid fragmentation into endemic clades. In a recombining species, individuals with many different, recombined genotypes would have to be able to disperse throughout the globe and mate to avoid establishing endemic populations. As mentioned above, many socially important fungi, including species of *Aspergillus*, *Coccidioides* and *Fusarium*, are morphologically mitosporic and can reproduce clonally. Alas, morphology is not necessarily an accurate means of assessing reproductive mode. In many cases, analyses of association of alleles at the same loci used to recognize species has been used to reject the hypothesis of exclusively clonal reproduction for these fungi (Burt *et al.*, 1996; Geiser *et al.*, 1998b) and, therefore, indicate that genetic recombination also is occurring. *A. fumigatus* is among those fungi that are morphologically mitosporic but whose population genetics give evidence of recombination (Paoletti *et al.*, 2005; Pringle *et al.*, 2005). Very recent research with *A. fumigatus* has discovered individuals of both mating types (Pöggeler, 2002; Dyer *et al.*, 2003), which are often in equal proportion at single geographic locations (Rydholm *et al.*, 2004; Paoletti *et al.*, 2005). This information, combined with the observation that many different genotypes of *A. fumigatus* can be found in the same geographic location (Pringle *et al.*, 2005), suggests that *A. fumigatus* does not maintain its global species by worldwide sweeps of a clonal genotype. Instead, almost any individual of *A. fumigatus* must be capable of very long-distance travel. A combination of clonality and recombination does not, however, explain the difference between *A. fumigatus* and other fungi as regards endemism. For example, *Coccidioides*, *Histoplasma* and at least some *Fusarium* species are capable of both clonal and recombining reproduction, but lack the global reach of *A. fumigatus*. The explanation for the difference in biogeography between *A. fumigatus* and the other species undoubtedly lies in their ecology, again a subject in need of additional research.

Challenging phylogenetic species recognition with biological species recognition

PSR relies on genetic isolation for species identification, but much of the discussion on species recognition has focused on biological species recognition (BSR) based on sexual compatibility within species and reproductive isolation between species. Much has been written about the relative merits of BSR and PSR but data comparing both methods in a single fungus were wanting. Unfortunately, in most cases where PSR had been used to recognize fungal species it was not been possible to compare PSR and BSR because the fungi were morphologically asexual or difficult to mate in the laboratory. Exceptions included the 'mating populations' of *Fusarium* species (O'Donnell *et al.*, 1998), species of *Ceratocystis* (Baker *et al.*, 2001, 2003; Engelbrecht & Harrington, 2005), and a number of Basidiomycota (Vilgalys & Sun, 1994; Petersen & Hughes, 1999; Burnett, 2003). Possibly the best fungus for a comparison of PSR and BSR is *Neurospora*, because it can complete its entire life cycle in cultivation, from meiospore to meiospore, in 14 days. This attribute caused it to be associated with elements of BSR from its description in 1927 (Shear & Dodge, 1927) and has made it the fungus most associated with BSR. Neurosporologists, principally Perkins and colleagues at Stanford University, have used mating testers to assign more than 6000 natural individuals to one of the five outbreeding (conidiating) species (*N. crassa, N. intermedia, N. sitophila, N. tetrasperma* and *N. discreta*) (Perkins & Turner, 1988; Turner *et al.*, 2001). Knowing that these five outbreeding species formed a monophyletic clade with respect to homothallic species (Pöggeler, 1999; Dettman *et al.*, 2001), PSR was applied to *c.* 150 individuals from all five outbreeding species, with the collection of individuals weighted toward what were thought to be the two most similar species, *N. crassa* and *N. intermedia.* With DNA sequences from four nuclear regions flanking microsatellites, and with three criteria for PSR (support from a majority of gene genealogies *or* significant support from one gene genealogy and no significant conflict from others *and* the rule that all individuals must belong to a species) eight species-level clades were defined, including three new phylogenetic species (Dettman *et al.*, 2003a).

The parallel and independent BSR study (Dettman *et al.*, 2003b), which employed almost half of the individuals in the PSR study, focused on *N. crassa, N. intermedia* and the nine individuals that were thought to be their hybrids. This collection included individuals that had previously been identified as *N. crassa, N. intermedia* or putative hybrids, but that proved to be members of the three new phylogenetic species. The many reciprocal matings (more than 1800 in all) were evaluated in a seven-level scale of reproductive success based on methods developed by Perkins and colleagues (Perkins & Turner, 1988; Turner *et al.*, 2001). When crosses were arranged to consolidate matings with the highest success (production of at least 50% viable ascospores), four biological species were recognized (Fig. 15.4), that is, all of those recognized by PSR, save one (Dettman *et al.*, 2003b). The remarkable agreement of PSR and BSR (Fig. 15.3) further supports the use of PSR in the many fungi for which BSR is either impossible or prohibitively arduous.

Speciation processes and the maintenance of species

We already have noted that a multilocus phylogenetic analysis of *Coccidioides* and *Uncinocarpus* species suggested that far more species are initiated than persist. Newly diverged species can be lost by chance events (especially if they originate as a numerically small population) or because individuals compete poorly with individuals from genetically similar sympatric populations, or because the two novel species reticulate. Newly or recently diverged species may not be kept separate if hybridization and introgression facilitate enough gene flow to reverse or stall genetic differentiation. As noted above, PSR for *Neurospora* included nine natural isolates thought to be hybrids because each of the individuals did not mate well with testers for any of the original five outbreeding species. Comparison of multiple gene genealogies can identify individuals as hybrids because hybrids will fall into different clades in different gene genealogies. In *Neurospora*, by PSR, all putative hybrids were found to be members of not more than one of the eight phylogenetic species by comparison of the four sequenced regions. Most putative hybrids proved to be members of new phylogenetic species (Fig. 15.3), but two belonged to *N. crassa* and another to *N. intermedia* (Dettman *et al.*, 2003b). The inability to recognize the true affinities of the putative hybrids when mating testers were available for only five species is easy to understand. We now know that they mate with other members of their species, and testers have been proposed for the new biological species. As regards hybridization, *Neurospora* stands in contrast to other genera associated with agriculture or forestry that are known to form hybrids, for example, *Fusarium* (O'Donnell *et al.*, 2000, 2004), *Botrytis* (Nielsen & Yohalem, 2001), *Neotyphodium* (Schardl & Craven, 2003; Moon *et al.*, 2004), *Ophiostoma* (Brasier *et al.*, 1998; Konrad *et al.*, 2002), *Melampsora* (Newcombe *et al.*, 2001) or *Heterobasidion* (Garbelotto *et al.*, 2004), and also to the clade of *Saccharomyces* yeasts whose ancestor arose from hybridization followed by allopolyploidy (Wolfe & Shields, 1997; Wong *et al.*, 2002).

Surprisingly, BSR provides more information about hybrid matings than it does about conspecific matings, because most of the crosses are

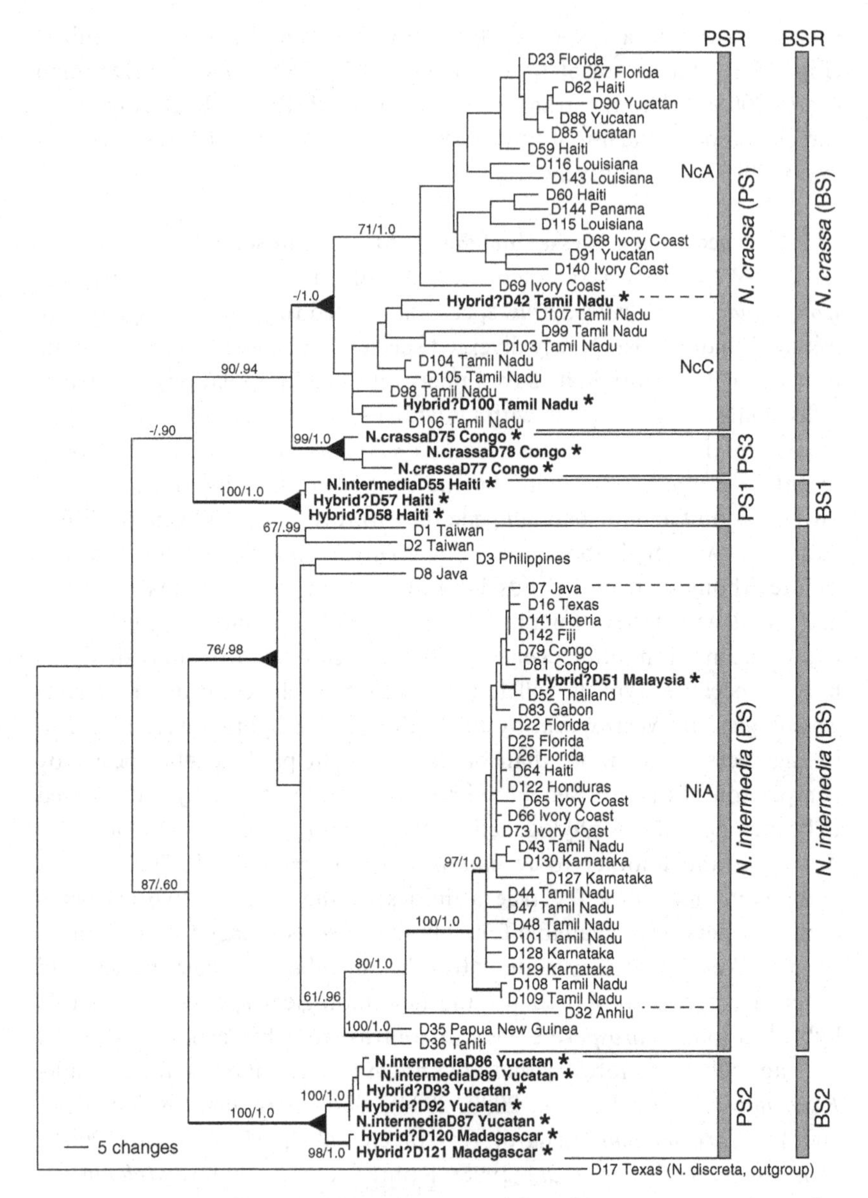

Fig. 15.3. Phylogenetic species recognition of *Neurospora* species and a graphic comparison to biological species recognized in the same species. Maximum parsimony (MP) phylogram produced from the combined analysis of DNA sequences from four anonymous nuclear loci (TMI, DMG, TML and QMA loci, a total of 2141 aligned nucleotides). Tree length = 916 steps. Consistency index = 0.651. Labels to the right of the phylogram indicate groups identified by phylogenetic species recognition and biological species recognition. Bold branches were concordantly

between parents from different species (Fig. 15.4). BSR focuses on the conspecific crosses that delimit species, but scrutiny of the hybrid or heterospecific crosses is informative about species maintenance. With *Neurospora*, there was a large variation in mating success (Fig. 15.4), from crosses that were almost as successful as conspecific matings (level 5, with more than 15% viable ascospores) to those at level 1 (abortion of incipient perithecial development) and those at level 0 (a complete absence of mating effort with no hint of perithecial development) (Dettman *et al.*, 2003b). Remarkably, this variation was correlated with the geographic origin of mating partners: hybrid matings of sympatric *Neurospora* pairs were significantly less successful than those of allopatric pairs (Fig. 15.4). In other words, species are kept reproductively isolated either by geographic distance in allopatry or by genetic reinforcement of the species barrier in sympatry. Our inability to find natural hybrids implies that hybrid matings do not occur or do not produce successful progeny under natural conditions. The laboratory observation that hybrid matings produce fewer viable progeny than conspecific matings, coupled with the aforementioned inability to find natural hybrids, suggest that there is selection against hybridization. Individuals that can recognize and abort hybrid matings would be favoured, and this would select for the reinforcement of reproductive isolation in sympatry over geologic time scales. However, when allopatric individuals are brought together over human time scales their unreinforced reproductive isolation can be overcome, particularly in disturbed agricultural settings that may be permissive to the survival of hybrid progeny. This scenario may explain the conflicting

---

Fig. 15.3. (cont.)
supported by the majority of the loci, or were well supported by at least one locus but not contradicted by any other locus. Triangles at nodes indicate that all taxa united by (or distal to) a node belong to the same phylogenetic species. Taxon labels indicate strain number and geographic source. If a strain was originally identified by traditional mating tests to a species that did not match the phylogenetic or biological species identification, the original species name is listed before the strain number and is followed by an asterisk, all in bold type. If the original species identification matched both the phylogenetic and biological species identification, no name appears before the strain number. Branch support values for major branches with significant support are indicated by numbers above or below branches (MP bootstrap proportions/Bayesian posterior probabilities). Figure and legend from Dettman *et al.* (2003b) with permission of the authors and publisher.

Phylogenetic species →					int	int	int	int	int	int	int	int	int	int*	int	int	int	int	int	int	int	int	int	cra	cra	cra	cra	cra	cra	cra	cra	cra	cra	cra	cra	cra	cra	PS3*	PS2*	PS2*	PS1*	PS1*
↓ Biological species →					int	int	int	int	int	int	int	int	int	int*	int	int	int	int	int	int	int	int	int	cra	cra	cra	cra	cra	cra	cra	cra	cra	cra	cra	cra	cra	cra	cra*	BS2*	BS2*	BS1*	PS1*
↓ Original sp. →					int	int	int	int	int	int	int	int	int	hyb*	int	int	int	int	int	int	int	int	int	cra	cra	cra	cra	cra	cra	cra	cra	cra	cra	cra	cra	cra	cra	cra*	int*	hyb*	hyb*	hyb*
↓ Region →					India	India	India	India	India	India	Carib	Carib	Carib	Asia	Asia	Asia	Asia	Asia	Asia	Asia	Africa	Africa	Africa	India	India	India	India	Carib	Carib	Carib	Carib	Carib	Carib	Carib	Carib	Africa	Africa	Africa	Carib	Africa	Carib	Carib
			↓	a↓A→	D44	D48	D101	D108	D128	D129	D26	D64	D122	D51	D142	D7	D32	D36	D52	D1	D141	D65	D83	D105	D107	D98	D99	D23	D27	D62	D90	D91	D115	D144	D143	D68	D140	D78	D87	D120	D57	D58
int	int	int	India	D43	6/6	6/6	6/6	6/6	6/6	6/6	6/2	6/6	6/6	1/0	6/0	5/0	6/6	6/6	6/0	5/5	2/2	6/1	6/6	1/2	1/3	2/2	0/2	1/2	1/2	3/1	1/2	1/2	0/0	0/2	1/2	0/2	1/1	0/2	1/5	1/0	0/0	0/0
int	int	int	India	D47	5/5	6/6	6/6	5/6	6/6	6/6	5/0	5/5	1/1	0/0	6/1	1/1	6/6	6/6	6/1	5/5	5/1	6/6	6/1	0/1	1/3	1/1	1/1	0/2	1/1	0/2	0/4	0/1	0/0	0/1	0/4	0/3	0/3	0/3	0/2	2/3	0/0	0/0
int	int	int	India	D109	6/6	6/6	6/6	6/6	6/6	6/6	6/6	5/5	6/0	0/0	6/0	4/1	6/6	6/6	0/0	5/5	4/1	6/0	6/6	0/1	1/3	0/1	0/1	0/2	2/2	0/2	1/2	0/2	0/0	0/1	0/2	1/3	0/3	0/1	0/5	2/3	0/0	0/0
int	int	int	India	D127	6/6	5/6	5/6	6/6	6/6	6/6	5/5	5/5	5/6	4/0	6/0	4/4	6/6	6/6	6/6	5/5	4/0	6/6	5/6	2/1	3/1	2/1	1/0	1/1	2/1	1/2	1/1	1/1	1/0	1/0	3/3	2/2	1/2	3/3	4/4	3/3	1/0	1/0
int	int	int	India	D130	5/6	4/6	6/6	6/6	6/6	6/6	6/6	6/6	6/1	6/6	6/5	5/0	6/6	6/6	6/6	5/5	5/5	6/6	6/6	2/2	2/2	1/2	1/0	3/1	3/1	2/1	3/2	3/2	3/1	3/1	2/2	2/1	1/1	2/2	0/1	0/3	1/0	1/0
int	int	int	Carib	D22	0/0	6/6	0/5	6/6	1/0	6/6	6/6	6/6	6/6	6/6	6/6	0/0	2/6	5/5	6/0	6/6	6/1	6/6	1/0	0/4	0/4	0/0	0/4	1/3	1/2	0/1	1/2	0/2	0/0	0/1	1/3	0/1	0/1	0/4	3/4	3/3	1/0	1/0
int	int	int	Carib	D16	6/6	6/6	6/6	6/6	6/6	6/6	6/6	5/1	4/6	0/0										3/3	3/4	0/1	0/5												1/4	3/3	0/0	0/0
int	int	int	Carib	D25	6/6	6/6	6/6	6/6	0/6	6/6	6/6	6/5	6/6	4/0		6/1								2/3	0/4	0/3	0/5												1/5	3/3	3/0	3/0
int	int	int	Africa	D66							0/0	0/0	0/6		0/0	6/0					0/5	0/6	0/0	0/3	0/3	0/0	0/5												0/4	0/3		
int	int	int	Africa	D81							6/0	0/0	6/6		0/0	0/0					6/6	6/6	0/0	1/3	0/4	0/0	0/1												1/4	0/3		
int	int	int	Africa	D73							6/6	0/0	6/6		6/6	6/0					6/6	6/6	6/6	0/4	0/4	0/0	0/5												0/4	1/4		
int	int	int	Africa	D79							6/6	6/0	6/6		6/6	6/0					6/6	6/6	6/6	4/1	1/4	0/0	1/1												1/4	3/3		
int	int	int	Asia	D2							6/6	6/0	6/6	1/2	5/0	5/0	6/6	6/6	6/0	5/6	6/0			3/3	3/3	0/0	4/4												5/5			
int	int	int	Asia	D3							0/0	0/0	0/5	5/5	4/4	5/5	6/6	6/6	6/6	6/6	1/1			1/4	3/3	0/0	4/4												3/3			
int	int	int	Asia	D35							6/1	5/1	1/5	1/0	5/0	1/1	6/6	6/6	6/0	5/5	5/2			3/3	3/3	0/0	4/1												4/4			
int	int	int	Asia	D8	2/1	2/5	2/2	2/1	1/1	2/4	4/0	3/0	1/1	2/0	4/4	1/3	5/5	5/5	5/0	2/2	2/5	5/5	5/5	1/3	2/4	0/0	1/1												1/4			
cra	cra	cra	India	D103							5/5	5/0	1/1	4/1										6/6	6/6	6/6	6/6	5/5	5/5	5/5	5/5	5/5	5/5	1/1	5/5				4/4	3/4	1/0	1/0
cra	cra	cra	India	D104							4/1	4/4	4/2	4/3										6/6	6/6	6/6	6/6	5/5	5/5	5/4	5/5	5/5	0/0	5/2	5/5	5/5	5/5	6/6	5/5	4/4	4/4	4/1
cra	cra	cra	India	D106							4/3	4/0	3/1	4/3										6/6	6/6	6/0	6/6	5/4	4/4	5/5	5/5	6/5	0/0	5/1	5/5	0/5	5/5	0/5	0/5	3/4	0/3	3/0
cra*	cra*	hyb*	India	D42	2/2	2/2	2/2	2/1	3/3	2/1	5/5	5/5	5/0	3/0	4/3						3/0	5/0	5/4	6/6	6/6	5/5	6/6	5/4	3/3	3/4	5/4	4/5	5/5	3/3	4/4	3/3	4/4	5/5	4/4	3/3	3/3	3/0
cra*	cra*	hyb*	India	D100	2/2	0/1	0/1	1/1	1/0	1/1	4/4	4/2	3/2	3/3	4/0						4/0	4/4	4/4	6/6	6/6	6/6	6/6	4/4	3/3	4/4	5/5	5/5	6/5	4/1	3/3	1/5	5/5	5/5	3/4	3/3	3/3	3/0
cra	cra	cra	Carib	D116	5/0	3/3	3/3	3/2	3/0	3/3	3/1	3/0	0/0	3/0	3/0	3/0	3/3	3/3	3/0	3/3	3/0	3/0	3/3	6/5	5/5	0/0	6/6	6/6	6/6	6/6	6/6	6/6	6/6	5/2	6/6	6/6	6/6	6/6	3/3	3/3	3/0	3/0
cra	cra	cra	Carib	D85	1/1	3/3	3/3	3/3	3/0	1/3	1/0	3/0	0/0	3/0	2/0	3/0	3/3	3/3	1/0	3/3	2/0	1/1	1/0	6/5	4/4	0/0	4/5	6/6	6/6	6/6	6/6	6/6	6/2	6/2	6/6	6/6	6/6	6/6	3/3	3/3	3/0	1/0
cra	cra	cra	Carib	D59	2/0	3/3	3/1	2/1	3/3	3/3	2/1	3/1	1/3	2/0	3/0	3/0	3/3	3/3	3/0	3/3	2/0	3/1	3/1	5/5	4/5	0/0	4/4	6/6	6/6	6/6	6/5	6/6	6/6	6/2	6/6	6/6	6/6	6/6	3/3	3/4	1/0	1/0
cra	cra	cra	Carib	D60	2/1	3/3	3/3	3/1	3/0	3/3	1/1	1/0	0/0	3/0	3/0	3/0	3/3	3/3	3/0	3/3	3/0	2/0	2/0	5/5	4/5	0/0	5/5	6/6	6/6	6/6	6/6	6/6	6/6	5/1	6/6	6/6	6/6	6/6	3/3	3/3	3/0	3/0
cra	cra	cra	Carib	D88	3/0	3/3	2/2	3/1	3/0	3/3	0/1	3/0	3/0	0/0	1/0	3/0	3/3	3/3	3/0	3/3	1/0	1/0	0/0	5/5	5/5	0/0	5/5	6/6	6/6	6/6	6/6	6/6	6/1	6/1	6/6	6/6	6/6	6/6	3/3	3/3	0/0	0/0
cra	cra	cra	Africa	D69							1/1	1/1	1/0	3/0										5/5	5/5	6/5	6/6								6/6	6/6	6/6	6/6	4/4	3/3	3/0	3/0
PS3*	cra*	cra*	Africa	D77							3/3	3/3	3/0	3/3										5/1	5/5	5/5	6/5								6/6	6/6	6/6	6/6	3/3	4/3	3/0	3/0
PS3*	cra*	cra*	Africa	D75							4/1	4/0	3/0	3/0							5/0	5/0	3/1	5/5	6/6	6/6	6/6								0/0	6/6	6/6	6/6	4/4	3/3	3/0	3/0
PS2*	BS2*	int*	Carib	D86	0/0	0/0	0/4	0/0	0/0	0/0	0/4	0/0	0/2	0/0							0/0	0/1	0/0	0/3	0/5	0/0	0/5								0/3	0/3	0/1	0/4	0/6	0/6	0/0	0/0
PS2*	BS2*	int*	Carib	D89	1/3	3/4	3/0	3/0	2/0	1/1	4/3	4/3	3/1	3/0										3/3	4/4	0/0	4/4								1/3			4/4	6/6	6/6	5/0	1/0
PS2*	BS2*	hyb*	Carib	D92	1/1	1/3	3/1	1/0	1/3	2/1	3/4	4/4	4/0	3/3	3/0						3/0	3/0	0/4	4/4	4/4	3/3	3/3	3/1	3/3	3/3	3/3	3/3	4/4	3/0	3/3	3/3	1/1	3/3	6/6	5/5	3/3	3/0
PS2*	BS2*	hyb*	Carib	D93	3/0	1/5	3/1	3/0	4/0	2/1	4/4	4/4	3/3	3/0	4/0						3/0	4/0	4/4	4/3	4/4	3/3	3/3	3/3	3/3	3/3	3/3	3/3	4/3	3/1	4/4	3/3	1/1	4/4	6/6	6/6	4/0	4/0
PS2*	BS2*	hyb*	Africa	D121	5/4	4/4	3/3	4/4	5/0	4/1	5/5	5/4	4/0	3/0	4/3						3/0	4/0	4/4	4/4	4/4	3/3	3/3	4/4	4/4	4/4	4/4	4/4	4/4	3/3	4/4	4/4	4/3	4/4	6/6	6/6	5/5	4/0
PS1*	BS1*	int*	Carib	D55	0/0	0/1	0/1	0/0	0/0	0/0	0/1	0/0	0/0	3/1										1/4	0/3	1/3	0/4								0/3				0/4	0/4	6/6	6/6

Category of Reproductive Success

6	>50% black ascospores
5	15–50% black ascospores
3 & 4	<1% black ascospores; & 1–15% black ascospores
2	perithecia developed ostioles, no ascospores ejected
0 &1	sterile, no perithecia produced; & barren perithecia, no ostiole developed

Fig. 15.4. Matrix displaying the reproductive success of 929 crosses (1858 matings) of outbreeding *Neurospora* species. Columns represent the 38 *mat A* strains and rows represent the 35 *mat a* strains, with strain numbers along the row and column headings of the matrix. Numbers within matrix cells indicate the reproductive success ratings (see inset) of the two reciprocal matings of the cross between the corresponding strains (*mat a* strain as the perithecial parent/*mat A* strain as the perithecial parent). For example, the top and leftmost cell of the matrix indicates that both matings of the cross between *mat a* strain D43 and *mat A* strain D44 received a rating of 6. The matrix cells have been shaded in proportion to the reproductive success of the best mating. A reproductive success rating of 6 was used to delineate biological species; and crosses satisfying that criterion have been filled black. Matrix cells without entries indicate that matings were not performed for that cross. Additional row and column headings indicate the phylogenetic species designation (see Fig. 15.3), biological species designation, original species designation and geographic source of the strains (int, *N. intermedia*; cra, *N. crassa*; PS, phylogenetic species; BS, biological species; hyb, possible hybrid between *N. crassa* and *N. intermedia*; Carib, Caribbean Basin). Asterisks indicate that the phylogenetic or biological species designation differed from the original species identification. Figure and legend from Dettman *et al.* (2003b) with permission of the authors and publisher.

observations of an absence of hybrids in *Neurospora* compared with the aforementioned cases of hybridization in agriculturally important fungi (Schardl & Craven, 2003).

Burnett has opined that the failure to produce hybrid progeny occurs late in the mating pathway of Ascomycota and early in the mating pathway in Basidiomycota (Burnett, 2003). In the ascomycete *Neurospora*, however, hybrid mating success varies across mated pairs and reproductive failure is not confined to the later stages (Dettman *et al.*, 2003b). In a comparison of hybrid or heterospecific mating success between allopatric and sympatric partners (Fig. 15.5) only 70% of both allopatric and sympatric heterospecific matings make any sort of perithecium, only 55% of allopatric and 25% of sympatric crosses proceeded to make well-developed perithecia, and only 48% of allopatric and 13% of sympatric matings produced any ascospores. Although it is true that mating success is indistinguishable between allopatric and sympatric heterospecific partners in the very earliest stages of mating, it is not true that the barriers to

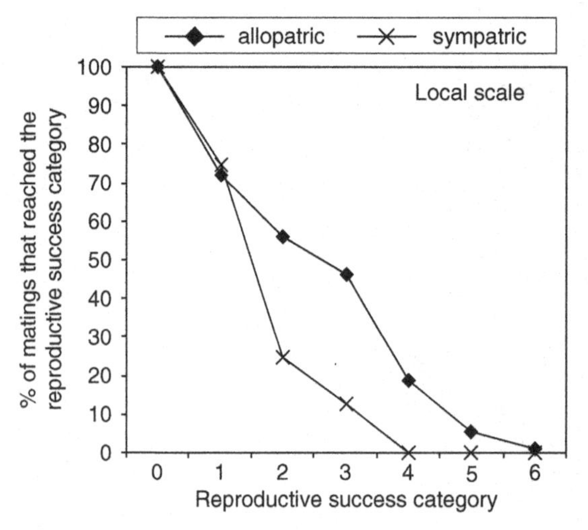

Fig. 15.5. Mating success for sympatric and allopatric heterospecific pairings: the percentage of heterospecific matings between allopatric or sympatric individuals that reached the successive categories of reproductive success. At all geographic scales (regional, sub-regional and local), allopatric matings were significantly more likely than sympatric matings to proceed through the consecutive stages of the sexual cycle. The discrepancy between the allopatric and sympatric curves increases as the scale of sympatry decreases and is most pronounced for local sympatry, shown here. Figure and legend modified from Dettman *et al.* (2003b) with permission of the authors and publisher.

success are all late in the process. The application of BSR in Basidiomycota typically assesses mating success only to the point where clamp connections are formed, for example as reported for the recent studies of the *Hebeloma crustuliniforme* complex (Aanen & Kuyper, 1999) and those of other mushrooms (Petersen & Hughes, 1999). There may also be significant reductions in mating success subsequent to the production of clamp connections; additional studies should be made with this fungus or another where the complete reproductive cycle could be assessed. *Schizophyllum commune*, in which population structure has recently been studied by enzyme electrophoresis (James *et al.*, 1999) and which completes the life cycle from meiospore to meiospore in cultivation, would seem to be an ideal candidate.

Selection

There is a debate about whether sympatric species need to exploit different niches to avoid competition for exactly the same resources and rapid extinction of one of the pair (Coyne & Orr, 2004). However, there is little debate that selection, along with drift, contributes to the genetic differentiation that supports the persistence of newly diverged species. Even when selection cannot act directly on reproductive isolation barriers (as in allopatry), selection can drive genetic divergence that results in genomic incompatibilities between new species, and also drive phenotypic differentiation that renders hybrids with intermediate phenotypes less fit than either parent.

The molecular signatures of directional and purifying selection, that is, the excess or deficit, respectively, of nucleotide substitutions causing amino acid replacements (non-synonymous substitutions, Ka) compared to those that do not (synonymous substitutions, Ks), are typically detected by comparing within- and between-species polymorphisms (McDonald & Kreitman, 1991). To pursue these analyses, *a priori* one must be able to place individuals in phylogenetic species. With knowledge of PSR, the effects of selection can be studied for any gene by using this and other population genetic approaches initially developed from studies of flies or mammals (McDonald & Kreitman, 1991; Yang & Bielawski, 2000; Yang & Nielsen, 2002). The first such studies in fungi focused on genes that, curiously enough, may be little affected by speciation, those genes responsible for mating recognition and regulation in basidiomycetes (May *et al.*, 1999). This is not to say that mating genes evolve slowly (Pöggeler, 1999; Swanson & Vacquier, 2002) or that specificity is not involved in recognizing potential mates (Casselton & Olesnicky, 1998). However, fungi transformed with the mating type allele of a close relative can mate with conspecific partners (Sharon *et al.*, 1996; Turgeon, 1998), suggesting that mating loci are not involved with reproductive isolation. May and colleagues (May *et al.*, 1999) showed that the diversity of mating alleles present early in the evolutionary history of a group of *Coprinus* species has been maintained through speciation events. These studies were followed by research on more genes with transspecies polymorphisms, i.e., vegetative self- and non-self-recognition genes in Ascomycota (Leslie, 1993). With these heterokaryon incompatibility loci, Glass and colleagues showed that more than two alleles are maintained in populations at a single locus by balancing selection (Wu *et al.*, 1998). Most recently, and still with genes that maintain diversity through speciation events, O'Donnell and colleagues found in a very thorough study of PSR in *Fusarium graminearum, sensu lato*, that clusters of genes responsible for synthesis of secondary products (trichothecene toxins) produced gene genealogies in strong conflict with those for six other regions (O'Donnell *et al.*, 2000). Subsequent analysis showed that individuals of a single phylogenetic species were capable of making one of three different trichothecenes found in the species because each individual possessed genes for only one of the three toxins (Ward *et al.*, 2002). Again, the maintenance of intraspecific diversity through speciation events could be explained by balancing or frequency-dependent selection (O'Donnell *et al.*, 2004).

Most genes, however, are not under balancing or frequency-dependent selection and their variation is very much related to the species in which they are found. Recently, selection was studied in genes from *Coccidioides* species that produce proteins that have been shown to elicit an immune response in mice. In the first of these studies, Johannesson and colleagues found that a fungal gene product that stimulates a protective acquired immune response in mice (that is, a good vaccine candidate) was under strong positive selection in contrast to housekeeping genes (Johannesson *et al.*, 2004). However, the selection did not correlate with pathogenicity or virulence because it could not be demonstrated to be stronger in *Coccidioides* species than in *Uncinocarpus* species, their close, non-pathogenic relatives (Johannesson *et al.*, 2004). In the second study, a *Coccidioides* gene known to produce a protein that is a poor vaccine candidate (because it elicits a non-protective immunological response from the host) was found to have different numbers of repeats that influence the host immune response. The evolution of the repeated regions was shown by phylogenetic analysis to be concerted (Johannesson *et al.*, 2005), as with ribosomal repeat units (Dover, 1982); there is a possibility that the diversity of repeat number is maintained in this species by a conflict between greater numbers of repeats promoting virulence and lesser numbers favouring the function of the protein as an adhesion factor (Hung *et al.*, 2002). The existing studies of selection for individual genes, and individual nucleotide positions therein, are merely a prelude to the coming wave of whole-genome examination of selection and of accelerated nucleotide substitution by comparative genomics (Clark *et al.*, 2003; Cliften *et al.*, 2003; Kellis *et al.*, 2003). Here, the many genomes sequenced for socially important fungi should prove extremely valuable. However, knowing which genes are under strong selection without understanding how the gene products relate to the natural history or ecology of the fungus will be less than satisfying and will leave us short of the goal of understanding ecologically relevant adaptation at the genetic level. Studies of adaptation also will require a thorough understanding of how to measure the fitness of filamentous fungi (Pringle & Taylor, 2002).

We have attempted to contribute to the discussion on identifying fungal species and to the discussion of the evolutionary processes whose study depends on accurate species recognition: species formation and persistence, species maintenance by geographic distance in allopatry and by reinforcement in sympatry, and natural selection. Like Brasier, we have left discussion of the genetic systems of mating compatibility to others (Glass & Kuldau, 1992; Fraser *et al.*, 2004; Fraser & Heitman, 2004). Most of our comments address the pattern of speciation, but the new challenge facing mycology is to better understand the processes of speciation and adaptation. Rapid speciation certainly can be achieved by a change in the mating system, for example, from homothallism to heterothallism or back (Geiser *et al.*, 1998a; Turgeon, 1998; Yun *et al.*, 1999; Lee *et al.*, 2003). Likewise, hybridization events, sometimes associated with polyploidy, as in yeast (Wong *et al.*, 2002; Dujon *et al.*, 2004), *Botrytis* (Nielsen & Yohalem, 2001) or the asexual *Epichloë* relatives in the genus *Neotyphodium* elegantly unravelled by Schardl and colleagues (Moon *et al.*, 2004) can produce instant speciation. Mycologists also accord species status to clonal lineages that probably emerge from sexual species (Summerbell, 2002; Kaszubiak *et al.*, 2004) and that are probably evolutionary dead ends (LoBuglio *et al.*, 1993), although the argument has been made that such lineages are better considered as individuals (Janzen, 1977). However, less well studied is the more common speciation process, whereby geographic distance, or adaptation to heterogenous environments in the same geographic location [which might have to very small to be considered sympatric for a fungus (Le Gac & Giraud, 2004)], contribute to a diminution of gene flow between populations of a single species and lead to speciation. We hope that studies of 'normal' speciation, of the adaptation of newly diverged species to different micro-habitats, and of reinforcement of the species barrier in sympatry, will become the subject of sustained study, at least in species for which phylogenetic species have been recognized.

Acknowledgements

Writing of this manuscript was supported by grants NSF DEB 0316710 and DEB 0516511 to JWT. We thank D. J. Jacobson and D. D. Perkins for comments on the draft manuscript.

References

Aanen, D. K. & Kuyper, T. W. (1999). Intercompatibility tests in the *Hebeloma crustuliniforme* complex in northwestern Europe. *Mycologia* **91**, 783–95.

Ainsworth, G. C., Kirk, P. M., Bisby, G. R., Cannon, P. F., David, J. C. & Stalpers, J. A. (2001). *Ainsworth and Bisby's Dictionary of the Fungi.* Wallingford: CABI Bioscience.

Avise, J. C. & Wollenberg, K. (1997). Phylogenetics and the origin of species. *Proceedings of the National Academy of Sciences of the USA* **94**, 7748–55.

Baker, C. J., Alfenas, A. C. & Harrington, T. C. (2001). Host specialization and cryptic species within *Ceratocystis fimbriata. Phytopathology* **91**, S4–S5.

Baker, C. J., Harrington, T. C., Krauss, U. & Alfenas, A. C. (2003). Genetic variability and host specialization in the Latin American clade of *Ceratocystis fimbriata. Phytopathology* **93**, 1274–84.

Brasier, C. M. (1997). Fungal species in practice: identifying species units in fungi. In *Species: the Units of Biodiversity*, ed. M. F. Claridge, H. A. Dawah & M. R. Wilson, pp. 135–70. London and New York: Chapman and Hall.

Brasier, C. M., Kirk, S. A., Pipe, N. D. & Buck, K. W. (1998). Rare interspecific hybrids in natural populations of the Dutch elm disease pathogens *Ophiostoma ulmi* and *Ophiostoma novo-ulmi. Mycological Research* **102**, 45–57.

Burnett, J. H. (1983). Presidential address: speciation in fungi. *Transactions of the British Mycological Society* **81**, 1–14.

Burnett, J. H. (2003). *Fungal Populations and Species.* Oxford: Oxford University Press.

Burt, A., Carter, D. A., Koenig, G. L., White, T. J. & Taylor, J. W. (1996). Molecular markers reveal cryptic sex in the human pathogen *Coccidioides immitis. Proceedings of the National Academy of Sciences of the USA* **93**, 770–3.

Burt, A., Koufopanou, V. & Taylor, J. W. (2000). Population genetics of human-pathogenic fungi. In *Molecular Epidemiology of Infectious Diseases*, ed. R. C. A. Thompson, pp. 229–44. London: Arnold.

Carbone, I. & Kohn, L. M. (2001a). A microbial population-species interface: Nested cladistic and coalescent inference with multilocus data. *Molecular Ecology* **10**, 947–64.

Carbone, I. & Kohn, L. M. (2001b). Multilocus nested haplotype networks extended with DNA fingerprints show common origin and fine-scale, ongoing genetic divergence in a wild microbial metapopulation. *Molecular Ecology* **10**, 2409–22.

Casselton, L. A. & Olesnicky, N. S. (1998). Molecular genetics of mating recognition in basidiomycete fungi. *Microbiology and Molecular Biology Reviews* **62**, 55–70.

Clark, A. G., Glanowski, S., Nielsen, R., Thomas, P. D., Kejariwal, A., Todd, M. A., Tanenbaum, D. M., Civello, D., Lu, F., Murphy, B., Ferriera, S., Wang, G., Zheng, X. G., White, T. J., Sninsky, J. J., Adams, M. D. & Cargill, M. (2003). Inferring nonneutral evolution from human-chimp-mouse orthologous gene trios. *Science* **302**, 1960–3.

Cliften, P., Sudarsanam, P., Desikan, A., Fulton, L., Fulton, B., Majors, J., Waterston, R., Cohen, B. A. & Johnston, M. (2003). Finding functional features in *Saccharomyces* genomes by phylogenetic footprinting. *Science* **301**, 71–6.

Couch, B. C. & Kohn, L. M. (2002). A multilocus gene genealogy concordant with host preference indicates segregation of a new species, *Magnaporthe oryzae*, from *M. grisea*. *Mycologia* **94**, 683–93.

Coyne, J. A. & Orr, H. A. (2004). *Speciation*. Sunderland, MA: Sinauer Associates.

Debeaupuis, J.-P., Sarfati, J., Chazalet, V. & Latge, J.-P. (1997). Genetic diversity among clinical and environmental isolates of *Aspergillus fumigatus*. *Infection and Immunity* **65**, 3080–5.

Dettman, J. R. & Taylor, J. W. (2004). Mutation and evolution of microsatellite loci in *Neurospora*. *Genetics* **168**, 1231–48.

Dettman, J. R., Harbinski, F. M. & Taylor, J. W. (2001). Ascospore morphology is a poor predictor of the phylogenetic relationships of *Neurospora* and *Gelasinospora*. *Fungal Genetics and Biology* **34**, 44–61.

Dettman, J. R., Jacobson, D. J. & Taylor, J. W. (2003a). A multilocus genealogical approach to phylogenetic species recognition in the model eukaryote *Neurospora*. *Evolution* **57**, 2703–20.

Dettman, J. R., Jacobson, D. J., Turner, E., Pringle, A. & Taylor, J. W. (2003b). Recognizing species under biological and phylogenetic species concepts: reproductive isolation versus phylogenetic divergence. *Evolution* **57**, 2721–41.

Dover, G. (1982). Molecular drive: a cohesive mode of species evolution. *Nature* **299**, 111–17.

Dujon, B., Sherman, D., Fischer, G., Durrens, P., Casaregola, S., Lafontaine, I., de Montigny, J., Marck, C., Neuveglise, C., Talla, E., Goffard, N., Frangeul, L., Aigle, M., Anthouard, V., Babour, A., Barbe, V., Barnay, S., Blanchin, S., Beckerich, J. M., Beyne, E., Bleykasten, C., Boisrame, A., Boyer, J., Cattolico, L., Confanioleri, F., de Daruvar, A., Despons, L., Fabre, E., Fairhead, C., Ferry-Dumazet, H., Groppi, A., Hantraye, F., Hennequin, C., Jauniaux, N., Joyet, P., Kachouri, R., Kerrest, A., Koszul, R., Lemaire, M., Lesur, I., Ma, L., Muller, H., Nicaud, J. M., Nikoloski, M., Oztas, S., Ozier-Kalogeropoulos, O., Pellenz, S., Potier, S., Richard, G. F., Straub, M. L., Suleau, A., Swennen, D., Tekaia, F., Wesolowski-Louvel, M., Westhof, E., Wirth, B., Zeniou-Meyer, M., Zivanovic, I., Bolotin-Fukuhara, M., Thierry, A., Bouchier, C., Caudron, B., Scarpelli, C., Gaillardin, C., Weissenbach, J., Wincker, P. & Souciet, J. L. (2004). Genome evolution in yeasts. *Nature* **430**, 35–44.

Dyer, P. S., Paoletti, M. & Archer, D. B. (2003). Genomics reveals sexual secrets of *Aspergillus*. *Microbiology* **149**, 2301–3.

Engelbrecht, C. J. B. & Harrington, T. C. (2005). Intersterility, morphology and taxonomy of *Ceratocystis fimbriata* on sweet potato, cacao and sycamore. *Mycologia* **97**, 57–69.

Falush, D., Stephens, M. & Pritchard, J. K. (2003). Inference of population structure using multilocus genotype data: linked loci and correlated allele frequencies. *Genetics* **164**, 1567–87.

Field, D. & Wills, C. (1996). Long, polymorphic microsatellites in simple organisms. *Proceedings of the Royal Society of London* **B263**, 209–15.

Field, D., Eggert, L., Metzger, D., Rose, R. & Wills, C. (1996). Use of polymorphic short and clustered coding-region microsatellite to distinguish strains of *Candida albicans*. *FEMS Immunology and Medical Microbiology* **15**, 73–9.

Fisher, M. C., Koenig, G., White, T. W. & Taylor, J. W. (2000). A test for concordance between the multilocus genealogies of genes and microsatellites in the pathogenic fungus *Coccidioides immitis*. *Molecular Biology and Evolution* **17**, 1164–74.

Fisher, M. C., Koenig, G. L. and White, T. J., San-Blas, G., Negroni, R., Alvarez, I. G., Wanke, B. & Taylor, J. W. (2001). Biogeographic range expansion into South America by *Coccidioides immitis* mirrors New World patterns of human migration. *Proceedings of the National Academy of Sciences of the USA* **98**, 4558–62.

Fisher, M. C., Koenig, G. L., White, T. J. & Taylor, J. T. (2002a). Molecular and phenotypic description of *Coccidioides posadasii* sp nov., previously recognized as the non-California population of *Coccidioides immitis*. *Mycologia* **94**, 73–84.

Fisher, M. C., Rannala, B., Chaturvedi, V. & Taylor, J. W. (2002b). Disease surveillance in recombining pathogens: multilocus genotypes identify sources of *Coccidioides* infections. *Proceedings of the National Academy of Sciences of the USA* **99**, 9067–71.

Fisher, M. C., Aanensen, D., de Hoog, S. & Vanittanakom, N. (2004). Multilocus microsatellite typing system for *Penicillium marneffei* reveals spatially structured populations. *Journal of Clinical Microbiology* **42**, 5065–9.

Fraser, J. A. & Heitman, J. (2004). Evolution of fungal sex chromosomes. *Molecular Microbiology* **51**, 299–306.

Fraser, J. A., Diezmann, S., Subaran, R. L., Allan, A., Lengeler, K. B., Dietrich, F. S. & Heitman, J. (2004). Convergent evolution of chromosomal sex-determining regions in the animal and fungal kingdoms. *PLOS Biology* **2**, 2243–55.

Garbelotto, M., Gonthier, P., Linzer, R., Nicolotti, G. & Otrosina, W. (2004). A shift in nuclear state as the result of natural interspecific hybridization between two North American taxa of the basidiomycete complex *Heterobasidion*. *Fungal Genetics and Biology* **41**, 1046–51.

Geiser, D. M., Frisvad, J. C. & Taylor, J. W. (1998a). Evolutionary relationships in *Aspergillus* section *Fumigata* inferred from partial beta-tubulin and hydrophobin DNA sequences. *Mycologia* **90**, 831–45.

Geiser, D. M., Pitt, J. I. & Taylor, J. W. (1998b). Cryptic speciation and recombination in the aflatoxin-producing fungus *Aspergillus flavus*. *Proceedings of the National Academy of Sciences of the USA* **95**, 388–93.

Geiser, D. M., Dorner, J. W., Horn, B. W. & Taylor, J. W. (2000). The phylogenetics of mycotoxin and sclerotium production in *Aspergillus flavus* and *Aspergillus oryzae*. *Fungal Genetics and Biology* **31**, 169–79.

Giraud, T., Fortini, D., Levis, C. *et al.* (1999). Two sibling species of the *Botrytis cinerea* complex, *transposa* and *vacuma*, are found in sympatry on numerous host plants. *Phytopathology* **89**, 967–73.

Glass, N. L. & Kuldau, G. A. (1992). Mating type and vegetative incompatibility in filamentous ascomycetes. *Annual Review of Phytopathology* **30**, 201–24.

Harrington, T. C. & Rizzo, D. M. (1999). Defining species in the fungi. In *Structure and Dynamics of Fungal Populations*, ed. J. J. Worrall, pp. 43–70. Dordrecht and Boston: Kluwer Academic Publishers.

Hibbett, D. S., Fukumasa-Nakai, Y., Tsuneda, A. & Donoghue, M. J. (1995). Phylogenetic diversity in shiitake inferred from nuclear ribosomal DNA sequences. *Mycologia* **87**, 618–38.

Hung, C. Y., Yu, J. J., Seshan, K. R., Reichard, U. & Cole, G. T. (2002). A parasitic phase-specific adhesin of *Coccidioides immitis* contributes to the virulence of this respiratory fungal pathogen. *Infection and Immunity* **70**, 3443–56.

James, T. Y., Porter, D., Hamrick, J. L. & Vilgalys, R. (1999). Evidence for limited intercontinental gene flow in the cosmopolitan mushroom, *Schizophyllum commune*. *Evolution* **53**, 1665–77.

Janzen, D. H. (1977). What are dandelions and aphids? *American Naturalist* **111**, 586–9.

Johannesson, H. & Stenlid, J. (2003). Molecular markers reveal genetic isolation and phylogeography of the S and F intersterility groups of the wood-decay fungus *Heterobasidion annosum*. *Molecular Phylogenetics and Evolution* **29**, 94–101.

Johannesson, H., Vidal, P., Guarro, J., Herr, R. A., Cole, G. T. & Taylor, J. W. (2004). Positive directional selection in the proline-rich antigen (PRA) gene among the human pathogenic fungi *Coccidioides immitis, C. posadasii* and their closest relatives. *Molecular Biology and Evolution* **21**, 1134–45.

Johannesson, H., Townsend, J. P., Hung, C.-Y., Cole, G. T. & Taylor, J. W. (2005). Concerted evolution in the repeats of an immunomodulating cell surface protein, SOWgp, of the human pathogenic fungi *Coccidioides immitis* and *C. posadasii. Genetics* **171**, 109–17.

Kasuga, T., White, T. J. & Taylor, J. W. (2002). Estimation of nucleotide substitution rates in eurotiomycete fungi. *Molecular Biology and Evolution* **19**, 2318–24.

Kasuga, T., White, T. J., Koenig, G., McEwen, J., A., R., Castaneda, E., Lacaz, C. d. S., Heins-Vaccari, E. M., de Freitas, R. S., Zancope-Oliveira, R. M., Qin, Z., Negroni, R., Carter, D. A., Mikami, Y., Tamura, M., Taylor, M. L., Miller, G. F., Poonwan, N. & Taylor, J. W. (2003). Phylogeography of the fungal pathogen *Histoplasma capsulatum. Molecular Ecology* **12**, 3383–401.

Kaszubiak, A., Klein, S., de Hoog, G. S. & Gräser, Y. (2004). Population structure and evolutionary origins of *Microsporum canis, M. ferrugineum* and *M. audouinii. Medical Mycology* **4**, 179–86.

Kellis, M., Patterson, N., Endrizzi, M., Birren, B. & Lander, E. S. (2003). Sequencing and comparison of yeast species to identify genes and regulatory elements. *Nature* **423**, 241–54.

Kohn, L. M. (2005). Mechanisms of fungal speciation. *Annual Review of phytopathology* **43**, 279–308. (This thorough review was published too late to be included in this chapter. It is highly recommended to readers.)

Kohn, L. M. & Carbone, I. (2001). Origins and epidemiology of *Sclerotinia sclerotiorum* genotypes on Southeastern US canola: The power of multilocus DNA sequence data, combined datasets and phylogeographic statistical approaches. *Phytopathology* **91**, S115.

Konrad, H., Kirisits, T., Riegler, M., Halmschlager, E. & Stauffer, C. (2002). Genetic evidence for natural hybridization between the Dutch elm disease pathogens *Ophiostoma novo-ulmi* ssp *novo-ulmi* and *O-novo-ulmi* ssp *americana. Plant Pathology* **51**, 78–84.

Koufopanou, V., Burt, A. & Taylor, J. W. (1997). Concordance of gene genealogies reveals reproductive isolation in the pathogenic fungus *Coccidioides immitis. Proceedings of the National Academy of Sciences of the USA* **94**, 5478–82.

Koufopanou, V., Burt, A., Szaro, T. & Taylor, J. W. (2001). Gene genealogies, cryptic species, and molecular evolution in the human pathogen *Coccidioides immitis* and relatives (Ascomycota, Onygenales). *Molecular Biology and Evolution* **18**, 1246–58.

Kurtzman, C. P. & Fell, J. W. (1998). *The Yeasts: a Taxonomic Study*. New York: Elsevier.

Kwon-Chung, K. J. & Bennett, J. E. (1992). *Medical Mycology*. Philadelphia: Lea & Febiger.

Le Gac, M. & Giraud, T. (2004). What is sympatric speciation in parasites? *Trends in Parasitology* **20**, 207–8.

Lee, J., Lee, T., Lee, Y. W., Yun, S. H. & Turgeon, B. G. (2003). Shifting fungal reproductive mode by manipulation of mating type genes: obligatory heterothallism of *Gibberella zeae. Molecular Microbiology* **50**, 145–52.

Leslie, J. F. (1993). Fungal vegetative compatibility. *Annual Review of Phytopathology* **31**, 127–50.

LoBuglio, K. F., Pitt, J. I. & Taylor, J. W. (1993). Phylogenetic analysis of two ribosomal DNA regions indicates multiple independent losses of a sexual *Talaromyces* state among asexual *Penicillium* species in sub genus *Biverticulum. Mycologia* **85**, 592–604.

Maiden, M. C. J., Bygraves, J. A., Feil, E., Morelli, G., Russell, J. E., Urwin, R., Zhang, Q., Zhou, J. J., Zurth, K., Caugant, D. A., Feavers, I. M., Achtman, M. & Spratt, B. G. (1998). Multilocus sequence typing: A portable approach to the identification of clones within populations of pathogenic microorganisms. *Proceedings of the National Academy of Sciences of the USA* **95**, 3140–5.

May, G., Shaw, F., Badrane, H. & Vekemans, X. (1999). The signature of balancing selection: Fungal mating compatibility gene evolution. *Proceedings of the National Academy of Sciences of the USA* **96**, 9172–7.
McDonald, J. H. & Kreitman, M. (1991). Adaptive protein evolution at the Adh locus in *Drosophila*. *Nature* **351**, 652–4.
Moon, C. D., Craven, K. D., Leuchtmann, A., Clement, S. L. & Schardl, C. L. (2004). Prevalence of interspecific hybrids amongst asexual fungal endophytes of grasses. *Molecular Ecology* **13**, 1455–67.
Natvig, D. O. & May, G. (1996). Fungal evolution and speciation. *Journal of Genetics* **75**, 441–52.
Newcombe, G., Stirling, B. & Bradshaw, H. D. (2001). Abundant pathogenic variation in the new hybrid rust *Melampsora* × *columbiana* on hybrid poplar. *Phytopathology* **91**, 981–5.
Nielsen, K. & Yohalem, D. S. (2001). Origin of a polyploid *Botrytis* pathogen through interspecific hybridization between *Botrytis aclada* and *B. byssoidea*. *Mycologia* **93**, 1064–71.
O'Donnell, K., Cigelnik, E. & Nirenberg, H. I. (1998). Molecular systematics and phylogeography of the *Gibberella fujikuroi* species complex. *Mycologia* **90**, 465–93.
O'Donnell, K., Kistler, H. C., Tacke, B. K. & Casper, H. H. (2000). Gene genealogies reveal global phylogeographic structure and reproductive isolation among lineages of *Fusarium graminearum*, the fungus causing wheat scab. *Proceedings of the National Academy of Sciences of the USA* **97**, 7905–10.
O'Donnell, K., Ward, T. J., Geiser, D. M., Kistler, H. C. & Aoki, T. (2004). Genealogical concordance between the mating type locus and seven other nuclear genes supports formal recognition of nine phylogenetically distinct species within the *Fusarium graminearum* clade. *Fungal Genetics and Biology* **41**, 600–23.
Paoletti, M., Rydholm, C., Schwier, E. U., Anderson, M. J., Szakacs, G., Lutzoni, F., Debeaupuis, J. P., Latgé, J. P., Denning, D. W. & Dyer, P. S. (2005). Evidence for sexuality in the opportunistic fungal pathogen *Aspergillus fumigatus*. *Current Biology* **15**, 1242–8.
Perkins, D. D. & Turner, B. C. (1988). *Neurospora* from natural populations: toward the population biology of a haploid eukaryote. *Experimental Mycology* **12**, 91–131.
Petersen, R. H. & Hughes, K. W. (1999). Species and speciation in mushrooms. *Bioscience* **49**, 440–52.
Phillips, D. V., Carbone, I., Gold, S. E. & Kohn, L. M. (2002). Phylogeography and genotype-symptom associations in early and late season infections of canola by *Sclerotinia sclerotiorum*. *Phytopathology* **92**, 785–93.
Pitt, J. I. (1979). *The genus Penicillium and its Teleomorphic States Eupenicillium and Talaromyces*. London: Academic Press.
Pöggeler, S. (1999). Phylogenetic relationships between mating-type sequences from homothallic and heterothallic ascomycetes. *Current Genetics* **36**, 222–31.
Pöggeler, S. (2002). Genomic evidence for mating abilities in the asexual pathogen *Aspergillus fumigatus*. *Current Genetics* **42**, 153–60.
Pringle, A. & Taylor, J. W. (2002). Fitness in filamentous fungi. *Trends in Microbiology* **10**, 474–81.
Pringle, A., Baker, D. M., Platt, J. L., Wares, J. P., Latgé, J.-P. & Taylor, J. W. (2005). Cryptic speciation in the cosmopolitan and clonal human pathogenic fungus *Aspergillus fumigatus*. *Evolution* **59**, 1886–99.
Pritchard, J. K. & Feldman, M. W. (1996). Statistics for microsatellite variation based on coalescence. *Theoretical Population Biology* **50**, 325–44.
Rannala, B. & Mountain, J. L. (1997). Detecting immigration by using multilocus genotypes. *Proceedings of the National Academy of Sciences of the USA* **94**, 9197–201.

Rydholm, C., Paoletti, M., Dyer, P. & Lutzoni, F. (2004). Recombination and mating loci in the "asexual" *Aspergillus fumigatus* and sexual *Neosartorya fischeri* species pair. *Inoculum* **55**, 33.

Schardl, C. L. & Craven, K. D. (2003). Interspecific hybridization in plant-associated fungi and oomycetes: a review. *Molecular Ecology* **12**, 2861–73.

Sharon, A., Yamaguchi, K., Christiansen, S., Horwitz, B. A., Yoder, O. C. & Turgeon, B. G. (1996). An asexual fungus has the potential for sexual development. *Molecular and General Genetics* **251**, 60–8.

Shear, C. L. & Dodge, B. O. (1927). Life histories and heterothallism of the red bread-mold fungi of the *Monila sitophila* group. *Journal of Agricultural Research* **34**, 1019–42.

Summerbell, R. C. (2002). What is the evolutionary and taxonomic status of asexual lineages in the dermatophytes? *Studies in Mycology* **47**, 97–101.

Swanson, W. J. & Vacquier, V. D. (2002). The rapid evolution of reproductive proteins. *Nature Reviews Genetics* **3**, 137–44.

Taylor, J. W. & Fisher, M. C. (2003). Fungal multilocus sequence typing – it's not just for bacteria. *Current Opinion in Microbiology* **6**, 351–6.

Taylor, J. W., Jacobson, D. J., Kroken, S., Kasuga, T., Geiser, D. M., Hibbett, D. S. & Fisher, M. C. (2000). Phylogenetic species recognition and species concepts in fungi. *Fungal Genetics and Biology* **31**, 21–32.

Tran-Dinh, N., Pitt, J. I. & Carter, D. A. (1999). Molecular genotype analysis of natural toxigenic and nontoxigenic isolates of *Aspergillus flavus* and *A. parasiticus*. *Mycological Research* **103**, 1485–90.

Turgeon, B. G. (1998). Application of mating type gene technology to problems in fungal biology. *Annual Review of Phytopathology* **36**, 115–37.

Turner, B. C., Perkins, D. D. & Fairfield, A. (2001). *Neurospora* from natural populations: a global study [Review]. *Fungal Genetics and Biology* **32**, 67–92.

Vilgalys, R. & Sun, B. L. (1994). Ancient and recent patterns of geographic speciation in the oyster mushroom *Pleurotus* revealed by phylogenetic analysis of ribosomal DNA sequences. *Proceedings of the National Academy of Sciences of the USA* **91**, 4599–603.

Ward, T. J., Bielawski, J. P., Kistler, H. C., Sullivan, E. & O'Donnell, K. (2002). Ancestral polymorphism and adaptive evolution in the trichothecene mycotoxin gene cluster of phytopathogenic *Fusarium*. *Proceedings of the National Academy of Sciences of the USA* **99**, 9278–83.

Wolfe, K. H. & Shields, D. C. (1997). Molecular evidence for an ancient duplication of the entire yeast genome. *Nature* **387**, 708–13.

Wong, S., Butler, G. & Wolfe, K. H. (2002). Gene order evolution and paleopolyploidy in hemiascomycete yeasts. *Proceedings of the National Academy of Sciences of the USA* **99**, 9272–7.

Worrall, J. J., Ed. (1999). *Structure and Dynamics of Fungal Populations*. Dordrecht: Kluwer.

Wu, J., Saupe, S. J. & Glass, N. L. (1998). Evidence for balancing selection operating at the *het*-c heterokaryon incompatibility locus in a group of filamentous fungi. *Proceedings of the National Academy of Sciences of the USA* **95**, 12 398–403.

Yang, Z. H. & Bielawski, J. P. (2000). Statistical methods for detecting molecular adaptation. *Trends in Ecology and Evolution* **15**, 496–503.

Yang, Z. H. & Nielsen, R. (2002). Codon-substitution models for detecting molecular adaptation at individual sites along specific lineages. *Molecular Biology and Evolution* **19**, 908–17.

Yun, S. H., Berbee, M. L., Yoder, O. C. & Turgeon, B. G. (1999). Evolution of the fungal self-fertile reproductive life style from self-sterile ancestors. *Proceedings of the National Academy of Sciences of the USA* **96**, 5592–7.

16

Multilocus sequence typing (MLST) and multilocus microsatellite typing (MLMT) in fungi

MATTHEW C. FISHER

Imperial College Faculty of Medicine, London

What is multilocus typing?

Origins of the technique

The characterization of genetic variation has revolutionized our understanding of fungal populations and species. Traditionally, advances have been most rapid in the fields of medical mycology and phytopathology (Taylor *et al.*, 1999b) owing to the need for effective molecular epidemiological tools. Epidemiological studies are typically concerned with disease outbreaks, the origin and spread of virulent strains, or the emergence of an interesting phenotype such as antibiotic resistance. Genetic variation in the genomes of pathogens provides a means by which isolates can be differentiated from one another (Taylor *et al.*, 1999b). The characterization of this molecular variation has given rise to the field of molecular epidemiology, whereby genetic variation is used to address questions about the biology and transmission of infectious diseases. However, the techniques developed for molecular epidemiology are not limited to medical fields, and there is huge potential to apply these methodologies to non-disease-causing organisms.

The power of molecular epidemiology as an analytical tool has led to a period of rapid development, resulting in many methods for indexing genetic variation, such as VNTRs (variable number tandem repeats), MLEE (multilocus enzyme electrophoresis), RFLPs (restriction fragment length polymorphisms), RAPDs (randomly amplified polymorphic repeats) and PFGE (pulse field gel electrophoresis), to name but a few (Taylor *et al.*, 1999b; McEwen *et al.*, 2000). Typically, laboratories have tended to develop in-house techniques that are specifically focused on a particular problem, and are usually a variant on the above. However,

Fungi in the Environment, ed. G. M. Gadd, S. C. Watkinson & P. S. Dyer. Published by Cambridge University Press.

although specific techniques are often well suited to the task for which they were developed, it is often problematic applying the protocols within another laboratory; reproducibility may not be high or the genetic variation may simply not be present in another population or species. This has led to a proliferation of techniques that are specific to the specialities and interests of a research group, and a situation has developed whereby there are almost as many typing systems as there are questions. Moreover, there has been a huge redundancy in research effort; genotypes that have been painstakingly scored from a carefully sampled cohort of isolates are presented once in a scientific publication, and often as a series of summary statistics within tables or figures. Without any means for other researchers to acquire the genotypes, research effort is effectively 'fossilized' upon publication, as it cannot be easily acquired by another researcher and used for re-analysis, or even applied to address other relevant biological questions.

Multilocus sequence typing (MLST)

Multilocus typing approaches were developed to address these questions and concerns. The technique was originally developed as an epidemiological tool by the bacterial research community (Maiden *et al.*, 1998), although nearly identical approaches were developed in parallel by mycologists (Taylor & Fisher, 2003). Specifically, a method for discriminating isolates was sought that obeyed the following criteria: (i) the technique should be 100% reproducible between laboratories; (ii) the technique should be able to discriminate between isolates yet remain stable over time (iii) the technique should accurately portray evolutionary relationships at a variety of scales; and (iv) the technique should be portable, and accessible, to the wider research community (Feil & Spratt, 2001).

MLST works by directly amplifying, by using the polymerase chain reaction (PCR), a 450–500 nucleotide fragment from each of 7–10 housekeeping genes for each isolate within a cohort (Enright & Spratt, 1999). This length of PCR product is chosen so that it can be sequenced in its entirety for both forward and reverse strands by using a single pair of primers. This ensures that all polymorphisms are scored twice, at the very least, and sequencing errors are therefore minimized as a consequence. All fragments are sequenced so that each stretch of *c.* 3500 nucleotides represents a series of seven independent samples of the amount of genetic variation present in the genome of each isolate. Once all isolates within the cohort have been sequenced at the chosen loci, they are used to build up a multilocus database (Fig. 16.1). Simply, each allele at a particular locus is given a unique integer code. Therefore, the concatenated string of all integers coding the alleles present at each locus represents the allelic profile, or sequence type (ST) of an isolate. Sequences that differ by a single nucleotide change are assigned as different alleles and will therefore be represented as a unique ST within the database. No weighting schemes are used, so a single point mutation carries as much weight as a recombinational exchange that translocates a large number of polymorphisms. This simplifies analyses; however, the raw sequence data are always accessible from the websites if weighted analyses are required.

Once isolates have been sequenced for the MLST loci, and STs assigned, then the data is uploaded to the Internet. STs are held online in an SQL server relational database that is configured to be accessible via a web interface located at http://www.mlst.org/. This site allows laboratories to upload and compare the STs of their isolates to those that are represented in the database, thus allowing the direct comparison of data with those

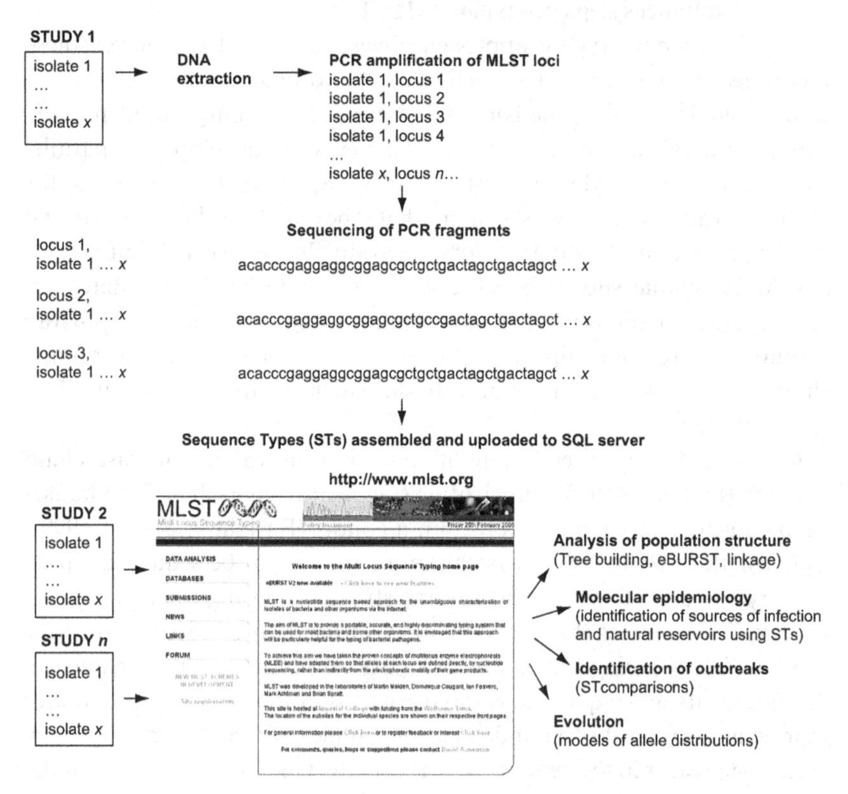

Fig. 16.1. The processes involved in developing an MLST scheme.

being generated by other laboratories. Once curated, STs and associated information can become a permanent feature of the online database. This increases the size of the database and therefore its power to accurately portray the genetic structure of the species.

MLST relies on indexing the genetic variation at housekeeping genes; the accumulation of nucleotide changes in these genes is a relatively slow process, ensuring that STs are stable over recent time scales (Maiden *et al.*, 1998). Owing to the large number of samples within a ST (*c.* 3500–4000), sufficient genetic variation is indexed to discriminate all but the most closely related of isolates. However, as mentioned above, MLST was developed for prokaryotes, in which levels of naturally occuring nucleotide polymorphisms are often an order of magnitude higher than those often observed in eukaryotes. Therefore MLST may run into problems when the species, or population, in question may be depauperate in genetic variation. This has been observed, for instance, in the South American population of the onygenalean fungus *Coccidioides posadasii*, where all isolates share an identical ST and no discrimination is achieved (Fisher *et al.*, 2001). In such a situation, more rapidly evolving loci need to be sought to develop a multilocus typing scheme. Such typing techniques (multilocus microsatellite typing, MLMT) will be described in a later section.

MLST in fungi

Despite the reduced levels of genetic variation found within eukaryotes, MLST schemes have proved very successful in a number of species of medically important fungi. The first fungal MLST scheme was developed by Koufopanou *et al.* (1997, 1998) by sequencing five genes (*CHS1*, *pyrG*, *tcrP*, *CTS2* and a serine proteinase) in *Coccidioides immitis*. Analysis of these genes showed that there was, on average, a 1.4% nucleotide diversity between isolates; the gene genealogies demonstrated the existence of two strongly supported clades, separated from one another by 11–12.8 million years of evolution (Koufopanou *et al.*, 1997; Fisher *et al.*, 2000). This evidence was used to name a new sister species to *Coccidioides immitis*, *C. posadasii* (Fisher *et al.*, 2002). The *Coccidioides* MLST scheme was further used to demonstrate that isolates of both species within North and Central America were composed of geographically separated, genetically recombining populations, and that geography covaried with distance in a linear manner. This feature of the population genetics of *Coccidioides* spp. was used to show that successful long-distance dispersal of spores across space is low, resulting in low effective gene-flow. However, isolates of the pathogen in South America comprised a genetically bottlenecked sub-set of the genetic diversity found in Texas, North America. This finding was used to argue that a southwards dispersal of the fungus followed the last Pleistocene glaciation event (Fisher *et al.*, 2001). This series of papers is a useful illustration of how MLST can be used in conjunction with MLMT as tools for for assessing global epidemiology at a variety of scales, in closely related species, and in populations with varying degrees of inter-isolate relatedness.

There has been extensive interest in, and work on, developing MLST as a tool for research into human candidiasis. Initially, the technique was developed by Bougnoux, *et al.* (2002) by sequencing fragments of six genes, *CaGLN4*, *CaADP1*, *CaRPN2*, *CaACC1*, *CaSYA1* and *CaVPS13*. Subsequently, this database was adapted by Tavanti *et al.* (2003) to become a seven-gene scheme comprising the following genes: *AAT1a*, *ACC1*, *ADP1*, *MPIb*, *SYA1*, *VPS13* and *ZWF1b*. This set has become the internationally agreed reference standard and can be accessed at http://calbicans.mlst.net/. Currently, the *C. albicans* database holds MLST genotypes for 279 isolates, of which 244 (87%) are unique STs (Tavanti *et al.*, 2003).

In parallel, there has been a development of an MLST scheme for *Candida glabrata* (Dodgson *et al.*, 2003) with a six-gene scheme; *FKS*, *LEU2*, *NMT1*, *TRP1*, *UGP1* and *URA3*. This is also accessible at the Imperial College, London, MLST website. As of early 2005, this scheme contained 109 isolates, comprising 30 STs. Dendrograms generated by the MLST scheme were used to show that there are five principal clusters. Comparison with a previously developed fingerprinting method based on Southern blot hybridization with a CG6/Cg12 probe (Lockhart *et al.*, 1997) showed that both methods defined the same five major clusters and that, interestingly, there was a geographical component to the population genetics of this pathogen. However, the CG6/Cg12 probe discriminated between all isolates within groups whereas the MLST scheme did not, showing that this MLST scheme suffers from low resolution. Therefore, when asking questions about microevolution and nosocomial transmission in *C. glabrata*, it will be necessary either to sequence more genes or to focus on more highly variable loci.

In addition to work on *Coccidioides* and *Candida* species, there has been much effort focused on generating multilocus genealogies from multiple genes in many other fungal species, for both pathogens and environmental species. Extensive gene genealogies have been generated for *Histoplasma capsulatum* (Kasuga *et al.*, 1999, 2003), *Cryptococcus neoformans* (Xu *et al.*, 2000), *Batrachochytrium dendrobatidis* (Morehouse *et al.*, 2003),

Saccharomyces sp. (Johnson *et al.*, 2004), *Fusarium* sp. (O'Donnell *et al.*, 2004), and lichenized ascomycetes such as *Letharia* (Kroken & Taylor, 2001). However, this list is by no means exhaustive and most intensively studied fungal species will have been sequenced at more than one locus for a sample of isolates. Therefore, it is only a matter of time before a set of genes with the required properties is chosen for each species and MLST schemes developed for incorporation into a web-accessible database.

Multilocus microsatellite typing (MLMT)

When developing a MLST scheme for a fungal species it is important to use pilot studies to determine (i) which loci are polymorphic enough to use and (ii) whether nucleotide diversity *per se* is high enough to justify developing an MLST scheme. If the answer to (ii) is no, then more highly polymorphic loci will need to be sought.

Microsatellites have become the marker of choice for many eukaryotic systems. These loci are short stretches of DNA composed of a repeated motif. The motif can be composed of di-, tri-, tetra- or pentanucleotide repeats and, owing to strand slippage mispairing during meiosis (Hancock, 1999), accumulates variation between lineages as length polymorphisms. The rates of mutation at microsatellite repeats are high, estimated in yeast to be 10^{-4}–10^{-5} mutations per generation compared with 10^{-9} for point mutations (Henderson & Petes, 1992; Weber & Wong, 1993). Therefore, microsatellites accumulate much higher levels of genetic diversity than do 'normal' housekeeping loci. This leads to their being used as markers for typing genetically depauperate species. Typing with microsatellites is similar to MLST in that an entire fragment of sequence, spanning a microsatellite region, is amplified by PCR. However, the two techniques differ in the subsequent analytical step. Instead of sequencing, genetic variation is assayed at microsatellites by electrophoresing the amplified loci through a high-resolution acrylamide gel and scoring the size products; bands with different mobilities are scored as distinct alleles. Modern sequencing technology and capillary electrophoresis are used to ensure that reproducibility is high between experiments and laboratories, and that alleles that differ in length by a single nucleotide can be reliably scored. Following genotyping, all alleles are coded with unique identifiers; the composite string of alleles defines the microsatellite type (MT) of the isolate. Therefore, the MT of an isolate is analogous to its ST and can be analysed in much the same manner.

The main motivation for developing an MLMT scheme is to uncover genetic variation within closely related populations. An MLST scheme for

Coccidioides immitis showed that all 14 isolates recovered from sources in South America were of the same ST. However, typing these isolates with a nine-locus MLMT system showed that 10 out of the 14 isolates had unique MTs, demonstrating the increased resolution that is made available by MLMT.

Similarly, analysis of 4955 nucleotides of the HIV-associated mycosis *Penicillium marneffei* in Thailand for four genes (*StlA*, *StuA*, *GasB*, *GasC*) showed that only seven nucleotides were polymorphic, corresponding to 0.141% sequence diversity (M. C. Fisher, unpublished observations). In this species, in Thailand, MLST is not sufficiently discriminatory to be used as a technique. Therefore, a MLMT scheme was developed, based on the amplification of 21 microsatellite-containing loci (Fisher *et al.*, 2004b). To minimize confusion between MLST and MLMT schemes, a separate website has been created to house MLMT schemes and can be accessed at http://www.multilocus.net/. Analysis of 21 isolates from the Centraalbureau voor Schimmelculture culture collection showed the existence of 19 unique MTs (http://pmarneffei.multilocus.net/). These were clustered into an 'eastern' and a 'western' clade (Fisher *et al.*, 2004a), demonstrating strong geographical structuring of the pathogen, a feature that is shared with *Coccidioides* spp. It is so far not known why endemic mycoses exhibit such highly genetically heterogenous distributions; however, it is thought that local adaptation of isolates may occur, therefore decreasing the probability of survival following aerial dispersal.

Considerations when using MLMT schemes

Despite the considerable advantages of using microsatellite-based systems owing to their high resolution, using microsatellites carries an associated penalty. Because of their stepwise mode of mutation (SMM), high mutation rates and constraints on the maximum, and minium, size of the microsatellite array, identically sized alleles can evolve. Therefore the reappearance of alleles that were previously lost from populations is theoretically possible and will counteract the effects of genetic drift by creating alleles that are identical in size but are not identical by descent (Hancock, 1999). The outcome of this is that microsatellite data sets inevitably have a degree of phylogenetic 'noise', termed homoplasy, where unrelated isolates can share an identical allele. This effect would therefore give a phylogenetic signal similar to that of a recombinational exchange between two distantly related isolates. An associated consideration is that, owing to constraints on the sizes of the repeat array, genetic distances at microsatellite loci reach an asymptote more rapidly than loci that obey an infinite-alleles model of mutation (Nauta & Weissing, 1996). Despite theory showing that genetic distances based on the SMM can maintain linearity over millions of years (Goldstein *et al.*, 1995), a comparative study between microsatellites and their flanking sequence genealogies in *C. immitis* and *C. posadasii* showed that microsatellites greatly under-estimated the split between the two species lineages by an order of magnitude (flanking sequence estimate, 12 800 000 years; microsatellite estimate, 760 000 years); (Fisher *et al.*, 2000). However, for populations that had only recently diverged from one another, dating with microsatellites and flanking sequence genealogies gave the same answer, *c.* 40 000 years (Fisher *et al.*, 2000). This analysis shows that, at this reduced temporal scale, genetic divergence is occuring in a linear manner for the microsatellite loci.

Although these problems are an unavoidable fact of MLMT schemes, as long as researchers are aware of them then they can be taken into consideration and analyses adjusted accordingly. For instance, to compensate for the effect of a certain proportion of alleles being identical by size, but not by descent, the solution is to use more loci. Construction of organismal phylogenies for *Coccidioides* spp. using both gene genealogies and microsatellite data showed that both methods produced congruent genealogies (Fisher *et al.*, 2000). In this study, the effects of homoplasy at the microsatellite loci did not impair the MLMT scheme's ability to recover the correct phylogenetic relationships, once enough loci were used.

Bioinformatic tools

Once a research community has decided upon a method and an appropriate set of loci, then there are a number of bioinformatic tools that can be used to analyse MLST and MLMT data.

Phylogenies and eBURST

Most fungi have complex life cycles that incorporate a mixture of meiotic and mitotic stages. More than half of the ascomycetes are mitosporic, and the meiosporic species exibit a continuum of homothallic and heterothallic behaviour. Therefore an approporiate model of fungal evolution needs to take into account these two reproductive modes (Taylor *et al.*, 1999a). Relationships between isolates can be analysed within a phylogenetic framework, by using distance, parsimony or maximum likelihood methodologies (Burt *et al.*, 1996; Koufopanou *et al.*,

1997). A consideration here is that, for species that exhibit even low levels of recombination, different gene genealogies will be incongruent and, if combined, will result in trees that have very low statistical support. This has the effect of obscuring relationships between groups, because the signal of recombination overwhelms the signal generated by clonal descent. However, MTs and STs can be usefully analysed within phylogenetic packages (such as PAUP*, MrBAYES and PHYLIP), bearing in mind these caveats.

The simplest model for the evolution of a fungal species that exhibits a mixture of mitosporic and meiosporic reproduction is that clonal complexes will emerge from a founding genotype and increase in frequency in the population as a consequence of genetic drift or a fitness advantage. Initially, variants of the founding genotype will be identical; however, variation will be introduced by either mutation or recombination, resulting in the generation of derived genotypes. These derived genotypes will have allelic profiles that differ from the founding genotype at only one of the MLST or MLMT loci, and are called single-locus variants, SLVs (Maiden *et al.*, 1998). Further mutation, or recombination, will cause diversification into double-locus variants (DLVs), then triple-locus variants (TLVs), etc. Although the resulting clonal complexes can be depicted by using dendrograms, relationships between isolates are arbitrary and there is little information on the identity of the founding genotype. To resolve this, Feil *et al.* (2004) designed a bioinformatics algorithm, termed eBURST. This program was designed to illuminate the short-term evolutionary history of a species by identifying non-overlapping groups of closely related STs (or MTs), termed clonal complexes. The program subsequently identifies the predicted founder of the clonal complex and then ascertains the most parsimonious line of descent of isolates from the founder. Therefore, eBURST is designed to specifically ignore all deep evolutionary structure within the data, and to focus on clonal diversification over short evolutionary time scales.

Analysis of the *C. albicans* MLST data set showed that there are 10 clonal complexes linking the 98 STs, and 146 singletons with no immediate SLV relationships. These relationships between isolates can be visualized by the use of 'stars at night' plots, where clonal complexes are linked by lines, and which provide a snapshot of the MLST database (Fig. 16.2). Once clonal complexes have been identified, further research can then focus on the specific mechanisms that may have caused particular clonal complexes to increase in frequency and dominate a population; mechanisms that are of interest are the relative fitnesses of the founding clones and the impact of neutral processes on population diversification (Fraser *et al.*, 2005).

The use of eBURST in conjunction with MLMT as an epidemiological tool is ably demonstrated by recent work on the host range of *P. marneffei.* This mycosis is found to infect a large proportion of bamboo rat species within its endemic range, and it has long been debated whether or not isolates that infect bamboo rats are infectious to humans. Data from case-controlled studies suggest that bamboo rat exposure is not a significant risk factor and indicates exposure to dust in a rural environment as a source of infectious spores (Chariyalertsak *et al.*, 1997). However, to rule out bamboo rats as a potential source of infection it is necessary to show that the multilocus genotypes of *P. marneffei* from bamboo rats and humans are mutually exclusive. To this end, MLMT was used to genotype *P. marneffei* isolates from bamboo rats and a patient with AIDS in the Manipur district of Assam, India; the relationships between MTs from this

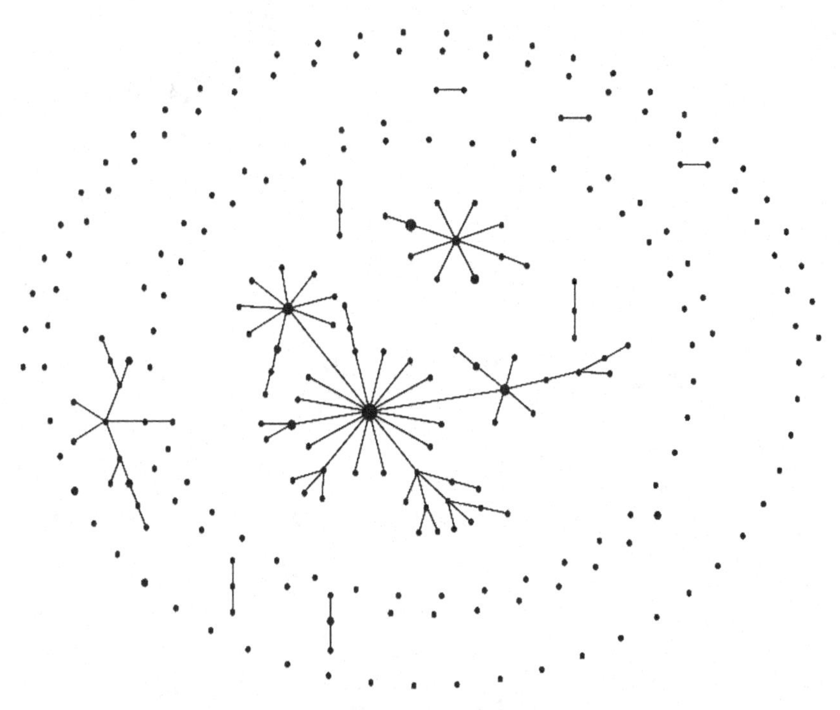

Fig. 16.2. Snapshot of the *Candida albicans* MLST database shown by a 'stars at night' plot, generated by eBURST. Clusters of linked isolates correspond to clonal complexes with the founders positioned centrally within the clusters. Unlinked isolates are known as 'singletons' with no known SLV relatives.

analysis are shown in Fig. 16.3. Analysis of the eBURST plot clearly shows that one of the MTs recovered from a Bay bamboo rat, *Cannomys badius*, is identical at all 21 loci to that recovered from the AIDS patient and provides strong evidence that rodent-associated *P. marneffei* can disperse, and cause disease, in humans. This finding has recently been corroborated in a Thailand study using a much larger cohort of both rodent and human isolates (M. C. Fisher, personal observations).

Multilocus typing and environmental mycology

MLST and MLMT have been developed primarily as molecular tools to investigate the epidemiology of infectious diseases. However, there is no reason why these approaches and techniques are not entirely relevant to studies on environmental fungi. For instance, Johnson *et al.* (2004) surveyed a 10 km^2 sampling area of southern England for wild isolates of

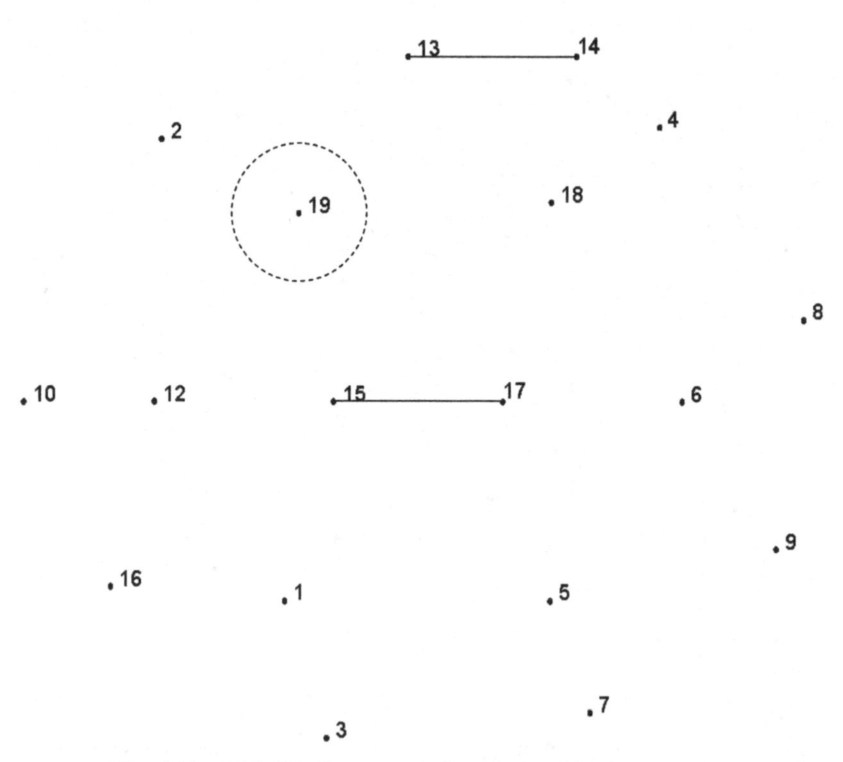

Fig. 16.3. eBURST diagram of the relationships between *P. marneffei* MTs recovered from bamboo rats in Manipur, India. The single MT (MT 19) that has been observed to co-infect humans as well as bamboo rats is circled by a dotted line.

Saccharomyces paradoxus. The resulting samples were sequenced at seven loci and the data used to show that clonal replication, inbreeding and outbreeding have all contributed to the population genetic structure of this species. Similarly, an extensive survey of the lichenized ascomycete *Letharia* 'gracilis' and *L.* 'lupina' in North America by using a 12-locus MLST scheme showed that whereas *L.* 'lupina' reproduces principally by asexual structures (soredia) the genetic structure of this species is as recombined in nature as the sexually reproducing *L.* 'gracilis' sister taxon (Kroken & Taylor, 2001).

Conclusions

By their very nature, MLST and MLMT promote an 'open-source' approach to fungal population genetics and epidemiology by ensuring that data are produced in a manner that can be accessed by all, and be developed to include further isolates as needed. The centralization of resources, and standardization of techniques, makes international collaborations simple; huge resources can be generated as a result. For instance, the current (2005) sizes of the principal bacterial databases are as follows: *Neisseria meningitidis*, 6414 isolates; *Staphylococcus aureus*, 1082 isolates; and *Streptococcus pneumoniae*, 2681 isolates. The creation of such data sets provides hugely powerful tools to analyse the evolutionary genomics of species, and to shed much-needed light on the processes that generate, and maintain, biodiversity in nature.

References

Bougnoux, M. E., Morand, S. & d'Enfert, C. (2002). Usefulness of multilocus sequence typing for characterization of clinical isolates of *Candida albicans*. *Journal of Clinical Microbiology* **40**, 1290–7.

Burt, A., Carter, D. A., Koenig, G. L., White, T. J. & Taylor, J. W. (1996). Molecular markers reveal cryptic sex in the human pathogen *Coccidioides immitis*. *Proceedings of the National Academy of Sciences of the USA* **93**, 770–3.

Chariyalertsak, S., Sirisanthana, T., Supparatpinyo, K., Praparattanapan, J. & Nelson, K. E. (1997). Case-control study of risk factors for *Penicillium marneffei* infection in human immunodeficiency virus-infected patients in northern Thailand. *Clinical Infectious Diseases* **24**, 1080–6.

Dodgson, A. R., Pujol, C., Denning, D. W., Soll, D. R. & Fox, A. J. (2003). Multilocus sequence typing of *Candida glabrata* reveals geographically enriched clades. *Journal of Clinical Microbiology* **41**, 5709–17.

Enright, M. C. & Spratt, B. G. (1999). Multilocus sequence typing. *Trends in Microbiology* **7**, 482–7.

Feil, E. J. & Spratt, B. G. (2001). Recombination and the population structures of bacterial pathogens. *Annual Review of Microbiology* **55**, 561–90.

Feil, E. J., Li, B. C., Aanensen, D. M., Hanage, W. P. & Spratt, B. G. (2004). eBURST: inferring patterns of evolutionary descent among clusters of related bacterial genotypes from multilocus sequence typing data. *Journal of Bacteriology* **186**, 1518–30.

Fisher, M. C., Koenig, G., White, T. J. & Taylor, J. W. (2000). A test for concordance between the multilocus genealogies of genes and microsatellites in the pathogenic fungus *Coccidioides immitis*. *Molecular Biology and Evolution* **17**, 1164–74.

Fisher, M. C., Koenig, G. L., White, T. J., San- Blas, G., Negroni, R., Alvarez, I. G., Wanke, B. & Taylor, J. W. (2001). Biogeographic range expansion into South America by *Coccidioides immitis* mirrors New World patterns of human migration. *Proceedings of the National Academy of Sciences of the USA* **98**, 4558–62.

Fisher, M., Koenig, G., White, T. & Taylor, J. (2002). Molecular and phenotypic description of *Coccidioides posadasii* sp. nov., previously recognized as the non-California population of *Coccidioides immitis*. *Mycologia* **94**, 73–84.

Fisher, M. C., Aanensen, D., de Hoog, S. & Vanittanakom, N. (2004a). Multilocus microsatellite typing system for *Penicillium marneffei* reveals spatially structured populations. *Journal of Clinical Microbiology* **42**, 5065–9.

Fisher, M. C., de Hoog, G. S. & Vannittanakom, N. (2004b). A highly discriminatory Multilocus Microsatellite Typing System (MLMT) for *Penicillium marneffei*. *Molecular Ecology Notes* **5**, 231–4.

Fraser, C., Hanage, W. P. & Spratt, B. G. (2005). Neutral microepidemic evolution of bacterial pathogens. *Proceedings of the National Academy of Sciences of the USA* **102**, 1968–73.

Goldstein, D. B., Ruiz Linares, A., Cavalli-Sforza, L. L. & Feldman, M. W. (1995). An evaluation of genetic distances for use with microsatellite loci. *Genetics* **139**, 463–71.

Hancock, J. M. (1999). Microsatellites and other simple sequences: genomic context and mutational mechanisms. In *Microsatellites*, ed. D. B. Goldstein & C. Schlotterer, pp. 1–6. Oxford: Oxford University Press.

Henderson, S. T. & Petes, T. D. (1992). Instability of simple sequence DNA in *Saccharomyces cerevisiae*. *Molecular and Cellular Biology* **12**, 2749–57.

Johnson, L. J., Koufopanou, V., Goddard, M. R., Hetherington, R., Schafer, S. M. & Burt, A. (2004). Population genetics of the wild yeast *Saccharomyces paradoxus*. *Genetics* **166**, 43–52.

Kasuga, T., Taylor, J. W. & White, T. J. (1999). Phylogenetic relationships of varieties and geographical groups of the human pathogenic fungus, *Histoplasma capsulatum* Darling. *Journal of Clinical Microbiology* **37**, 653–63.

Kasuga, T., White, T. J., Koenig, G., McEwen, J., Restrepo, A., Castaneda, E., Da Silva Lacaz, C., Heins-Vaccari, E. M., De Freitas, R. S., Zancope-Oliveira, R. M., Qin, Z., Negroni, R., Carter, D. A., Mikami, Y., Tamura, M., Taylor, M. L., Miller, G. F., Poonwan, N. & Taylor, J. W. (2003). Phylogenetic relationships of varieties and geographical groups of the human pathogenic fungus *Histoplasma capsulatum* Darling. *Molecular Ecology* **12**, 3383–401.

Koufopanou, V., Burt, A. & Taylor, J. W. (1997). Concordance of gene genealogies reveals reproductive isolation in the pathogenic fungus *Coccidioides immitis*. *Proceedings of the National Academy of Sciences of the USA* **94**, 5478–82.

Koufopanou, V., Burt, A. & Taylor, J. W. (1998). Concordance of gene genealogies reveals reproductive isolation in the pathogenic fungus *Coccidioides immitis*. *Proceedings of the National Academy of Sciences of the USA* **95**, 8414.

Kroken, S. & Taylor, J. W. (2001). Outcrossing and recombination in the lichenized fungus *Letharia*. *Fungal Genetics and Biology* **34**, 83–92.

Lockhart, S. R., Joly, S., Pujol, C., Sobel, J. D., Pfaller, M. A. & Soll, D. R. (1997). Development and verification of fingerprinting probes for *Candida glabrata*. *Microbiology* **143**, 3733–46.

Maiden, M. C., Bygraves, J. A., Feil, E., Morelli, G., Russell, J. E., Urwin, R., Zhang, Q., Zhou, J., Zurth, K., Caugant, D. A., Feavers, I. M., Achtman, M. & Spratt, B. G. (1998). Multilocus sequence typing: a portable approach to the identification of clones within populations of pathogenic microorganisms. *Proceedings of the National Academy of Sciences of the USA* **95**, 3140–5.

McEwen, J. G., Taylor, J. W., Carter, D., Xu, J., Felipe, M. S., Vilgalys, R., Mitchell, T. G., Kasuga, T., White, T., Bui, T. & Soares, C. M. (2000). Molecular typing of pathogenic fungi. *Medical Mycology*, **38** (suppl. 1), 189–97.

Morehouse, E. A., James, T. Y., Ganley, A. R., Vilgalys, R., Berger, L., Murphy, P. J. & Longcore, J. E. (2003). Multilocus sequence typing suggests the chytrid pathogen of amphibians is a recently emerged clone. *Molecular Ecology* **12**, 395–403.

Nauta, M. J. & Weissing, F. J. (1996). Constraints on allele size at microsatellite loci: implications for genetic differentiation. *Genetics* **143**, 1021–32.

O'Donnell, K., Ward, T. J., Geiser, D. M., Corby Kistler, H. & Aoki, T. (2004). Genealogical concordance between the mating type locus and seven other nuclear genes supports formal recognition of nine phylogenetically distinct species within the *Fusarium graminearum* clade. *Fungal Genetics and Biology* **41**, 600–23.

Tavanti, A., Gow, N. A., Senesi, S., Maiden, M. C. & Odds, F. C. (2003). Optimization and validation of multilocus sequence typing for *Candida albicans*. *Journal of Clinical Microbiology* **41**, 3765–76.

Taylor, J., Jacobson, D. & Fisher, M. (1999a). The evolution of asexual fungi: reproduction, speciation and classification. *Annual Review of Phytopathology* **37**, 197–246.

Taylor, J. W. & Fisher, M. C. (2003). Fungal multilocus sequence typing – it's not just for bacteria. *Current Opinions in Microbiology* **6**, 351–6.

Taylor, J. W., Geiser, D. M., Burt, A. & Koufopanou, V. (1999b). The evolutionary biology and population genetics underlying fungal strain typing. *Clinical Microbiological Reviews* **12**, 126–46.

Weber, J. L. & Wong, C. (1993). Mutation of human short tandem repeats. *Human Molecular Genetics* **2**, 1123–8.

Xu, J., Vilgalys, R. & Mitchell, T. G. (2000). Multiple gene genealogies reveal recent dispersion and hybridization in the human pathogenic fungus *Cryptococcus neoformans*. *Molecular Ecology* **9**, 1471–81.

VI

Molecular ecology of fungi in the environment

17

Fungi in the hidden environment: the gut of beetles

MEREDITH BLACKWELL, SUNG-OUI SUH

Department of Biological Sciences, Louisiana State University

JAMES B. NARDI

Department of Entomology, University of Illinois

Introduction

Over the past century the recognition of the presence of endosymbionts in a variety of arthropods has become well established (Buchner, 1965). Intense interest in the rickettsial endosymbionts, widespread among insects (van Meer *et al.*, 1999), led to the discovery that they may induce sterility of the host, and increased rates of speciation have been attributed to their presence (Shoemaker *et al.*, 1999). Bacteria also have long been known for their nutritional contributions to insects, but more recently indigenous gut bacteria have been recognized for their ability to prevent colonization of non-indigenous microbes. In fact the insect gut is considered a 'hot spot' of bacterial gene exchange and bacterial adaptation (Dillon & Dillon, 2004). Thus, important attributes that affect speciation, habitat utilization, and survival are provided by prokaryotic symbionts. By contrast, although there were a number of early reports of fungal endosymbionts of insects, fewer of them were substantiated after the original reports (Buchner, 1965). More recent work, however, indicates that insect–yeast interactions abound in nature, although the exact nature of many of the interactions is less well understood (Suh & Blackwell, 2005; Vega & Dowd, 2005).

Yeasts and yeast-like fungi from the guts of a small group of planthoppers (Homoptera) and beetles in three families (Coleoptera: Anobiidae, Cerambycidae and Scolytidae) have been studied most extensively. These fungi have a yeast growth form with single cells and asexual reproduction by budding, the hallmark of the 'yeast habit', although many yeasts actually have filamentous growth as well. The so-called true yeasts lack fruiting bodies and constitute a monophyletic group (Phylum Ascomycota: Subphylum Saccharomycotina). In addition to the true yeasts, several other

Fungi in the Environment, ed. G. M. Gadd, S. C. Watkinson & P. S. Dyer. Published by Cambridge University Press.

ascomycetes, including *Symbiotaphrina* and unnamed planthopper yeast-like endosymbionts (YLSs) (Sub-phylum Pezizomycotina), have secondarily assumed the yeast habit (Jones & Blackwell, 1996; Jones *et al.*, 1999; Suh *et al.*, 2001). Some eukaryotic endosymbionts play important roles in insect evolution; the examples of *Symbiotaphrina* in detoxification of food resources for anobiid beetles (Dowd, 1989, 1991), and the contribution of YLSs to planthopper nutrition (Noda & Koizumi, 2003) support this contention. Besides numerous other examples of presumed endosymbiotic relationships including termite associations with yeasts (Prillinger *et al.*, 1996; Schäfer *et al.*, 1996; Vega & Dowd, 2005), another yeast system, although not endosymbiotic, the cactophilic yeast–*Drosophila* association studied by Starmer and his colleagues (Starmer *et al.*, 1991), provides an additional example of the influence that yeasts have in ameliorating the nutritional resources of insects.

Our study of gut yeasts began with a preliminary investigation of the occurrence of the fungi in the gut of basidioma-feeding beetles. There were several reasons for the selection of yeasts in the beetle gut habitat, including (1) previous reports of ballistosporic yeasts in mushrooms (Prillinger, 1987), (2) the exclusive use of basidiomata as nutritional resources by a wide range of beetles that could be targeted for re-sampling, and (3) the availability of a dense database of large-subunit ribosomal RNA gene sequences (rDNA) that includes all known yeasts (Kurtzman & Robnett, 1998). Furthermore, the research focus of the participants was already on fungi associated with insects, particularly mushroom-feeding beetles infected with Laboulbeniales, the objects of another study (Weir & Blackwell, 2001). It was economical, therefore, to investigate the same beetles for the presence of yeasts.

Mushroom-feeding beetles

Initially, the study was designed to discover new yeast species and reveal associations between yeasts and the beetles they inhabited. In addition there were many other questions that had bearing on determining the degree of the interactions between the organisms: Is there specificity in the yeast–insect association? Have arthropod-associated yeasts diverged many times or are they acquired repeatedly from the habitat? Do distantly related beetles using the same group of basidiomata as nutritional resources have similar yeasts? Lawrence (1973, 1989) suggested that certain groups of ciid beetles utilized a restricted selection of basidiomata, and he correlated the groups of basidiomata by 'hardness' and texture. Could such a host distribution be due to the presence, not only of specialized mouthparts, but also of functionally similar yeasts to provide digestive enzymes? Do predaceous beetles that feed on fungus-feeding arthropods in the basidiocarp habitat also contain gut yeasts? How widespread are yeast associations within beetle clades? Most important, are yeasts involved in rapid radiation of beetle lineages because they perform an essential service for the beetles, or, conversely, do beetle lineages diverge rapidly when freed of gut yeasts? Is it possible that only the yeast benefits from such associations as it is bathed in a soup of undigested nutrients? Later, questions were asked about whether mycophagous beetles more commonly associate with gut yeasts than other groups of beetles or not, and whether microbes that cannot be cultured also were present in the gut of beetles.

Yeasts from the gut of beetles: species, specificity and expansion of clades

Several years ago, a proposal to fund the isolation of yeasts from the gut of beetles was justified by predicting that more than 50 undescribed species would be discovered. The results, however, have far surpassed those high expectations with significant increases in unexpected microbial species diversity (Boekhout, 2005; Suh *et al.*, 2005). The major findings on beetles isolated primarily from the basidioma habitat are summarized below.

- Six hundred and fifty yeasts were isolated from the digestive tracts of more than 90% of all of the beetles dissected (Boekhout, 2005; Suh *et al.*, 2005).
- Two hundred and ninety yeast genotypes representing more than 200 undescribed taxa were discovered among the isolates (Suh & Blackwell, 2005; Suh *et al.*, 2005).
- Insect gut yeasts were distributed in clusters throughout the yeast phylogenetic tree based on analysis of the small subunit (SSU) and large subunit (LSU) rDNA (Suh *et al.*, 2005).
- Several previously unknown, entirely insect-associated yeast clades have been discovered, such as the *Candida tanzawaensis* clade, which contained 30% of all the gut yeasts collected (Suh *et al.*, 2004b).
- One dominant yeast usually was present in a beetle gut, based on cloning methods (Zhang *et al.*, 2003).
- Yeast–beetle specificity was observed between certain yeasts and beetles across broad geographical ranges and multiple developmental stages of some beetles (Suh *et al.*, 2004b).

- Because almost all dissected beetles in Cucujoidea and Tenebrionoidea bore yeasts, untapped beetle diversity lends support to high estimates of undiscovered yeasts (Suh *et al.*, 2004b, 2005; Boekhout, 2005).
- The discovery of almost 200 undescribed yeasts gains greater significance with the realization that fewer than 700 species of ascomycete yeast have been described previously from all of the Earth's habitats (Boekhout, 2005).

Additional questions

Prompted by the preliminary findings on beetles collected from basidiomata, we became interested in asking two additional questions about insect gut fungi: Are all gut yeasts recovered in culture? Do yeasts occur in the gut of beetles that rely on nutritional resources other than fungi? Gene cloning (Zhang *et al.*, 2003) and microscopy techniques (Nardi *et al.*, 2006) were used to address these questions.

In attempts to isolate gut yeasts on yeast malt agar, we typically recovered large numbers of colony-forming units of a single species. When a second species of yeast was present it was usually recovered in low numbers, indicating that rare yeasts might have been overlooked in the cultures. In addition, it was considered possible that microbes that cannot be cultured by using our methods might be present. In a search for such nonculturable microbes from the gut of beetles, LSU and SSU rDNA genes were cloned from the beetle guts (Zhang *et al.*, 2003) and were compared with cloned DNAs by restriction enzyme patterns and sequencing. Clones were sequenced from the LSU rDNA clone libraries of each of five beetle species collected in the Baton Rouge, Louisiana. The beetles were *Triplax* sp. (Erotylidae, ex *Pleurotus* sp.), *Platydema* sp. (Tenebrionidae, ex *Ganoderma* sp.), *Neomida* sp. (Tenebrionidae, ex *Ganoderma* sp. and *Fomitella supina*), *Ceracis curtus* (Ciidae, ex *Fomitella supina*), and *Odontotaenius disjunctus* (Passalidae, ex rotten log) (Table 17.1). The yeasts cloned from the gut of several species indicated that our culture methods recovered most ascomycete yeasts. For example, a yeast near *C. ambrosiae* was cloned from the gut of both *Triplax* sp. and *Platyderma* sp., and similar yeasts had been isolated in culture from both beetle species. Although previously we had obtained a few yeast isolates from the gut of *Ceracis curtus*, no yeasts were cloned from that species; our number of attempts, however, was low. The gut of the wood-ingesting beetle *Odontotaenius disjunctus* was of special interest when it was

Table 17.1. *Closest known taxa to selected sequenced clones of LSU rDNA from BLAST searches*

Highlighted taxa are similar to cultured yeasts recovered from the same beetle species. Filamentous fungi were observed in cultural studies but not studied further because they appeared to be substrate contaminants ingested by the beetles. The basidiomycete yeasts *Malassezia* spp. were common in sequenced clones but were not isolated in cultural studies, most likely because special media were not used. Several protist (*) and insect (**) sequences were recovered as well as a microsporidian sequence. No identical sequence matches were discovered, and some sequences are significantly different from the closest sequence in the BLAST searches

Beetle	Closest known GenBank sequence to clone sequences
Erotylidae: *Triplax*	***Candida ambrosiae*** (Saccharomycotina) *Malassezia restricta* (Ustilaginomycetes)
Tenebrionidae: *Platydema* sp.	***Candida ambrosiae*** (Saccharomycotina)
Tenebrionidae: *Neomida* sp.	***Candida pyralidae*** (Saccharomycotina) ***Blastobotrys elegans*** (Saccharomycotina) ***Stephanoascus farinosus*** (Saccharomycotina) ***Zygozyma smithiae*** (Saccharomycotina) *Cercophora newfieldiana* (Pezizomycotina) *Capronia coronata* (Pezizomycotina) *Malassezia restricta* (Ustilaginomycetes) *Nosema* sp. (Microsporidia)
Ciidae: *Ceracis curtus*	(no yeasts cloned, although two species were cultured) *Capronia coronata* (Pezizomycotina) *Dissophora decumbens* (Mortierellales) *Gromia oviformis* (Cercozoa*) *Leptocarabus truncaticollis* (Carabidae**)
Passalidae: *Odontotaenius disjunctus*	***Pichia stipitis*** (Saccharomycetales) *Rhodotorula* sp. CBS 8885 (Sporidiobolales) *Chalara* sp. CL157 (Pezizomycotina) *Pentatrichomonas hominis* (Parabasalida*) *Carpophilus* sp. MAL-2003 (Nitidulidae**)

Source: After Zhang *et al.* (2003).

compared with that of the fungus-feeding beetles. The wood-ingesting beetle consistently yielded clones of a yeast similar to *Pichia stipitis* (Suh *et al.*, 2003) (Table 17.1). It is important to note that there were no sequences identical to those we cloned; the sequences discovered in BLAST (Altschul *et al.*, 1990) searches were always somewhat different and in several cases of very low similarity to known sequences. This result indicates that DNA databases are still low in numbers of many microbial sequences.

Pichia stipitis clade: associations with *Odontotaenius disjunctus* (Passalidae) and other wood-boring beetles

In comparing gut microbes of mushroom-feeding beetles with those using another nutritional resource, we examined the gut inhabitants of the widespread wood-boring beetle *Odontotaenius disjunctus* (Passalidae) by using not only cultures and gene cloning but also microscopy. Although we had discovered a high diversity of undescribed yeasts in the gut of fungus-feeding beetles, there was low diversity of other organisms. Conversely, the gut yeast diversity of passalid beetles is low and previously reported (Lichtwardt *et al.*, 1999; Suh *et al.*, 2003, 2004c); the gut community of passalids, however, appears to be overall more diverse and distinct from that of the fungus-feeding beetles.

The presence of physiologically different yeasts in the wood-ingesting beetle was expected because of the differences between the nutritional resources; a single yeast taxon was isolated from the gut of more than 400 individuals of *O. disjunctus* that were examined. This yeast was similar to *Pichia stipitis*, a xylose-fermenting and -assimilating yeast, rare attributes among all known species of the Saccharomycetes (Jeffries & Kurtzman, 1994) (Fig. 17.1). Other members of the *P. stipitis* clade also have been isolated from associations with wood-boring beetles (Table 17.2). In addition to the *P. stipitis* clade of about six taxa, only five relatively distantly related taxa outside of the *P. stipitis* clade are known to ferment and assimilate xylose. There is a high degree of correlation between the occurrence of certain yeasts that process and utilize xylose and wood-ingesting beetles in several distantly related families; this finding is significant because xylose subunits form the backbone of the hemicellulose component of the plant cell walls making up wood. It is important to note, however, that not all yeasts associated with wood-boring beetles have xylose fermentation and assimilation properties.

Compartmentalization of microbes in the hindgut of *Odontotaenius disjunctus*

The gut of insects comprises three main divisions: the foregut, midgut and hindgut. In many insects that we have examined the foregut and midgut contain yeasts. For example, certain species of green lacewing in the genus *Chrysoperla* (Neuroptera: Chrysopidae) have several closely related yeast taxa in the crop, a chamber of the foregut. These yeasts form

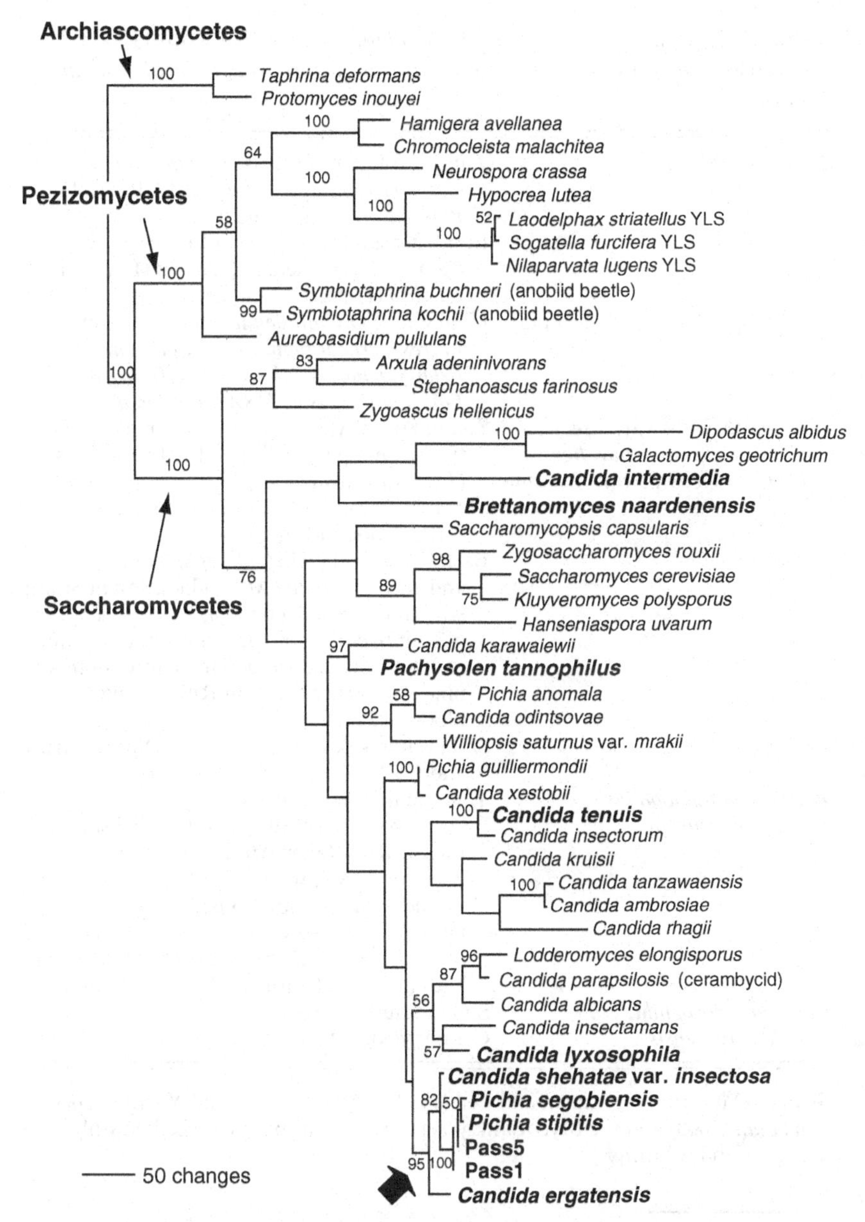

Fig. 17.1. Most parsimonious tree based on maximum parsimony analysis of SSU and LSU rDNA (2400 bp) depicts relationships of all known yeasts with the rare attributes of xylose fermentation and assimilation or xylan hydrolysis in the case of *Candida ergatensis* (**bold typeface**). Among xylose-fermenting yeasts only the *Pichia stipitis* clade (block arrow) and *C. tenuis* have been isolated from associations with

Table 17.2. *Xylose-fermenting and -assimilating yeasts (in* Candida ergatensis, *hydrolysing xylan) isolated from natural and several clinical isolations*

Pichia segobiensis	Tunnel of larva of *Calcophora mariana massiliensis* (Buprestidae) under bark of *Pinus silvestris* in Spain
Pichia stipitis	Insect larva on fruit trees in France; *Cetonia* (Scarabaeidae); *Dorcus* (Lucanidae); *Laphria* (Diptera: Asilidae)
Pichia stipitis-like (Pass 1 and 5)	Hind gut of *Odontotaenius disjunctus* and *Verres sternbergianus (Passalidae) in Pennsylvania, South Carolina, Georgia, Louisiana, Kansas, USA, and Panama*
Candida shehatae var. *insectosa*, *Candida shehatae* var. *lignose*, *Candida shehatae* var. *shehatae*	Soil in South Africa; dead insect-invaded pine tree; rose hip in British Columbia; beetles (*Leptura maculicornis, L. cerambyciformis, Spondylis buprestoides* (Cerambycidae)); rotten wood in Chile
Candida tenuis	Bark beetle in the USA; *Rhagium bifasciatum* and *R. sycophanta* (Anobiidae); soil in South Africa; larva of beetle *Harpium mordax* (Cerambycidae); in bark of birch tree; liquid sweetener in The Netherlands; litter of mixed pine and spruce forest in Russia; human blood in Peru
Brettanomyces naardenensis	Soda water, soft drinks in Europe, South Africa and USA
Pachysolen tannophilus	Tanning fluid in France
Candida intermedia	Feces of patient infected with tropical sprue, Puerto Rico; fallen *Musanga smithii* (Cecropiaceae); washed beer bottles in Sweden; human isolations, Norway, Germany; sea water off coast of USA; grape in Brazil; soil and beer in South Africa; brine bath in cheese factory in The Netherlands
Candida lyxosophila	Soil in South Africa
Candida ergatensis	Cerambycidae, Spain, Peru

Source: Data from Kurtzman and Fell (1998), Barnett, Payne and Yarrow (2000), Suh *et al.*, (2003) and the MycoBank yeast database (http://www.cbs.knaw.nl/databases/index.htm).

Fig. 17.1. (cont.)
wood-boring beetles (*). Tree length = 3589; consistency index (CI) = 0.4266; homoplasy index (HI) = 0.5734; CI excluding uninformative characters = 0.3891; HI excluding uninformative characters = 0.6109; retention index (RI) = 0.6450; and rescaled consistency index (RC) = 0.2751. YLS = yeast-like endosymbiont. Pass 1 and 5 genotypes from *O. disjunctus*. (After Suh *et al.*, 2003).

a small clade of closely related taxa that have a broad geographic range across the southern USA from Arizona to northeastern Mississippi (Woolfolk & Inglis, 2003; Suh *et al.*, 2004a; Nhu *et al.*, 2006).

When the position of the yeasts in the gut of fungus-eating beetles is known, they have been present in the midgut, sometimes localized in pockets at the anterior end of the midgut, such as the gastric caeca of *Megalodacne heros* (McHugh *et al.*, 1997). The yeasts of passalid beetles, however, occur in a different location. They are positioned along a furrow extending the length of the relatively undifferentiated posterior two-thirds of the hindgut (Fig. 17.2) (Lichtwardt *et al.*, 1999; Suh *et al.*, 2003, 2004c; Nardi *et al.*, 2006). In addition to its hindgut localization, the *P. stipitis*-like yeast is unusual for its physiological competence in handling xylose, and it is one of the few yeasts known to have a holdfast structure by which its branched filaments are attached to the gut cuticle. It is of interest that many of the microbes of wood-ingesting detritivores such as termites and wood roaches are also restricted to the hindgut.

Comparison with the adult hindgut of *O. disjunctus* reveals that the hindgut of the larva is much smaller and less differentiated morphologically. In addition, the larval hindgut is sparsely populated by yeasts and bacteria compared with that of adults, perhaps a factor involved in the evolution of parental care in which the adults feed a mixture of frass and chewed wood to the larvae (Nardi *et al.*, 2006). Several earlier studies reviewed the life history and associations of the passalid beetles. Adults occupy the large galleries in white-rotted logs. They feed the larvae that share the galleries with a mixture of chewed wood and frass, which is often plastered on the gallery walls (Pearse *et al.*, 1936; Gray, 1946). We have isolated yeasts as well as several species of bacterium from the material on the gallery walls; this material may be an initial source of microbes for the larvae. Because the hindgut cuticle is shed when the beetle moults, the microbes must be replenished between larval moults and after the metamorphosis of the non-feeding pupa to the adult.

To date we have looked only at the hindgut of *O. disjunctus* among all wood-boring beetles and found it rich in microbial diversity. In addition to the *P. stipitis*-like yeasts occupying the posterior part of the hindgut, other microbes form a complex, albeit highly compartmentalized, hindgut community. Eccrinid trichomycetes and bacteria have been reported from the gut of passalid beetles for more than 100 years (Leidy, 1849, 1853; Lichtwardt *et al.*, 1999). In addition to these organisms, we have discovered amoeboid protists and parabasalids in differentiated hindgut

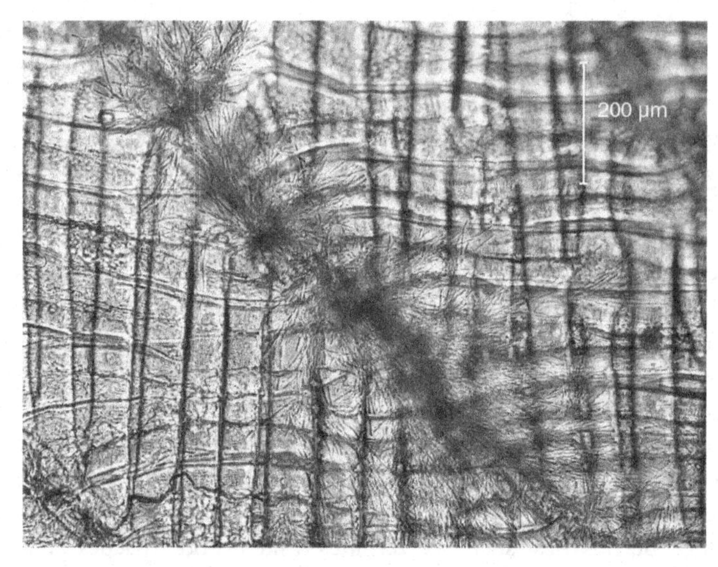

Fig. 17.2. *Pichia stipitis*-like yeast attached to a furrow in the lumen of the posterior two-thirds of the hindgut of *O. disjunctus* collected in Baton Rouge, Louisiana. The filamentous yeast is attached by a holdfast when observed *in situ*.

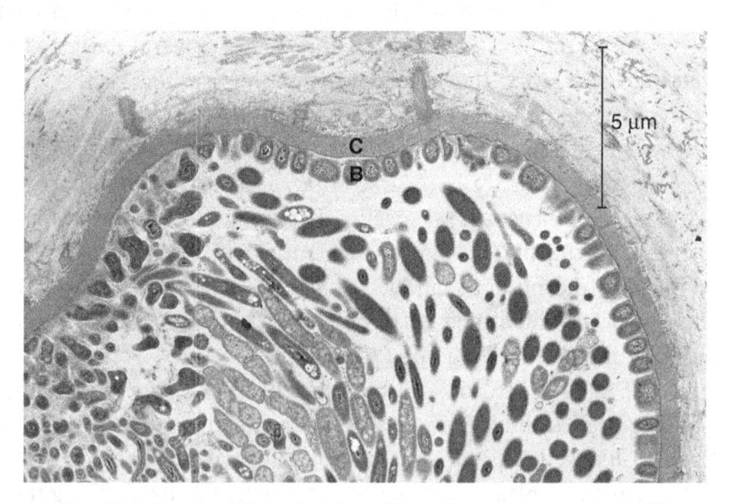

Fig. 17.3. Surface film of unidentified filamentous bacteria (B) attached to the cuticle (C) in a highly differentiated anterior chamber of the hindgut of *O. disjunctus* collected in Baton Rouge, Louisiana.

compartments. The most spectacular of the hindgut compartments based on morphology is the anterior third of the hindgut, a region with extreme expansion of the internal surface area. Not only is the region lobed, but it also has hundreds of spines projecting into the gut lumen. A surface film of morphologically diverse bacteria covers the greatly expanded surface area of this part of the gut (Fig. 17.3).

The presence of the yeasts as well as other microbes and the basis for their association with the insect in the gut region have not been explained, but we suspect functional parallels between these yeasts and yeasts in other wood-ingesting organisms such as termites (Prillinger *et al.*, 1996). In fact, the morphologically distinct compartments of the passalid hindgut segregate several classes of microbes that may be similar to those of termites, and we hypothesize that the passalid microbes will be shown to have similar functions. Suggestions of essential microbial function in the physiology and nutrition of wood-boring insects include cellulose, hemicellulose and lignin degradation, pheromone production, detoxification of foodstuffs, nitrogen fixation and vitamin synthesis (Martin, 1987; Suh & Blackwell, 2005). One needs to use care in making such assumptions, however, because a number of studies have shown that arthropods, including termites and wood-boring beetles, produce endogenous cellulases (Watanabe *et al.*, 1998; Lee *et al.*, 2004).

Conclusion

Since the studies of Leidy (1849, 1853) there has been increasing interest in gut microbes. Current intensified interest is exemplified by publications in two widely circulated general scientific journals. In the same week (25 March 2005) that *Nature* included an article in its News and Views section, 'Gut feeling for yeasts' (Boekhout, 2005), which discussed the large number of new yeast species being discovered in the gut of insects (Suh *et al.*, 2005), a special issue of *Science* entitled 'The gut: the inner tube of life' featured articles emphasizing the importance of microbes in the digestive system (see, for example, Bäckhed *et al.*, 2005; Simpson *et al.*, 2005). It is possible that many of these microbes will have similar functions in a wide variety of digestive tracts, and a community-level approach will be essential to understand the role of the gut inhabitants.

Acknowledgements

LSU undergraduate students Cennet Erbil and Nhu H. Nguyen provided their expert assistance throughout much of the study reported here. Our collaborator Joseph V. McHugh helped to collect and identify some of the beetles. Charles Mark Bee, Imaging Technology Group, Beckman Institute for Advanced Science and Technology, University of Illinois, and Lou Ann Miller, Center for Microscopy and Imaging, College of Veterinary Medicine, University of Illinois, graciously provided microscopic and photographic expertise. We acknowledge the curators and culture collections that preserve the germplasm and data derived from and used in our studies: NRRL, CBS and the NCBI databases. This work was supported in part by the National Science Foundation, Biodiversity Surveys and Inventories Program (NSF DEB-0417180). MB thanks John Peberdy and the British Mycological Society for providing travel funds to attend a memorable meeting in Nottingham.

References

Altschul, S. F., Gish, W., Miller, W., Myers, E.W. & Lipman, D.J. (1990). Basic local alignment search tool. *Journal of Molecular Biology* **215**, 403–10.

Bäckhed, F., Ley, R. E., Sonnenburg, J. L., Peterson D. A. & Gordon, J. I. (2005). Host-bacterial mutualism in the human intestine. *Science* **307**, 1915–20.

Barnett, J. A., Payne, R. W. & Yarrow, D. (2000). *Yeasts: Characteristics and Identification*, 3rd edn. Cambridge: Cambridge University Press.

Boekhout, T. (2005). Gut feeling for yeasts. *Nature* **434**, 449–50.

Buchner, P. (1965). *Endosymbiosis of Animals with Plant Microorganisms*. New York: John Wiley & Sons.

Dillon, R. J. & Dillon, V. M. (2004). The gut bacteria of insects: non-pathogenic interactions. *Annual Review of Entomology* **49**, 71–92.

Dowd, P. F. (1989). *In situ* production of hydrolytic detoxifying enzymes by symbiotic yeasts of cigarette beetle (Coleoptera: Anobiidae). *Journal of Economic Entomology* **82**, 396–400.

Dowd, P. F. (1991). Symbiont-mediated detoxification in insect herbivores. In *Microbial Mediation of Plant-Herbivore Interactions*, ed. P. Barbosa, V. A. Krischik & C. G. Jones, pp. 411–40. New York: John Wiley & Sons.

Gray, I. E. (1946). Observations on the life history of the horned *Passalus*. *American Midland Naturalist* **35**, 728–46.

Jeffries, T. W. & Kurtzman, C. P. (1994). Strain selection, taxonomy, and genetics of xylose-fermenting yeasts. *Enzyme and Microbial Technology* **16**, 922–32.

Jones, K. G. & Blackwell, M. (1996). Ribosomal DNA sequence analysis excludes *Symbiotaphrina* from the major lineages of ascomycete yeasts. *Mycologia* **88**, 212–18.

Jones, K. G., Dowd, P. F. & Blackwell, M. (1999). Polyphyletic origins of yeast-like endocytobionts from anobiid and cerambycid beetles. *Mycological Research* **103**, 542–6.

Kurtzman, C. P. & Fell, J. W. (1998). *The Yeasts, a Taxonomic Study*, 4th edn. Amsterdam: Elsevier.

Kurtzman, C. P. & Robnett, C. J. (1998). Identification and phylogeny of ascomycetous yeasts from analysis of nuclear large subunit (26S) ribosomal DNA partial sequences. *Antonie van Leeuwenhoek International Journal of General and Molecular Microbiology*, **73**, 331–71.

Lawrence, J. F. (1973). Host preference in ciid beetles (Coleoptera: Ciidae) inhabiting the fruiting bodies of basidiomycetes in North America. *Bulletin of the Museum of Comparative Zoology at Harvard University* **145**, 163–212.

Lawrence, J. F. (1989). Mycophagy in the Coleoptera: feeding strategies and morphological adaptations. In *Insect-Fungus Interactions*, ed. N. Wilding, N. M. Collins, P. M. Hammond & J. F. Webber, pp. 1–23. San Diego, CA: Academic Press.

Lee, S. J., Kim, S. R., Yoon, H. J., Kim, I., Lee, K. S., Je, Y. H., Lee, S. M., Seo, S. J., Sohn, H. D. & Jin, B. R. (2004). cDNA cloning, expression, and enzymatic activity of a cellulase from the mulberry longicorn beetle, *Apriona germari*. *Comparative Biochemistry and Physiology* **B139**, 107–16.

Leidy, J. (1849). On the existence of entophyta in healthy animals, as a natural condition. *Proceedings of the National Academy of Sciences of the USA* **4**, 225–33.

Leidy, J. (1853). A flora and fauna within living animals. *Smithsonian Contributions to Knowledge* **5**, 1–67.

Lichtwardt, R. W., White, M. M., Cafaro, M. J. & Misra, J. K. (1999). Fungi associated with passalid beetles and their mites. *Mycologia* **91**, 694–702.

Martin, M. M. (1987). *Invertebrate-Microbe Interactions*. Ithaca, NY: Cornell University Press.

McHugh, J. V., Marshall, C. J. & Fawcett, F. L. (1997). A study of adult morphology in *Megalodacne heros* (Coleoptera: Erotylidae). *Transactions of the American Entomological Society* **123**, 167–223.

Nardi, J. B., Bee, C. M., Miller, L. A.,Nguyen, N. H., Suh, S.-O. & Blackwell, M. (2006). Communities of microbes that inhabit the changing hind gut landscape of a subsocial beetle. *Arthropod Structure and Development* **35**, 57–68.

Nguyen, N. H., Suh, S.-O., Erbil, C. K. & Blackwell, M. (2006). *Metschnikowia noctiluminum* sp. nov., *Metschnikowia corniflorae* sp. nov., and *Candida chrysomelidarum* sp. nov., isolated from green lacewings and beetles. *Mycological Research* **110**, 346–56.

Noda, H. & Koizumi, Y. (2003). Sterol biosynthesis by symbiotes: cytochrome P450 sterol C-22 desaturase genes from yeastlike symbiotes of rice planthoppers and anobiid beetles. *Insect Biochemistry and Molecular Biology* **33**, 649–58.

Pearse, A. S., Patterson, M. T., Rankin, J. S. & Wharton, G. W. (1936). The ecology of *Passalus cornutus* Fabricius, a beetle which lives in rotting logs. *Ecological Monographs* **6**, 455–90.

Prillinger, H. (1987). Are there yeasts in Homobasidiomycetes? *Studies in Mycology* **30**, 33–59.

Prillinger, H., Messner, R., König, H., Bauer, R., Lopandic, K., Molnar, O., Dangel, P., Wergang, F., Kirisits, T., Nakase, T. & Sigler, L. (1996). Yeasts associated with termites: a phenotypic and genotypic characterization and use of coevolution for dating evolutionary radiations in asco- and basidiomycetes. *Systematic and Applied Microbiology* **19**, 265–83.

Schäfer, A., Konrad, R., Kuhnigk, T., Kampfer, P., Hertel, H. & Konig, H. (1996). Hemicellulose degrading bacteria and yeasts from the termite gut. *Journal of Applied Bacteriology* **80**, 471–8.

Shoemaker, D. D., Katju, V. & Jaenike, J. (1999). *Wolbachia* and the evolution of reproductive isolation between *Drosophilla recens* and *Drosophila subquinaria*. *Evolution* **53**, 1157–64.

Simpson, S., Ash, C., Pennisi, E. & Travis, J. (2005). The gut: inside out. *Science* **307**, 1895.

Starmer, W. T., Fogleman, J. C. & Lachance, M. A. (1991). The yeast community of cacti. In *Microbial Ecology of Leaves*, ed. J. H. Andrews & S. S. Hriano, pp. 158–78. New York: Springer.

Suh, S.-O. & Blackwell, M. (2005). Beetles as hosts for undescribed yeasts. In *Insect Fungal Associations: Ecology and Evolution*, ed. F. E. Vega & M. Blackwell, pp. 244–56. New York: Oxford University Press.

Suh, S.-O., Noda, H. & Blackwell, M. (2001). Insect symbiosis: derivation of yeast-like endosymbionts within an entomopathogenic lineage. *Molecular Biology and Evolution* **18**, 995–1000.

Suh, S.-O., Marshall, C., McHugh, J. V. & Blackwell, M. (2003). Wood ingestion by passalid beetles in the presence of xylose-fermenting gut yeasts. *Molecular Ecology* **12**, 3137–45.

Suh, S.-O., Gibson, C. M. & Blackwell, M. (2004a). *Metschnikowia chrysoperlae* sp. nov., *Candida picachoensis* sp. nov. and *Candida pimensis* sp. nov., isolated from the green lacewings *Chrysoperla comanche* and *Chrysoperla carnea* (Neuroptera: Chrysopidae). *International Journal of Systematics and Evolutionary Microbiology* **54**, 1883–90.

Suh, S.-O., McHugh, J. V. & Blackwell, M. (2004b). Expansion of the *Candida tanzawaensis* yeast clade: 16 novel *Candida* species from basidiocarp-feeding beetles. *International Journal of Systematics and Evolutionary Microbiology* **54**, 2409–29.

Suh, S.-O., White, M. M., Nguyen, N. H. & Blackwell, M. (2004c). The identification of *Enteroramus dimorphus*: a xylose-fermenting yeast attached to the gut of beetles. *Mycologia* **96**, 756–60.

Suh, S.-O., McHugh, J. V., Pollock, D. & Blackwell, M. (2005). Massive biodiversity of yeasts from the gut of basidiocarp-feeding beetles. *Mycological Research* **109**, 261–5.

Tatusova, T. A. & Madden, T. L. (1999). Blast 2 sequences – a new tool for comparing protein and nucleotide sequences. *FEMS Microbiology Letters* **174**, 247–50.

van Meer, M. M. M., Witteveldt, J. & Stouthamer, R. (1999). Phylogeny of the arthropod endosymbiont *Wolbachia* based on the *wsp* gene. *Insect Molecular Biology* **8**, 399–408.

Vega, F. E. & Dowd, P. F. (2005). The role of yeasts as insect endosymbionts. In *Insect Fungal Associations: Ecology and Evolution*, ed. F. E. Vega & M. Blackwell, pp. 211–43. New York: Oxford University Press.

Watanabe, H., Noda, H., Tokuda, G. & Lo, N. (1998). A cellulase gene of termite origin. *Nature* **394**, 330–1.

Weir, A. & Blackwell M. (2001). Molecular data support the Laboulbeniales as a separate class of Ascomycota, Laboulbeniomycetes. *Mycological Research* **105**, 715–22.

Woolfolk, S. W. & Inglis, G. D. (2003). Microorganisms with field-collected *Chrysoperla rufilabris* (Neuroptera: Chrysopidae) adults with emphasis on yeast symbionts. *Biological Control* **29**, 155–68.

Zhang, N., Suh, S.-O. & Blackwell, M. (2003). Microorganisms in the gut of beetles: evidence from molecular cloning. *Journal of Invertebrate Pathology* **84**, 226–33.

18

A saltmarsh decomposition system and its ascomycetous laccase genes

STEVEN Y. NEWELL, JUSTINE I. LYONS AND MARY ANN MORAN

The University of Georgia – Marine Institute

Introduction

The saltmarshes of the Georgia, USA, Atlantic coast are expansive and highly productive. The marshes form the intertidal ecosystem 5–10 km wide extending from the barrier-island chain to the mainland. The predominant macrophyte of the marshes is smooth cordgrass (*Spartina alterniflora* Loisel.). Cross-marsh average annual production of smooth cordgrass shoots in Georgia has been measured at approximately 1.3 kg m^{-2} of marsh (Newell, 2001a, from Dai & Wiegert, 1996). Like most grasses, smooth cordgrass does not abscise its leaf blades; they remain attached to the leaf sheath after senescence and death (Newell, 1993, and references therein). As new blades are produced at the apex of shoots, the bottom blades senesce and die, until the whole shoot dies after flowering. Therefore, a large crop of standing-dead litter is available to microbes for decomposition for much of the year (for leaf blades alone, up to 538 g dry mass m^{-2}) (Newell *et al.*, 1998).

Because smooth cordgrass is produced in an intertidal marsh, one might suspect that tidal flooding would be a major wetting phenomenon for the standing-dead cordgrass leaves. However, the grass shoots extend above the flooding-tidal level most of the time: it is estimated that most of the dead-blade mass is wetted by tides only about 10% of the time on an annual-average basis (Newell *et al.*, 1998). The most important wetting phenomenon is dewset, and the combination of tidal, rain, and dew wetting results in the leaves being wet roughly half of the time (Newell *et al.*, 1989; Newell, 2007).

Smooth-cordgrass ascomycetes are well adapted to alternating wet and dry periods, losing little or no biomass after aestivating and reviving (Newell, 1995). Putting this fact together with the pervasive capabilities

Fungi in the Environment, ed. G. M. Gadd, S. C. Watkinson & P. S. Dyer. Published by Cambridge University Press.

of ascomycetes in solid substrates, by way of turgor-pressure-driven drilling and solid-resistance reduction through external-enzymatic lysis, one can easily understand the fact that cordgrass ascomycetes thoroughly integrate their mycelia into standing-dead cordgrass. Inside the dead shoots, the ascomycetes can carry out their digestive work free from prokaryotic competition or pressure from micropredators such as amoebae (Newell *et al.*, 1996; Newell, 2001a, 2002). Other factors that might favour fungal production in standing-decaying smooth cordgrass are: (a) potential formation of consortia with dinitrogen-fixing bacteria on the decaying blades (Newell & Porter, 2000); (b) potential synergistic relationships with green microalgae on the blade surfaces (Newell & Porter, 2000); (c) potential uptake of nitrogen from flooding-tidal waters (White & Howes, 1994; Newell, 1996, 2001a, 2002); and (d) potential synergistic exchange of gases and water vapour between standing-decaying and living parts of shoots, which are connected by aerenchyma channels (Newell & Porter, 2000).

A major fate of the large quantities of fungal mass produced within smooth-cordgrass marshes (rough estimate: over $0.5\,kg\,m^{-2}\,yr^{-1}$) is the same as might be expected for forest-floor ecosystems: it flows to litter-comminuting invertebrates at the base of the marsh food web (gastropods, arthropods) (Graça *et al.*, 2000). Cordgrass ascomycetes are also prodigious expellers of ascospores: a conservative estimate is 5.9 billion spores $m^{-2}\,yr^{-1}$ (Newell, 2001b).

A potential problem for smooth-cordgrass ascomycetes in their activities as microbial producers at the base of the marsh food web is that standing-dead smooth cordgrass is composed largely of lignocellulose (70–75%) (Hodson *et al.*, 1984; Newell & Porter, 2000). Early results in testing of cordgrass-fungal ability to decompose lignocellulose (LC) from smooth cordgrass suggested that cordgrass ascomycetes were weak degraders of LC (less than $0.1\%\,d^{-1}$ mineralization), but this was likely to have been a consequence of the fact that the microcosm-method used was not favourable for ascomycetous LC-lysis (especially violent shaking and absence of non-LC carbohydrates) (Newell *et al.*, 1996). Subsequent tests with static incubation and presence of malt and yeast extract revealed that a major member of the cordgrass-ascomycete community (*Phaeosphaeria spartinicola* Leuchtmann) (Newell, 2001b), which had been tested in the earlier work, was indeed capable of mineralization of LC at more than $0.8\%\,d^{-1}$ (Bergbauer & Newell, 1992). More recent testing of a combination of the three predominant cordgrass-blade ascomycetes (*P. spartinicola, Phaeosphaeria halima* [Johnson] Shoemaker & Babcock, and

Mycosphaerella sp.2 [Kohlmeyer & Kohlmeyer, 1979]) (Newell, 2001b), incubated statically with wet but not submerged, radiolabelled, extracted, ground cordgrass LC and a small piece of pre-sterilized whole cordgrass leaf blade, showed that mineralization of LC by cordgrass ascomycetes can be more than 20% after 6 weeks at about 23 °C (and the rate of mineralization was not declining at that time point (Lyons, 2002)). Direct electron microscopic examination of changes in LC of smooth-cordgrass fibre tissue in shoots undergoing natural ascomycetous decay (Newell *et al.*, 1996) supported the work *in vitro*: cordgrass ascomycetes clearly brought about both type 1 and type 2 soft rot (Nilsson *et al.*, 1989). Type 2 soft rot is very similar to the white rot of basidiomycetes.

The major members of the smooth-cordgrass shoot-decomposer mycocommunity (*P. spartinicola*, *P. halima* and *Mycosphaerella* sp.2) were identified by direct observation of ascomata and by capture of expelled ascospores from standing-decaying blades (Newell, 2001a, b, and references therein). Percentage area of decaying blades occupied by at least one of the three predominant species of ascomycetes was greater than 90 at sites with low grazing pressure from periwinkle snails (Newell, 2001b). A regularly occurring species encountered at 40% or more of the area of blades examined was *Buergenerula spartinae* Kohlm. & Gessner. Other species observed at much lower percentages of blade area (less than 1%) were *Hydropisphaera erubescens* (Desm.) Rossman & Samuels and *Koorchaloma spartinicola* Sarma, Newell & Hyde, along with several species lacking published descriptions (Newell, 2001b; Sarma *et al.*, 2001; Buchan *et al.*, 2002; see also Kohlmeyer & Volkmann-Kohlmeyer, 2003, and references therein). Adaptation of DNA technology (rDNA/ITS PCR with ascomycete-selective primers) to identifying the members of the smooth-cordgrass ascomycetous community (Buchan *et al.*, 2002, 2003) revealed the same predominant species of ascomycetes as had direct microscopy. Two species more rarely encountered in the ITS clone libraries and by T-RFLP were *H. erubescens* and an undescribed ascomycete nicknamed '4clt' (capsule description in Buchan *et al.*, 2002).

Enzymatic lignocellulolysis

It is very clear from empirical testing, field sampling and transmission electron microscopy that the ubiquitous community of cordgrass ascomycetes brings about the breakdown of smooth-cordgrass lignocellulose during natural decay (see Introduction). Because these fungal species would not be expected to possess lignin or manganese peroxidases (the enzymes best known for their lignolytic capabilities) (Sinsabaugh &

Liptak, 1997), how do *Phaeosphaeria spartinicola* and its associates carry out lignocellulolysis? Ascomycetes have long been known to possess laccase, a copper-containing enzyme that is capable of oxidation of phenolic molecules, but it was believed that laccase did not have the power to oxidize the non-phenolic portions of lignin, thereby severely limiting lignolysis (Thurston, 1994). A recent discovery has altered this viewpoint. If a redox-mediator molecule is present along with laccase, non-phenolic portions of lignin can be oxidized and lignin can therefore be extensively broken down. Eggert *et al.* (1997) discovered that a basidiomycete that had no manganese or lignin peroxidase could carry out strong lignolysis. The species had this capability because of its ability to manufacture its own redox-mediator molecules to laccase (3-hydroxy anthranilate; HAA): the species could thus utilize laccase as its sole lignin-oxidizing enzyme (see also Johannes & Majcherczyk, 2000). Temp & Eggert (1999) have further shown that interactions of cellobiose dehydrogenases with the laccase-mediator molecules can probably allow cellulose lysis to go hand-in-hand with lignin oxidation when laccase is involved in lignocellulose breakdown (see also Daniel *et al.*, 1992). Thus it was clearly a reasonable hypothesis that at least some smooth-cordgrass ascomycetes possessed laccase, which they used to effectively attack cordgrass lignocellulose.

To give this hypothesis a preliminary test, we grew eight strains of the most common ascomycetes of smooth cordgrass (see Introduction) on agar containing ground leaves of smooth cordgrass, and flooded the plates with syringaldazine solution. All eight of these strains immediately presented a pink colour after the flooding, indicating that all eight produced laccase activity (Lyons *et al.*, 2003).

The foregoing result prompted us to move forward and apply DNA technology to enhance our understanding of ascomycetous laccases in the smooth-cordgrass decay system. We began by aligning and examining 20 published amino acid sequences for laccase genes of ascomycetes and basidiomycetes, and for bacteria, a plant and a nematode (Lyons *et al.*, 2003). We targeted conserved sequences within copper-binding sites II and III and designed degenerate forward and reverse primers such that they would be likely to be specific for fungal laccases and incapable of amplifying non-fungal laccases. Basidiomycetous laccases were similar enough to the ascomycetous laccases that our degenerate primers would be likely to potentially amplify partial basidiomycetous laccase genes, but mycelial basidiomycetes are not members of the smooth-cordgrass fungal community as determined by direct microscopy (Newell, 2001b).

Before applying our laccase-gene primers to saltmarsh microbes or cordgrass samples, we used them with DNA extracted from positive-control species of ascomycetes (*Aspergillus nidulans*, *Cryphonectria parasitica*, *Podospora* sp. and *Saccharomyces cerevisiae*). PCR with our primers produced amplification products of the expected size (about 900 base pairs) in all of the positive controls, and subsequent sequencing of the products confirmed their identities as portions of laccase genes (Lyons *et al.*, 2003).

When we extracted DNA from ascomycetes of the smooth-cordgrass standing-decay community (4–5 strains of the three predominant species [see Introduction] and 1–3 strains of six less frequent to rare species), and applied our laccase-gene primers to it in PCR/base-pair sequencing, we found that all species possessed laccase genes in at least one strain (Lyons *et al.*, 2003). All strains of the two major *Phaeosphaeria* species gave laccase PCR products; between them, they exhibited five different types of partial laccase gene (i.e. partial genes that were less than 96% similar in amino acid sequences) and as many as three different types within one strain. For all of the strains tested, 13 distinct sequence types were obtained.

We also extracted DNA from early and late stages of standing-decaying blades of smooth cordgrass, and used our laccase-gene primers in PCR/base-pair sequencing (Lyons *et al.*, 2003). We cloned and sequenced 26 partial laccase genes from the decaying blades from the marsh. Only two of these clones were not identifiable with the gene sequences that we had identified from the three major members (two *Phaeosphaeria* spp. and a *Mycosphaerella*; see Introduction) of the ascomycetous decay community. Thus there was a broad spectrum of types of laccase-activity potential among the cordgrass-decay ascomycetes (previous paragraph), and the major species identified by direct microscopy of ascospore production and by presence of rRNA genes (Newell, 2001b; Buchan *et al.*, 2003) were the same species producing potentially lignolytic laccase genes within the standing-decaying cordgrass blades.

Conclusions

Our findings with regard to laccase-gene occurrence in ascomycetes of the smooth-cordgrass standing-decay community go a long way towards explaining the very clear lysis of lignocellulose that takes place in the fibre cells of naturally decaying shoots of smooth cordgrass (Newell *et al.*, 1996). Laccase genes of several types are potentially expressible by all of the ascomycetes of naturally decaying smooth cordgrass, and it has been recently discovered that laccases can be used by fungi for extensive lignin breakdown without the assistance of other lignin oxidases. A variety of interesting questions are raised by our laccase-gene findings. For example: (1) What sorts of oxidation mediators (if any) are produced by cordgrass ascomycetes for action alongside their laccases? (2) Is there synergy between or among laccases produced by the three major species of cordgrass ascomycetes? (3) Which of the ascomycetous laccases are active at different points in the decay process? (4) Are there major differences in the products of laccase lignolysis among the laccases of the cordgrass ascomycetes?

References

Bergbauer, M. & Newell, S. Y. (1992). Contribution to lignocellulose degradation and DOC formation from a salt marsh macrophyte by the ascomycete *Phaeosphaeria spartinicola*. *FEMS Microbiology Ecology* **86**, 341–8.

Buchan, A., Newell, S. Y., Moreta, J. I. L. & Moran, M. A. (2002). Analysis of internal transcribed spacer (ITS) regions of rRNA genes in fungal communities in a southeastern U.S. salt marsh. *Microbial Ecology* **43**, 329–40.

Buchan, A., Newell, S. Y., Butler, M., Biers, E. J., Hollibaugh, J. T. & Moran, M. A. (2003). Dynamics of bacterial and fungal communities on decaying salt marsh grass. *Applied and Environmental Microbiology* **69**, 6676–87.

Dai, T. & Wiegert, R. G. (1996). Ramet population dynamics and net aerial primary productivity of *Spartina alterniflora*. *Ecology* **77**, 276–88.

Daniel, G., Volc, J., Kubatova, E. & Nilsson, T. (1992). Ultrastructural and immunocytochemical studies of the H_2O_2-producing enzyme pyranose oxidase in *Phanerochaete chrysosporium* grown under liquid culture conditions. *Applied and Environmental Microbiology* **58**, 3667–76.

Eggert, C., Temp, U. & Eriksson, K.-E. (1997). Laccase is essential for lignin degradation by the white-rot fungus *Pycnoporus cinnabarinus*. *FEBS Letters* **407**, 89–92.

Graça, M. A., Newell, S. Y. & Kneib, R. T. (2000). Grazing rates of organic matter and living fungal biomass of decaying *Spartina alterniflora* by three species of salt-marsh invertebrates. *Marine Biology* **136**, 281–9.

Hodson, R. E., Christian, R. R. & Maccubbin, A. E. (1984). Lignocellulose and lignin in the salt marsh grass *Spartina alterniflora*: initial concentrations and short-term, post-depositional changes in detrital matter. *Marine Biology* **81**, 1–7.

Johannes, C. & Majcherczyk, A. (2000). Natural mediators in the oxidation of polycyclic aromatic hydrocarbons by laccase mediator systems. *Applied and Environmental Microbiology* **66**, 524–8.

Kohlmeyer, J. & Kohlmeyer, E. (1979). *Marine Mycology. The Higher Fungi*. New York: Academic Press.

Kohlmeyer, J. & Volkmann-Kohlmeyer, B. (2003). *Octopodotus stupendus* gen. & sp. nov. and *Phyllachora paludicola* sp. nov., two marine fungi from *Spartina alterniflora*. *Mycologia* **95**, 117–23.

Lyons, J. I. (2002). Diversity of ascomycete laccase sequences and contributions of bacteria and ascomycetous fungi to lignocellulose degradation in a southeastern U.S. salt marsh. Master's Thesis, University of Georgia, Athens, GA, USA.

Lyons, J. I., Newell, S. Y., Buchan, A. & Moran, M. A. (2003). Diversity of ascomycete laccase gene sequences in a southeastern US salt marsh. *Microbial Ecology* **45**, 207–81.

Newell, S. Y. (1993). Decomposition of shoots of a saltmarsh grass. *Advances in Microbial Ecology* **13**, 301–26.

Newell, S. Y. (1995). Minimizing ergosterol loss during preanalytical handling and shipping of samples of plant litter. *Applied and Environmental Microbiology* **61**, 2794–7.

Newell, S. Y. (1996). Established and potential impacts of eukaryotic mycelial decomposers in marine/terrestrial ecotones. *Journal of Experimental Marine Biology and Ecology* **200**, 187–206.

Newell, S. Y. (2001a). Multiyear patterns of fungal biomass dynamics and productivity within naturally decaying smooth cordgrass shoots. *Limnology and Oceanography* **46**, 573–83.

Newell, S. Y. (2001b). Spore-expulsion rates and extents of blade occupation by ascomycetes of the smooth-cordgrass standing-decay system. *Botanica Marina* **44**, 277–85.

Newell, S. Y. (2002). Fungi in marine/estuarine waters. In *The Encyclopedia of Environmental Microbiology*, ed. G. Bitton, pp. 1394–400. New York: Wiley.

Newell, S. Y. (2007). Evolution at the Marine Institute of the story of the fate of smooth cordgrass shoots, 1979–2002. In *The University of Georgia Marine Institute – Five Decades of Research on Georgia's Coast*, ed. R. T. Kneib, in press. Athens, GA: University of Georgia Press.

Newell, S. Y. & Porter, D. (2000). Microbial secondary production from saltmarsh-grass shoots, and its known and potential fates. In *Concepts and Controversies in Tidal Marsh Ecology*, ed. M. P. Weinstein & D. A. Kreeger, pp. 159–85. Dordrecht: Kluwer Academic.

Newell, S. Y., Fallon, R. D. & Miller, J. D. (1989). Decomposition and microbial dynamics for standing, naturally positioned leaves of the salt-marsh grass *Spartina alterniflora*. *Marine Biology* **101**, 471–81.

Newell, S. Y., Porter, D. & Lingle, W. L. (1996). Lignocellulolysis by ascomycetes (fungi) of a saltmarsh grass (smooth cordgrass). *Microscopy Research and Technique* **33**, 32–46.

Newell, S. Y., Arsuffi, T. L. & Palm, L. A. (1998). Seasonal and vertical demography of dead portions of shoots of smooth cordgrass in a south-temperate saltmarsh. *Aquatic Botany* **60**, 325–35.

Nilsson, T., Daniel, G., Kirk, T. K. & Obst, J. R. (1989). Chemistry and microscopy of wood decay by some higher ascomycetes. *Holzforschung* **43**, 11–18.

Sarma, V. V., Newell, S. Y. & Hyde, K. D. (2001). *Koorchaloma spartinicola* sp. nov., a new sporodochial fungus from *Spartina alterniflora*. *Botanica Marina* **44**, 321–6.

Sinsabaugh, R. L. & Liptak, M. A. (1997). Enzymatic conversion of plant biomass. In *The Mycota*, vol. IV, *Environmental and Microbial Relationships*, ed. D. T. Wicklow & B. Söderström, pp. 347–57. Berlin: Springer-Verlag.

Temp, U. & Eggert, C. (1999). Novel interaction between laccase and cellobiose dehydrogenase during pigment synthesis in the white rot fungus *Pycnoporus cinnabarinus*. *Applied and Environmental Microbiology* **65**, 389–95.

Thurston, C. F. (1994). The structure and function of fungal laccases. *Microbiology* **140**, 19–26.

White, D. S. & Howes, B. L. (1994). Nitrogen incorporation into decomposing litter of *Spartina alterniflora*. *Limnology and Oceanography* **39**, 133–40.

Index